The Museum Environment

The Museum Environment

Second Edition

Garry Thomson CBE
Scientific Adviser, The National Gallery, London (1960–1985)

Routledge
Taylor & Francis Group
LONDON AND NEW YORK

First published by Elsevier
This edition published 2011 by Routledge
2 Park Square, Milton Park, Abingdon, Oxfordshire OX14 4RN
711 Third Avenue, New York, NY 10017

Routledge is an imprint of the Taylor and Francis Group, an informa business

First issued in hardback 2015

First edition 1978
Second edition 1986
Paperback edition 1994

Notice
No responsibility is assumed by the publisher for any injury and/or damage to persons or property as a matter of products liability, negligence or otherwise, or from any use or operation of any methods, products, instructions or ideas contained in the material herein. Because of rapid advances in the medical sciences, in particular, independent verification of diagnoses and drug dosages should be made

British Library Cataloguing in Publication Data
Thomson, Garry
The museum environment – 2nd ed.
1. Arts objects – Conservation and restoration
2. Museum conservation methods
I. Title II. International Institute for Conservation of Historic and Artistic Works
069.5′3 N8560

Library of Congress Cataloging-in-Publication Data
Thomson, Garry
The museum environment
Bibliography: p
Includes index.
ISBN 978-0-7506-2041-3 (pbk)
ISBN 978-1-138-13203-0 (hbk)
1. Museum buildings – environmental engineering
2. Museum conservation methods. 1. Title II. Series
TH6037.M87T46 069.5′3 85-25446

ISBN: 978-0-7506-2041-3

Transferred to Digital Print 2010

Series Editors' Preface

The conservation of artefacts and buildings has a long history, but the positive emergence of conservation as a profession can be said to date from the foundation of the International Institute for the Conservation of Museum Objects (IIC) in 1950 (the last two words of the title being later changed to Historic and Artistic Works) and the appearance soon after in 1952 of its journal *Studies in Conservation*. The role of the conservator as distinct from those of the restorer and the scientist had been emerging during the 1930s with a focal point in the Fogg Art Museum, Harvard University, which published the precursor to *Studies in Conservation*, *Technical Studies in the Field of the Fine Arts* (1932–42).

UNESCO, through its Cultural Heritage Division and its publications, had always taken a positive role in conservation and the foundation, under its auspices, of the International Centre for the Study of the Preservation and the Restoration of Cultural Property (ICCROM), in Rome, was a further advance. The Centre was established in 1959 with the aims of advising internationally on conservation problems, co-ordinating conservation activities and establishing standards and training courses.

A significant confirmation of professional progress was the transformation at New York in 1966 of the two committees of the International Council of Museums (ICOM), one curatorial on the Care of Paintings (founded in 1949) and the other mainly scientific (founded in the mid-1950s) into the ICOM Committee for Conservation.

Following the Second International Congress of Architects in Venice in 1964 when the Venice Charter was promulgated, the International Council of Monuments and Sites (ICOMOS) was set up in 1965 to deal with archaeological, architectural and town planning questions, to schedule monuments and sites and to monitor relevant legislation.

From the early 1960s onwards, international congresses (and the literature emerging from them) held by IIC, ICOM, ICOMOS and ICCROM not only advanced the subject in its various technical specializations but also emphasized the cohesion of conservators and their subject as an interdisciplinary profession.

The use of the term *Conservation* in the title of this series refers to the whole subject of the care and treatment of valuable artefacts both movable and immovable, but within the discipline conservation has a meaning which is distinct from that of restoration. *Conservation* used in this specialized sense has two aspects: firstly, the control of the environment to minimize the decay of artefacts and materials; and, secondly, their treatment to arrest decay and to stabilize them where possible against further deterioration. Restoration is the continuation of the latter process, when conservation treatment is thought to be insufficient, to the extent of reinstating an object, without falsification, to a condition in which it can be exhibited.

In the field of conservation conflicts of values on aesthetic, historical or technical grounds, are often inevitable. Rival attitudes and methods inevitably arise in a subject which is still developing and at the core of these differences there is often a deficiency of technical knowledge. That is one of the principal *raisons d'etre* of this series. In most of these matters ethical principles are the subject of much discussion, and generalizations cannot easily cover (say) buildings, furniture, easel paintings and waterlogged wooden objects.

A rigid, universally agreed principle is that all treatment should be adequately documented. There is also general agreement that structural and decorative falsification should be avoided. In addition there are three other principles which, unless there are over-riding objections, it is generally agreed should be followed.

The first is the principle of the reversibility of processes, which states that a treatment should normally be such that the artefact can, if desired, be returned to its pre-treatment condition even after a long lapse of time. This principle is impossible to apply in some cases, for example where the survival of an artefact may depend upon an irreversible process. The second, intrinsic to the whole subject, is that as far as possible decayed parts of an artefact should be conserved and not replaced. The third is that the consequences of the ageing of the original materials (for example 'patina') should not normally be disguised or removed. This includes a secondary proviso that later accretions should not be retained under the false guise of natural patina.

The authors of the volumes in this series give their views on these matters, where relevant, with reference to the types of material within their scope. They take into account the differences in approach to artefacts of essentially artistic significance and to those in which the interest is primarily historical or archaeological.

The volumes are unified by a systematic and balanced presentation of theoretical and practical materials with, where necessary, an objective comparison of different methods and approaches. A balance has also been maintained between the fine (and decorative) arts, achaeology and architecture in those cases where the respective branches of the subject have common ground, for example in the treatment of stone and glass and in the control of the museum environment. Since the publication of the first volume it has been decided to include within the series related monographs and technical studies. To reflect this enlargement of its scope the series has been renamed the Butterworths Series in Conservation and Museology.

Though necessarily different in details of organisation and treatment (to fit the particular requirements of the subject) each volume has the same general standard which is that of such training courses as those of the University of London Institute of Archaeology, The Victoria and Albert Museum, the Conservation Center, New York University, the Institute of Advanced Architectural Studies, York and ICCROM.

The authors have been chosen from among the acknowledged experts in each field, but as a result of the wide areas of knowledge and technique covered even by the specialized volumes in this series, in many instances multi-authorship has been necessary.

With the existence of IIC, ICOM, ICOMOS and ICCROM, the principles and practice of conservation have become as internationalized as the problems. The collaboration of two Consultant Editors, W T Chase, Head Conservator, Freer Gallery of Art, Smithsonian Institution, Washington, and Bernard M Feilden, Former Director of ICCROM will help to ensure that the practices discussed in this series will be applicable throughout the world.

Acknowledgement

ICCROM

The cost of drawing the line diagrams and printing the colour plates in this book was contributed by the International Centre for the Study of the Preservation and the Restoration of Cultural Property, Rome (ICCROM).

Preface to First Edition

This book has a double purpose, and so is divided into two parts. The first part is intended as a textbook for conservators and curators of museums concerning the damaging effects on exhibits of light, humidity and air pollution, and what to do to minimize this damage. The scientific background needed for this first part is kept to a minimum. The second part is meant for workers in the field of conservation research and summarises information which up to now has been widely scattered and sometimes difficult of access. I assume in the second part some familiarity with basic science.

I must admit to feeling a little uncertain about the level of technical knowledge that ought to be demanded in the first part. This is not yet a subject with a syllabus, to which students come prepared up to a certain level. While the explanations may seem laboured to some, they may be a struggle to many others in positions of importance in museums. Curators and restorers are today demanding a basic knowledge of what is required in the museum to safeguard its contents. They come highly trained in their own profession but often lacking a basic scientific background. Therefore I have tried in Part I to keep clear as far as possible of explanations which require such a background (these will be found in Part II), and yet to include all that need be known in order to act, and as much of the reasoning behind the recommendations as possible. Indeed in some of the most interesting aspects of the argument, particularly on lighting, some of my readers will be better qualified as curators than myself, a scientist.

No doubt in the future the domains of Parts I and II will find themselves in separate books and be the better for it.

This book is the first of a series, subsequent volumes of which will deal with the behaviour and conservation of the various classes of museum exhibit. The more general character of this work makes it no less practical, but certainly indicates a bias towards setting up a generalised framework of knowledge from which particular problems can be solved, and towards treating the environment rather than the exhibits. However it will inevitably reflect my special interests, while revealing the gaps in my experience.

No one who reads this book will fail to end with a realisation of our general ignorance. We have a very uneven knowledge of how fast things in the museum change and what causes these changes, and yet we have to erect this framework of preventive conservation before rather than after our research has reached a dignified level of completion.

The reader from outside the museum professions comes with the expectation that all technical books about antiquities must concern hunting for either treasures or fakes. After all this is the story that is fed to him by the popular press. Those who maintain that the main objective of museum science is the prevention of deterioration, as I do, are a little tired of this endless diet. I hope most strongly, but with due modesty, that this book may do a small amount to correct the aim of science in conservation.

Thus the book is based on the growing need for a summary of the 'preventive medicine' of conservation. This has meant that information from many fields has had to be brought under one cover, and I am therefore dependent on a large number of people. To many of these I owe personal gratitude, but to none more than to Michael Levey, Director of the National Gallery for his powerful and enthusiastic support of its laboratory; to all the members of that laboratory including Joyce Plesters and one who knows more about colour than I ever shall, Linda Bullock; to a conservator who has done as much as anyone in the history of the profession to raise its standards, Norman Brommelle; and to two very eminent physicists who have both made major contributions to colour science, Professor David Wright and Dr Brian Crawford.

G.T.

Preface to Second Edition

The first edition of this book was by chance timely: interest in preventive conservation was growing strongly. Since that time many studies have appeared, ranging from the elementary, exemplified by the excellent U.K. National Trust Manual of Housekeeping, to such advanced work as that of the Canadian Conservation Institute's Environment and Deterioration Research Division, to quote just two examples.

The changes in this new edition have mainly been on electronic hygrometry, new fluorescent lamps, buffered cases, air conditioning systems, data logging and control in historic buildings, though the latter subject remains mainly within the covers of Bernard Feilden's book, Conservation of Historic Buildings, in this series.

It is hoped that a summary of museum specifications for conservation in an Appendix will be useful for reference.

Several colleagues have given much help, none more so than Tim Padfield with his witty and cogent suggestions. I have tried to comply.

G. T.

Contents

Light Part I	Surface deterioration	2
	Light and heat energy	3
	The spectrum	4
	The basic light sources	7
	Colours and materials which change	10
	Damage caused by UV and visible radiation	16
	UV radiation and how to deal with it	17
	Measuring UV and visible radiation	19
	The reciprocity law	21
	Controlling visible radiation	22
	Reducing illuminance	22
	50 lux – artificial light	25
	Diffusion of light	27
	200 lux – daylight and artificial light	28
	Conservation lighting specifications	33
	Treatment of windows	34
	Angle at which light falls on exhibits	35
	Reducing time of exposure	37
	A suite of exhibition rooms	39
	Heat	43
	Control of temperature	45
	Lighting for professional photography, television and restoration	46
	Electronic flash	47
	Colour rendering	48
	The measurement of colour	55
	The lighting situation and the process of seeing	61
Humidity Part I	The importance of humidity	66
	Measuring the humidity in the air	67
	The wet-and-dry-bulb hygrometer	70
	Electronic hygrometers	76
	Non-mechanical hygrometers	76
	Understanding the hygrometric chart	77
	Response of museum material to RH	82

	Best RH for moisture-containing absorbent materials	86
	Climate inside and outside the museum	87
	Condensation and the dew point	92
	Humidity control	93
	RH control in a room	93
	The humidistat	94
	Humidifying equipment	95
	Dehumidifying equipment	97
	Room RH control: maintenance and air circulation	101
	Packaged air-conditioning units	104
	Ducted air conditioning	104
	RH control in a closed case – buffers	106
	Silica gel in packing cases	109
	Exhibition cases	110
	The buffered case: towards a practical solution	112
	RH control in a closed case – use of salts	113
	Mechanical RH stabilisation in cases	114
	Future development of exhibition case stabilisation	115
	RH is often a matter of compromise	116
	Historic buildings closed in winter and churches	119
	Improvisation and RH control	120
	Humidity control in archaeology	124
Air Pollution Part I	The problem	130
	Particulates	131
	Particulate concentrations today	132
	New concrete buildings	133
	Removal of particulates	133
	Electrostatic precipitators (electro-filters)	135
	Gaseous pollution	135
	Sulphur dioxide (SO_2)	136
	Damage caused by sulphur dioxide	138
	Glass and sulphur dioxide	147
	Effects of sulphur dioxide on lichens and mosses	148
	Ozone	149
	Effects of ozone	150
	Nitrogen oxides	151
	Effects of nitrogen dioxide	152
	Levels of ozone and nitrogen dioxide likely to be encountered	152
	Chlorides	153
	Pollution through storage conditions	154
	Removal of gaseous pollutants	156
	Fire extinguishers	159
	Sound and vibration	160

Light Part II	Spectral curves	164
	Sun and sky	167
	Lamps and control equipment	171
	Measuring UV	174
	Luminous efficiency and the light meter	175
	Some basic light units	177
	Visual performance	180
	Luminance and subjective brightness	181
	The Blue Wool standards	182
	Damage versus wavelength	184
	Heat radiated from light sources	186
	Activation energy	188
	The primary photochemical reaction	193
	Placing a colour on the CIE Chromaticity Chart	195
	The colour rendering calculation	199
	Colour rendering and the black body convention	206
	Choosing a fluorescent lamp	206
	Dimming	207
Humidity Part II	The standard hygrometric (psychrometric) chart	210
	The classical air-conditioning operation	214
	A museum air-conditioning system	218
	Control	219
	Heating and cooling loads	221
	Sensors	221
	External design conditions	221
	Dimensional changes caused by RH variation	222
	Outdoor climate and response of objects indoors	226
	Does constant RH keep dimensions unchanged at all temperatures?	228
	Effect of people on RH and temperature	230
	Use of the air moisture-content scale	232
	The closed and buffered museum case	232
	Hygrometric half-time	232
	Materials useful as buffers	236
	Penetration of oxygen and water vapour through plastic films	237
Air Pollution Part II	Plotting the size distribution of particulates	244
	Choice of particulate filter	252
	Efficiency of activated carbon filters	254
	Room air cleaners	254
	Measuring concentrations of pollutants in museums	255
	The fate of sulphur dioxide in the atmosphere	256
	The formation of ozone	258
	Computers in environment control	262
	Data logging	263

Future trends in environmental control 265
Appendix: Summary of specifications 268
References 271
Index 286

List of Colour Plates

Plate 1 The wavelength of ultraviolet, visible and infrared radiation

Plate 2 Green and purple water-colours mixed to produce a warm grey

Plate 3 The two effects of ultraviolet and visible radiation: colour change and weakening

Plate 4 The effect of light on a tapestry

Plate 5 Detail from a Turner water-colour which has been protected at the edge by the frame from damage by light

Plate 6 *Sunset Landscape* by Giorgione

Plate 7 *The Good Samaritan* by Jacopo Bassano

Plate 8 Cross-section of a tiny fragment of paint and ground from one of the apparently brown stripes on the scarf lying across the Samaritan's knee

Plate 9 *The Oise at Auvers* by Van Gogh

Plate 10 *Landscape with Sportsmen and Game* by Adam Pynacker

Plates 11 and 12 Lighting as in *Figure 14*, *Plate 11* direct and *Plate 12* diffuse

Plates 13, 14 and 15 The camera and colour rendering

Plate 16 Uccello's *S. George and the Dragon*

Light
Part One

Light

SURFACE DETERIORATION

No one needs to be told that light can change colours and rot materials, though with the fastness of modern dyes and the affluent tendency to throw away rather than repair this common knowledge has no longer the practical importance that it once had – except in the museum.

Light can only damage what it reaches and since most objects are opaque to light its major effect is on surface deterioration. But the surface is the very essence of many exhibits, above all of paintings and drawings.

We can say that all organic material is at risk under light. The term 'organic material' includes all things which originated in animals or plants – for example, paper, cotton, linen, wood, parchment, leather, silk, wool, feathers, hair, dyes, oils, glues, gums and resins — and in addition, because of similarities in chemical structure, almost all synthetic dyes and plastics.

It must be remembered that light can cause not only colour change but strength change, as in the weakening of textiles and the destruction of paint medium.

Some information on the materials affected by light is to be found on p. 10 and *Table 1* (p. 11).

Stone, metal, glass and ceramics, with some exceptions, are not affected by light, and we need not worry too much about wood, bone or ivory if their surface colour is not important. But this section on light is concerned to a greater or lesser extent with just about every other kind of museum material.

To reduce surface deterioration to a minimum we must control the lighting. But before we deal with ways and means it will help to examine the nature of light.

LIGHT AND HEAT ENERGY

The ultimate nature of energy is not understood. But what science can do is to define and quantify physical systems and the energy changes that they undergo. Heat, light and motion are forms of energy. Energy is so defined that it is conserved, which means that the total amount of energy in a closed system always remains unchanged. This is easily arranged by a convention which turns any energy which seems to disappear during a change into 'potential' energy of some kind. And indeed the potential energy locked up in each molecule of TNT is easily imagined as real.

A chemical reaction, which for us means deterioration, may absorb energy or may actually release some of the potential energy held in the molecules. But in either case a certain definite quantity of energy must be supplied in order to start the reaction. This is known as the *activation energy* of the reaction.

In the museum the activation energy may be brought to an object by heating it or illuminating it. To us a chemical change such as deterioration seems to be a gradual thing, but this is because many millions of molecules are involved. Each molecule changes in jumps, not gradually. The energy for a change must be delivered to a molecule all at once. Energy is always on the move. One might suppose that the delivery of energy in a chemical reaction is like applying our strength to break a stick: we contract our muscles more and more strongly until the stick breaks. But this is not so at the level of individual molecules. Quantum theory tells us that energy is delivered in separate *quanta* or bundles, a definite quantity in each, and it also shows us that the quanta of heat and light act in different ways. The following analogy might help to make this clear.

Many people spend part of every day of their working lives as commuters, pushing or being pushed into trains and buses. This jostling, though repeated thousands of times, causes no damage – at least to the body! It can be compared to the effect of ordinary room-temperature heat quanta on the molecules of an exhibit. Heat quanta are continually exchanged between molecules by, as it were, jostling each other, but the jostling causes no chemical damage. Very rarely someone in the rush-hour may hit out or trample on his fellow-passengers, but this remote possibility is not a thing we allow ourselves to worry about. However a crowd can become an ugly thing if the 'temperature is raised', and there can be widespread casualties, and in the same way chemical reactions begin to happen at an appreciable rate when we heat an object.

Quanta of light energy are called *photons* and most of them are very much more potent than room-temperature heat quanta. Turning again to our crowd we might suppose that a riot control squad now starts to loose off projectiles of various kinds, some of them lethal, on the crowd. These are analogous to photons. Some, equivalent to photons of red light, cause very little physical damage, while those equivalent to photons of blue and violet light would be the worst. However, of shorter wavelength than the violet and invisible to the eye, ultraviolet radiation photons are the most damaging of all.

Summary

Deterioration needs energy – either light or heat. Light is much more potent than heat in the museum.

THE SPECTRUM

We now look at the spectrum of colours formed when light passes through a prism (*Plate 1*). Light is that form of radiation which we can see. Since we are also concerned with 'colours' invisible to the human eye – the ultraviolet and the infrared – we will from now on frequently use the term radiation instead of light.

The radiation from white-light sources, such as daylight, tungsten and fluorescent lamps, can be split by a prism into all the colours of the rainbow, as shown in *Plate 1*, with wavelength shortest in the violet and longest in the red. Beyond the visible at the short end lies the ultraviolet (UV), and beyond the red at the long-wavelength end lies the infrared (IR). All these are emitted to various extents by white-light sources.

We now assign wavelengths so that we can deal in numbers instead of colours. It has just been explained that energy arrives in separate packets, which, for all forms of radiation, as well as for light, are called photons. But now we are talking of waves. We have come up against a central conundrum of modern physics: light behaves like particles but also like waves. The figures work out beautifully, but they seem to have no objective reality*. We must be content to talk in terms either of waves or photons, whichever is the most convenient.

At the long-wavelength end of the visible spectrum, the longest wavelength red which we can see is at about 760 nm, though red beyond 700 nm is barely visible (*Plate 1*). Beyond

*'The extraordinary paradox of a physical quantity that seems to depend upon the state of mind of the physicist is not to be shrugged off without comment.'[1]

this the radiation is called IR. All radiation, if it is absorbed by any material, causes a rise in temperature, so we would be quite correct to call light a form of radiant heat. But in practice the term 'radiant heat' is very often confined to IR radiation, because we can regard heating as its *only* effect. There is no definite wavelength limit to the IR radiation in museum light sources, since it falls off gradually as the wavelength increases.

There is nothing more that need be said about IR at this stage, but there is much that we have to consider about UV. Unlike IR this is confined within a band, in the light-sources we are considering, stretching between 300 and 400 nm. From the exhibit's point of view there is no line dividing UV from violet (or IR from red) – the 400 nm division is made entirely by our eyes' inability to see anything shorter. The 300 nm limit is set by the daylight spectrum. Wavelengths below 300 nm cannot penetrate the atmosphere. The radiation then has to pass through the glass of the museum window (or the glass envelope round a tungsten or fluorescent lamp). This removes some more UV, so that the effective limit is near 325 nm. Thus glass removes some UV, but the band between 325 and 400 nm gets through (*Figure 1*).

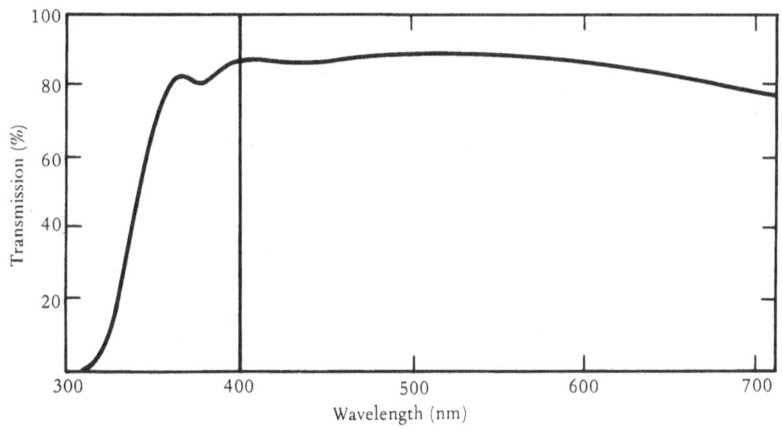

Figure 1
Transmission of a typical window glass (Pilkington 6 mm Float Glass). The height of the curve above the base line at any wavelength is a measure of the proportion of radiation transmitted at that wavelength. Thus no radiation gets through the glass at 300 nm, whereas throughout the visible range between 80 and 90% of the light is transmitted, except at the extreme red end, where transmission falls to just under 75%

Wave-length (nm)	*Trans-mission* (%)								
300	0	400	87	500	88.5	600	86	700	78.5
10	0.5	10	86.5	10	88.5	10	85	10	77.5
20	3.5	20	86	20	89	20	84.5	20	76.5
30	19	30	86	30	89	30	84	30	76
40	43	40	86.5	40	88.5	40	83	40	75
350	68	450	87	550	88	650	82.5	750	74.5
60	79.5	60	87	60	88	60	81.5	60	73.5
70	83	70	87.5	70	87.5	70	81		
80	81	80	88	80	87	80	80		
90	85	90	88	90	86.5	90	79.5		

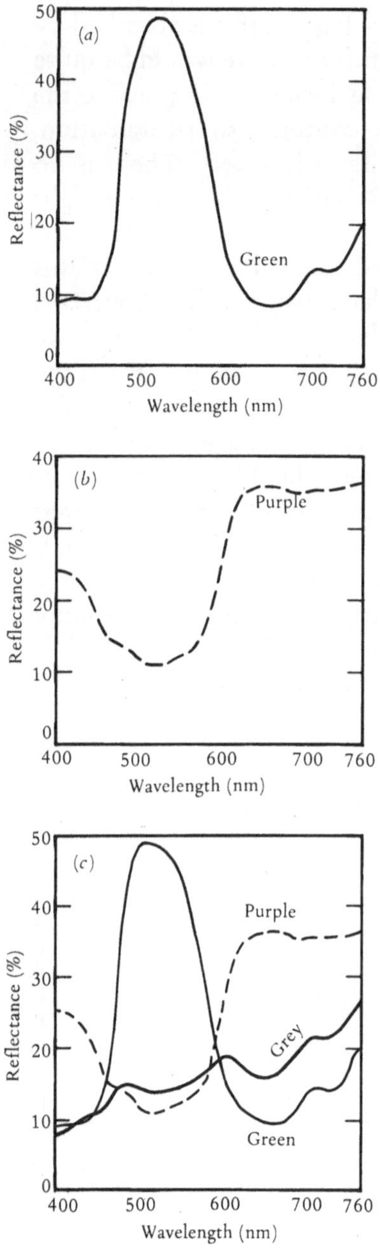

Figure 2
Reflectance curves of the three watercolours in *Plate 2*

We have now divided radiation from light sources into three regions, the UV, the visible and the IR, and we have added a wavelength scale. This scale has already been used to show the transmission of window glass, and it has many other useful functions.

The vertical scale can be assigned to measure any quantity which varies with wavelength. Consider the colour of light reflected from a brushstroke of green paint (*Plate 2*). White light falls on the brushstroke and green light is reflected. In practice all colours are reflected, green merely more than others. The light which is not reflected is absorbed and thereby heats up the paint by a small amount. A very tiny proportion of this absorbed light may even be used up in causing a chemical change to take place in the pigment. Green is in the middle of the spectrum, so less of the light at either end, blue and red, is reflected.

If we measure the amount of light reflected at each wavelength as a percentage of the light falling on the brushstroke at that wavelength, we can plot our figures in the form of a 'reflectance curve' (*Figure 2(a)*). This states quantitatively what we have already described.

Purple colours do the opposite to green: they reflect light predominantly at both blue and red ends of the spectrum, with less reflected in the middle (*Figure 2(b)*). Consequently, if the two pigments, green and purple, are mixed, light from all parts of the spectrum will be partly reflected and partly absorbed, so that one could get a fairly neutral grey (*Figure 2(c)*). For this reason green and purple are said to be complementary colours.

A white-light source emits energy throughout the spectrum: there are wavelengths of all colours. Following a similar procedure to our reflectance spectrum, if we now plot the amount of energy emitted by some particular white-light source at each wavelength through the spectrum we might expect to get a more or less level line. *Figure 59* (p. 165) shows that this is far from being the case. There are whites and whites, and the eye is amazingly adaptable to them*.

'White' is a term with no absolute meaning to us. Our eyes have the power to 'adapt' to any light source which emits throughout the spectrum, and even to some sources with large

*The figure in Part II referred to (*Figure 59*) actually plots power, not energy, against wavelength. We could if we wished plot energy, but we would then have to measure the energy emitted over a stated or assumed period of time. We are back to power, because energy per unit time is power. For example we measure the power consumed by an electric light bulb in watts, which are joules (energy units) per second (unit time).

gaps in the spectrum. In a room lit by one type of light both it and 'white' objects illuminated by it will appear white. Once at dawn I entered a hotel bathroom, flicked the light on and noticed a blue patch on the wall. A second glance showed this to be a frosted glass window lit by weak overcast daylight. My eyes were adapted to the tungsten light. After breakfast the situation was reversed. Now the bathroom was lit predominantly by daylight, to which my eyes were adapted. The window looked white, and any area shadowed from daylight but lit by the tungsten bathroom light looked brown-yellow.

The whiteness of white is further discussed under 'Colour rendering' on p. 55.

Summary

The spectrum of radiation from museum light sources (daylight, fluorescent and tungsten lamps, etc.) can be divided into three regions, by wavelength: ultraviolet radiation (300–400 nm), light or visible radiation (400–760 nm) and infrared radiation (beyond 760 nm).

THE BASIC LIGHT SOURCES

At the time of writing there are three types of light source suitable for general lighting in museums: tungsten, fluorescent and metal halide lamps.

At one time artificial light could only be produced by burning, as in a candle. But non-inflammable materials can be made to give out light if they are heated strongly enough. The ordinary domestic electric light bulb is referred to as the *tungsten* or incandescent lamp because it gives out light from a coiled tungsten filament heated to about 2700 °C by passing an electric current through it. The tungsten lamp comes in a great variety of forms and powers, some with clear glass and some with opal glass envelopes, some with built-in reflectors (*Figure 3. See also Table 17* p. 168). A useful variant, called by some manufacturers the 'Coolbeam', has a so-called dichroic reflector with the property of reflecting visible light forwards but allowing IR radiation to pass through to the back of the lamp.

Most of the electricity which passes through an ordinary tungsten lamp is converted into heat (94 per cent for a 100 watt lamp) not light. If we could heat the tungsten to a higher temperature its efficiency would increase but it would also evaporate too fast. The filament would thin out and break and

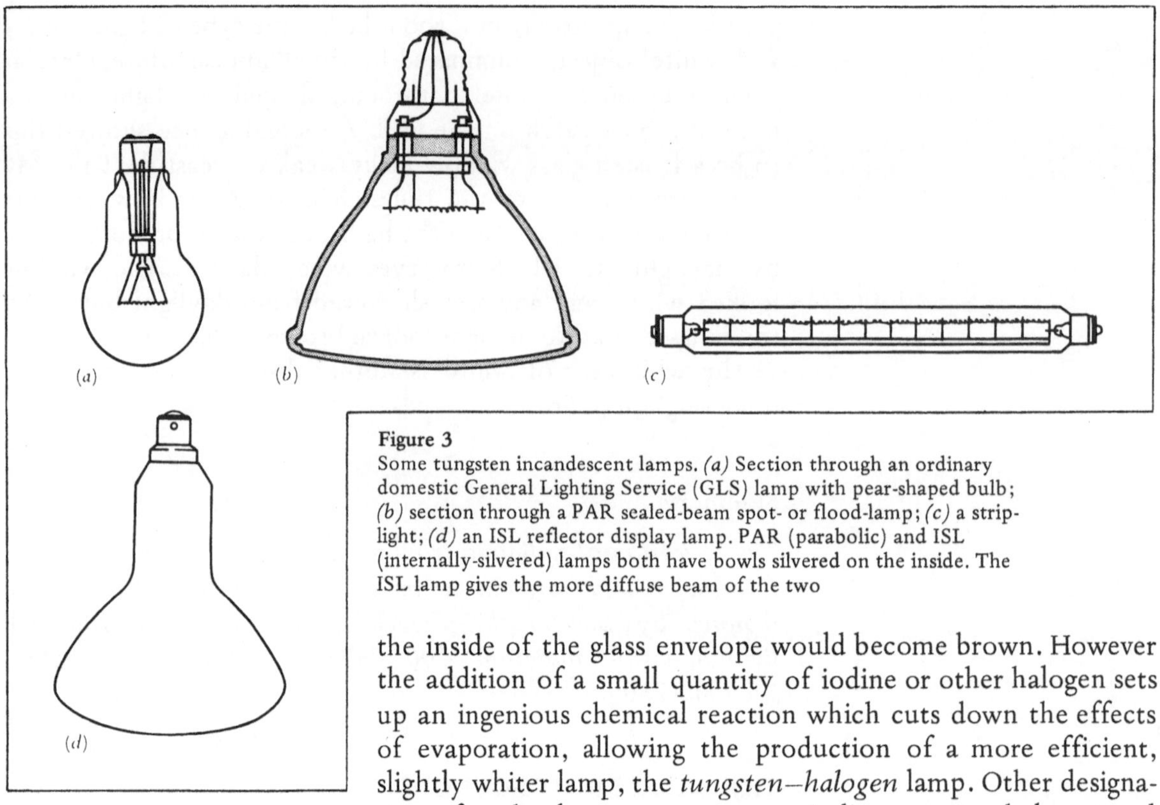

Figure 3
Some tungsten incandescent lamps. *(a)* Section through an ordinary domestic General Lighting Service (GLS) lamp with pear-shaped bulb; *(b)* section through a PAR sealed-beam spot- or flood-lamp; *(c)* a strip-light; *(d)* an ISL reflector display lamp. PAR (parabolic) and ISL (internally-silvered) lamps both have bowls silvered on the inside. The ISL lamp gives the more diffuse beam of the two

the inside of the glass envelope would become brown. However the addition of a small quantity of iodine or other halogen sets up an ingenious chemical reaction which cuts down the effects of evaporation, allowing the production of a more efficient, slightly whiter lamp, the *tungsten–halogen* lamp. Other designations for the lamp are tungsten–iodine, quartz–halogen and quartz–iodine, but in this book the term tungsten–halogen will be used. Because of the high operating temperature, quartz must be used in place of glass for the envelope. But quartz is transparent to the emitted UV, and a small amount of this is of wavelength less than 300 nm. This is very potent and must be eliminated. Fortunately ordinary glass, transparent to longer-wavelength UV (*Figure 1*), completely blocks this extra-short emission. Therefore tungsten–halogen lamps must have ordinary glass filters in front of them, but these must be so fitted as to allow free air ventilation and so prevent overheating (*Figure 4*).

Light can also be produced by passing electricity through a gas such as mercury vapour or sodium vapour or neon. Such lamps are widely used in street lighting and advertising but they are not sufficiently white to be used for lighting exhibits. However if a tubular lamp containing mercury vapour has the inside of its glass coated with a mixture of powders capable of fluorescing in the radiation emitted by the mercury vapour we get a *fluorescent* lamp (*Figure 5*). A substance is said to fluoresce if it absorbs radiation and re-emits it at a longer wavelength. In this case the UV radiation emitted by the mercury is absorbed by the fluorescent powders and re-emitted as visible light. Some fluorescent lamps are good at colour rendering, but others are bad (*see* p. 54).

Fluorescent lamps are cool enough to touch at the centre of the tube but not at either end where the heated filaments are. However quite a lot of heat also comes from the control equipment which must be used with them, for fluorescent lamps cannot be connected directly to mains electricity supply as can tungsten lamps. Such extra heating should be avoided in exhibition cases. The lamp unit can be on the case, but separated from it by glass. Control equipment can often be separated from the lamp and placed elsewhere.

Recently there has been much development of a third type of lamp, the *metal–halide* lamp. This is basically a modification of the high-pressure mercury bulb lamp to improve its colour. In certain cases the improvement is so great that the lamps, which are economical in electricity, become suitable for museum use. The colour-rendering improvements are obtained by adding

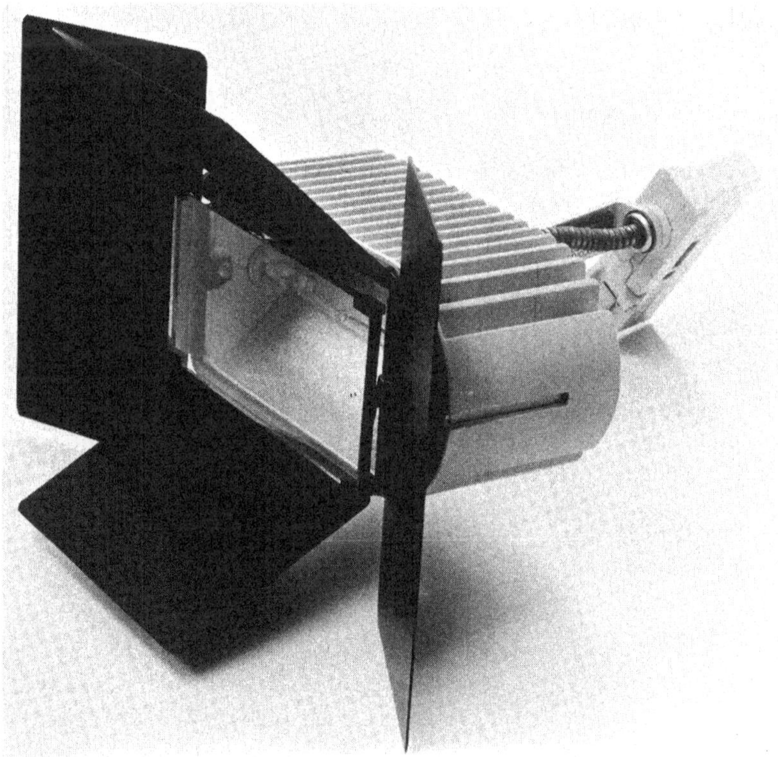

Figure 4
300-watt tungsten-halogen lamp in a Concord Lytespan fitting. The radiation emitted of wavelengths shorter than 300 nm makes it important to use a (heat-resistant) glass as a filter. Note the gap all round the glass for ventilation

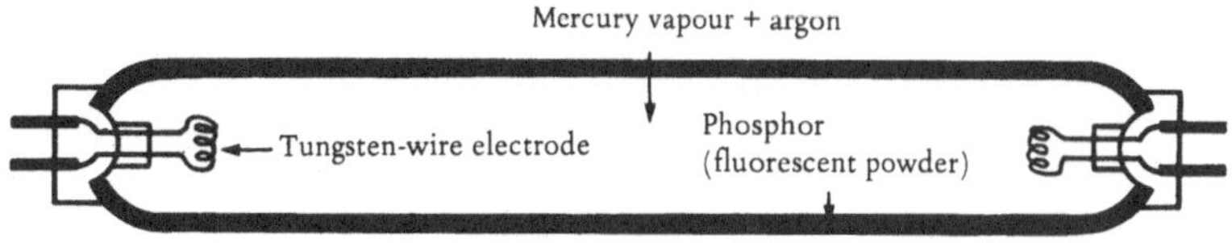

Figure 5
Section through a typical fluorescent lamp

small amounts of metal halide to the mercury. Colour can be further modified by surrounding the basic lamp with an envelope coated with fluorescent powder, as in the Thorn Kolorarc. We can expect an increasing use of metal–halide lamps in the museum, though at the time of writing few are made in sizes below 250 watts. The Osram/Wotan HMI lamp is, however, very popular for film and television. Emitting much less heat than an equivalent tungsten lamp, it is preferable for conservation.

The xenon lamp, though of good colour rendering, has not been included as a museum lamp because of its high UV output and bulky control gear. It is also generally too powerful for museum use. However as a flash lamp it can be recommended in photography (p. 47).

More lamp details are to be found in Part II p. 171.

Summary

Tungsten lamps give out light from a coil of tungsten wire heated electrically. The tungsten–iodine lamp is a variation designed for higher efficiency: a plain glass filter should be used in front of it. Fluorescent lamps emit light both from mercury vapour and from fluorescent powders; some are good at colour rendering. Metal–halide lamps are mercury lamps modified with additions which give some of them good colour rendering.

COLOURS AND MATERIALS WHICH CHANGE

It would be out of the question to compile a comprehensive list of the dyed fabrics found in museums, whether textile or tapestry, which change under the action of light, since a great variety were used, often difficult to identify and often prepared in significantly different ways, and there are none which are completely immune[2–4]. It is simplest to regard all natural textiles, with the obvious exception of those made of metal thread, as subject to the action of light[5–13]. Examination of parts masked from light, such as the backs of carpets, will soon show that even colours which might not appear to have faded have indeed changed. In many cases, with and without fading, the textile will have been weakened by light (*Plates 3* and *4*)[14–17].

TABLE 1. Lightfastness Categories for Some Common Pigments, Old and New

Chemical description	*Pigment names*
EXTREMELY PERMANENT COLOURS	
Arsenic sulphide (As_2S_3)	Orpiment
Calcium carbonate	Whiting, Lime White
Carbon blacks	Bone Black, Charcoal Grey, Chinese and Indian Ink, Ivory Black, Lamp Black
Chromium sesquioxide, hydrated	Viridian
Chromium sesquioxide, anhydrous	Chromium Oxide Green
Cobalt aluminate (often + phosphate)	Cobalt Blue, Thénard's Blue
Cobalt phosphate/arsenate	Cobalt Violet
Cobaltous stannate	Cerulean Blue
Copper carbonates (basic)	Azurite and Malachite
Iron oxides	Yellow and Gold Ochres, Raw and Burnt Sienas and Umbers, Indian Red, Light Red, the syn. Mars colours, Terre Verte, Venetian Red
Lead antimonate	Naples Yellow
Lead stannate	Lead-tin Yellow
Ultramarine (natural and synthetic)	Ultramarine
White pigments:	
Lead carbonate (basic)	Flake White, Silver White
Titanium oxide (Rutile)	Titanium White, Permanent White
Zinc oxide	Zinc White
DURABLE COLOURS	
Arsenic disulphide	Realgar
Barium chromate	Barium Yellow, Lemon Yellow
Barium manganate on sulphate	Manganese Blue
Cadmium sulphides and selenides	Cadmium Orange, Red and Yellow (Cadmium Yellow Pale is rather less durable)
Ferric ferrocyanide (and related compounds)	Prussian Blue
Mercury sulphide	Vermilion, Cinnabar
Zinc chromate	Zinc Yellow
Natural organics:	
Alizarin + derivatives	Alizarin Crimson, Madder Lake, Rose Madder
Synthetic organics:	
Copper phthalocyanine	Monastral Blue
Copper phthalocyanine, chlorinated	Monastral Green
Alizarin	
Arylamides	
Hansa Yellow	
Para Red	(Marketed under various trade names)
Quinacridone reds and violets	
Tartrazine Yellow	
Toluidine Red (Harrison Red)	
MODERATELY DURABLE COLOURS	
Copper acetate	Verdigris
Copper resinate	Verdigris or other copper compound dissolved in oil/resin
Lead chromate, neutral	Chrome Yellow
Lead chromate, basic	Chrome Red
Smalt	A cobalt glass
Natural organics:	
Gamboge	
Indigo	

(continued)

TABLE 1 *(continued)*

FUGITIVE COLOURS
Vandyke Brown and other natural bituminous mixtures (e.g. Asphaltum, Bistre). Red and yellow lakes of plant and animal origin, other than those mentioned specifically
Synthetic lakes:
Eosine
Perkin's Mauve
Rhodamine
Triphenyl methane derivatives (e.g. Magenta)
(all in this Fugitive category are organics)

Notes
1. This list is concerned with lightfastness only. The following pigments are affected by air pollutants to an extent greatly influenced by thickness and type of medium in which they are embedded.
By hydrogen sulphide: all the lead pigments (basic carbonate, chromates, and antimonate).
By acid gases (e.g. SO_2): Ultramarine, Lime White.
By alkalis: Prussian Blue.
Another cause of impermanence is pigment-medium interaction, usually in the presence of light. Thus Zinc White and Titanium White (even the preferred form, Rutile) may cause chalking. In a different category is the blue glass, Smalt, which because of its high alkalinity can become seriously discolored by interaction with oil, but appears highly permanent in watercolour.
2. A natural caution has prevented any of the modern organic synthetics from entering the highest category, though the phthalocyanines and quinacridones at least probably merit top marks for durability, and are certainly a great deal more permanent than indigo or alizarin, natural or synthetic.
3. The four grades, Extremely Permanent, Durable, Moderately Durable, and Fugitive are those used by Winsor and Newton in their valuable *Notes on the Composition and Permanence of Artists' Colours*, but the grading itself has been coordinated from a number of sources and with essential advice from Joyce Plesters. While at least four grades seem necessary since fastness varies so enormously, the placing of pigments in the two middle grades cannot be done with certainty. These two middle grades would probably comprise Feller's Class B (p. 175), though some of the modern organics might merit Class A. His Class A pigments are Extremely Permanent and Class C are Fugitive.
4. A colour which fades towards transparency, like a red lake, will obviously disappear visually more quickly in a thin wash of watercolour than in an oil impasto. Colour change in most cases involves fading, but Vermilion blackens and Copper Resinate turns from transparent green to opaque brown. Vermilion is here classified under Durable Colours because, although it is susceptible to discolouration by conversion to the black form of mercuric sulphide on exposure to light, this occurs comparatively rarely in paintings. The dyestuff Indigo varies in lightfastness with method of use. As a pigment for painting it is only moderately durable; mordanted on wool or silk it shows good lightfastness (as witness its survival in old tapestries), but on cotton and other cellulosics it is comparatively fugitive.

The colours in miniatures, in watercolours (*Plate 5*) and in other forms of art on paper are more susceptible to fading than when the same pigments are used in oils or tempera. The sensitivity of Japanese prints has recently been emphasised by Feller[18]. It is conceivable that locking pigment up in a thick medium may protect it from traces of pollutant gas (certainly not significantly from oxygen[19]), but the main cause of the difference between watercolours and oils lies surely in the amount of light received throughout the paint. In true watercolours (those with no white pigment) all the pigment particles are fully exposed to the light. This may also be true of, for example, the red-lake glazes used for flesh tones in oil paintings and these are notorious for fading. But on oil paintings there will usually be many areas with pigment in reserve, that is to say shielded or partly shielded from the light. The Impressionists tended to use (a) a single layer of paint, (b) applied thicker than strictly necessary, and (c) often mixed with white. For all these three reasons, even if the top layer of colour disappears there will be unaffected layers below, though these may be masked by the faded top layer.

There is no good modern review on the deterioration of pigments under light for the simple reason that the majority of them are too stable for measurements to be made in a reasonable time in the laboratory. Accelerated tests usually distort results beyond usefulness. Among the older sources, Russell and Abney's classical experiments published in 1888 on the action of light on watercolours can best be digested through an article by Brommelle[20]. *Table 1* summarises experience in testing and use. From this it can be seen that perhaps the majority of the pigments used in miniatures, watercolours and prints are more or less stable to light, though an important number are not.

Colour changes in oil paintings are particularly interesting just because with this supremely important group one can say that most of the colours are fast. Nevertheless the widespread use of the pigment known as 'copper resinate' for the green of leaves and grass in Italian and Flemish paintings has affected a high proportion of the paintings in national collections. Copper resinate can be made by heating almost any copper compound, usually verdigris, with resin or a mixture of oil and resin. The copper compound actually dissolves to form a clear transparent bluish green, turning yellowish if the heating is prolonged. This was used as a glaze. Exposure to light turns it from a transparent green to an opaque brown.

There are some precise examples where copper resinate has been protected at the edge of a painting by the frame and therefore has remained green while the exposed areas have become opaque brown. In certain cases the top layer which is attacked first has, by itself becoming opaque to light, protected the lower layer from change (*Plates 7 and 8*). Unchanged green has remained, though invisible to the eye, underneath the brown.

Other colours particularly susceptible to change are the red lakes, such as madder, which were used for flesh tones (*Plate 9*). A spectrometric study has recently been made of the fading of alizarin by Johnston-Feller[21,22].

The flower pieces popular in 17th-century Holland contain a variety of greens to suit differences in leaf colour. Unfortunately,

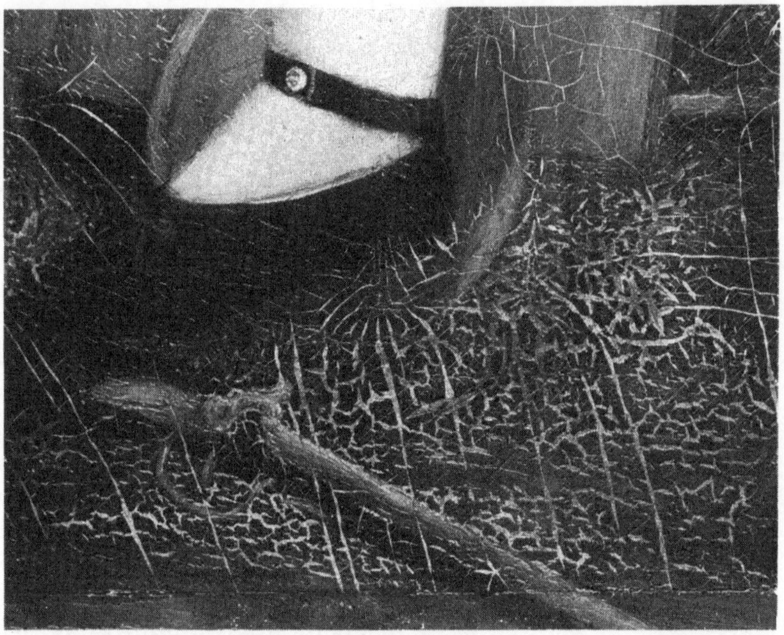

Figure 6
An effect of light on the paint medium. In this detail from a 19-century painting on canvas the frame has protected the bottom edge of the painting from developing drying cracks[23]

one system was to glaze a blue underpaint with a yellow lake such as quercitron. Light may bleach or destroy this glaze leaving a bright blue leaf (*Plate 10*).

It is difficult to assess the extent to which light has damaged the paint medium: linseed oil, egg tempera, gum or glue or whatever it happens to be (*Figure 6*). We know that light does eventually destroy all such organic media, and we know plenty of paintings which have 'worn thin'. But other explanations prevail, a common one being to blame the last restorer for being too heavy handed. We shall not know the answer until we devise and use methods for measuring the rate of attack by light on mature paint in museums[24–26].

It is perhaps not often appreciated that the paint on a car roof parked out of doors in Europe gets exposed to the same amount of light in three or four years that an old master has taken 500 years to receive, and this does not take into account the important extra exposure to UV that the car gets*.

Natural history specimens also need protection from the effects of light, which makes fur, feathers, scales and skin fade and weaken. We tend to forget the constant replacement of the outer coatings of all animals, including ourselves, which goes on during life but stops at death.

It has been emphasised on the very first page that all natural

*Average annual exposure in London, Stockholm or Leningrad 100 Mlx h, cut down to 75 Mlx h per year to allow for part-shading. 500 years at ½ Mlx h = 250 Mlx h. (*See Figure 61*, p. 162.)

organic material (and all synthetic organic material, but sometimes to a much smaller extent) is affected by light. There is no need to compile a long list, but one other immensely important material – paper – must be picked out. Paper is sensitive particularly to UV radiation[17,27], as are the other writing materials of antiquity other than stone and clay. Carbon black in its various forms is the only permanent black ink.

Summary

All textiles are subject to damage by light, as are many of the colours used in miniatures, watercolours and art on paper. Natural history specimens are also sensitive. Oil paintings change more slowly, but in important ways. The paint medium, whether oil, egg, gum or glue, is certainly damaged by light, but the extent of damage is difficult to assess.

DAMAGE CAUSED BY UV AND VISIBLE RADIATION

If we had a tunable light source which we could adjust to emit its radiant energy at any wavelength, and if we wanted to fade a colour as quickly as possible we would in the average case tune it to emit all its energy in the UV. This is because damage is related to wavelength. UV radiation causes more damage in general than the same amount of blue radiation, which causes more damage than yellow radiation. One can confidently assume that in the museum red light never causes any photochemical damage (damage due to chemical change by radiation).

But there is much less UV than visible radiation in all light sources, even in daylight. Balancing these two factors, the extra potency of UV against the smaller quantity of UV, it is still not possible to answer with any confidence the question, 'Which causes more damage, visible or UV radiation?' because many materials are only faded by radiation at certain wavelengths. However it is safe to assume that, under unfiltered daylight through glass, more damage will be caused in general by the small quantity of UV than by the whole of the visible radiation.

One may also generalise along the following lines. A very fugitive material will be damaged by either visible or UV radiation, and, since visible radiation is more plentiful, most of the damage will be done by the visible. A material which is fairly fast but nevertheless susceptible in the long run may be secure against most of the visible spectrum, and therefore will be changed mainly or wholly by UV radiation. This has been shown true of a large number of dyes (*Figure 66*).

A colourless material such as a varnish or paint medium, by virtue of its colourlessness, hardly absorbs any visible radiation but may absorb UV quite strongly. In such cases deterioration is also likely to be caused mainly by UV.

Thus, for the more sensitive dyes and pigments, though UV protection is important, this measure will not get us very far. Of much greater importance is the reduction of visible radiation. In contrast the more stable exhibits, such as oil paintings, stand to benefit very greatly by elimination of UV radiation, though visible radiation should obviously be controlled to reasonable levels, especially since the occasional fugitive material is to be found in them.

Summary

In the museum we must deal with both UV and visible radiation.

UV RADIATION AND HOW TO DEAL WITH IT

An important bit of information to be gleaned from *Figure 59*, is that there is much higher proportion of UV energy in daylight radiaion through glass than in tungsten radiation, characteristically about six times as much.

If an exhibit is lit during the day by daylight and after dark by tungsten light to the same brightness, it will be receiving about six times as much UV energy during an hour of daylight as it does in an hour of tungsten light. Tungsten lamps emit too little UV to require a filter.

Nearly all fluorescent lamps emit more UV than does a tungsten lamp, but none – except for a special high-UV lamp called 'Artificial Daylight' – emit as high a proportion as does daylight.

Therefore for UV protection we tackle first the daylight then the fluorescent lamps. We remove UV radiation by passing the light, before it reaches the exhibits, through a material transparent to visible light but opaque to UV.

The ideal UV-absorbing filter will prevent all the UV down to 400 nm from passing through, but will not hinder the passage of any visible light (*Figure 7(a)*). This is a lot to ask, and there is no filter made only of glass which will do this job well. However several kinds of plastic UV-absorbing filter of very satisfactory quality are available (*Figure 7(b) and (c)*). The filters

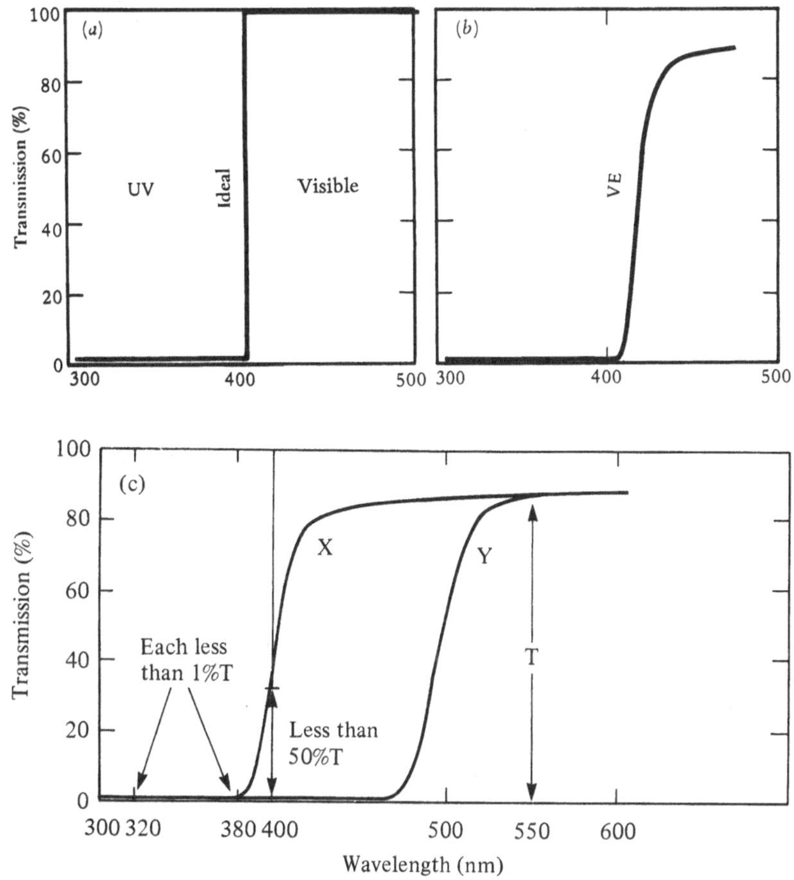

Figure 7

(a) The ideal UV-absorbing filter would transmit no radiation of wavelength shorter than 400 nm, while being completely transparent to all visible radiation. *(b)* ICE VE Perspex, an acrylic filter which eliminates UV radiation. Because it also removes a little violet light it has a slight yellow tint. *(c)* Curve X is of a typical satisfactory UV filter. This passes a suggested specification: Transmission at 400 nm should be less than half of transmission in the middle of the visible range: at 550 nm. Transmission at 320 and at 380 nm should be less than 1/100th of that at 550 nm. Y is of a filter which removes all UV and also all blue light. It can be used to give extra protection to delicate material whose colour is not important (e.g. black on white documents): it has a strong yellow colour.
Note. CIE proposals to define two UV bands, UV-A from 380–315 nm and UV-B from 315–280 nm, are useful in emphasising that short-wavelength UV is more damaging than long. But a good UV-absorbing filter should cut off at 400 not 380 nm. The residual sensitivity of the eye between 380 and 400 nm is too small to affect colour rendering in adapted siuations

available vary from country to country. A list for the U.K. is available from the Museums Association[28].

These filters must extend completely across windows, skylights or light fittings so that all light passes through them. They are manufactured in three forms:

1 Self-supporting acrylic or polycarbonate sheet 3–6 mm thick, either clear or diffusing, which can be used in place of glass in windows and skylights (if fire regulations allow) and for fluorescent lamp diffusers.
2 Thin foil, usually acetate, which can be cut to shape with knife or scissors and laid on glass, or adhered to glass.
3 Varnish.

Another useful form, but less easily available, is a sandwich of glass with a plastic interlayer. The interlayer contains the UV absorber. The glass can be made in security grades. Glass is less easily damaged by cleaning, and also the sealing of the plastic interlayer from air should increase its permanence.

Acrylic sheet is a loose term for polymethyl methacrylate, known variously under the trade names Perspex, Plexiglas, Oroglas, etc. In its normal grade this polymer may have weaker UV-absorbing properties than glass, and so it is important to note the special forms made for UV-absorption: Perspex VE, Oroglas UF3, Plexiglas 201.

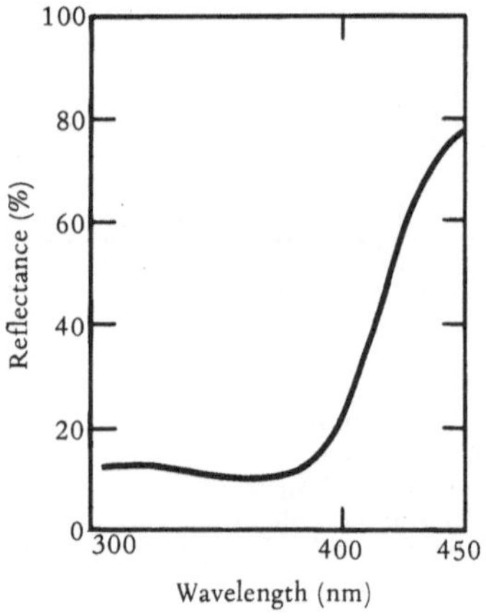

Figure 8
UV reflectance of a Rutile Titanium Dioxide paint

Acrylic and most other plastic sheets are liable to acquire an electrostatic charge if rubbed, and will then attract dust from the air, or even pigment from a pastel or charcoal sketch if this is very close to the sheet. To avoid this, an antistatic varnish should be applied to both sides of the acrylic sheet[29].

Polycarbonate sheet is harder than acrylic and much less affected by static, but again a special UV-absorbing grade is required.

Varnish is often the best answer for side windows. It can be made invisible if it is flowed on by the suppliers, using a special tool. Brush or spray is not satisfactory since thickness cannot be controlled and marks of flow will show.

It is useful to know that white paint, which contains titanium dioxide, is itself a fairly good UV absorber (*Figure 8*). In some cases, especially in the Tropics, it may be possible to ensure that all light entering a room is reflected at least once from a white wall. This will solve the UV problem. Titanium dioxide is the usual white pigment, but lead and zinc white are also good absorbers. Whitewash (chalk) is not effective.

The acrylic and polycarbonate sheets seem to be of high permanence, and can be expected to last at least ten years, even in moderately exposed situations. Foil and varnish filters, to err on the safe side, should be regarded as having the same sort of permanence as the paint on the walls, and should therefore be checked at not more than five-year intervals, using a UV monitor (*see* p. 21).

It has been mentioned that most fluorescent lamps also emit worrying amounts of UV radiation, though less strongly than daylight. The present situation is that all fluorescent lamps except the Philips 93 & 94 require UV filters. Some of the diffusing plastics used in fluorescent-lamp fittings act as reasonable UV filters. The only way to check whether this is so in any particular case is to use a UV monitor (p. 21).

Occasionally UV radiation is actually used in museums for fluorescent effects: to show for example the fluorescence of minerals in a geological museum or to brighten pigments in some modern works of art. The advantages in these special cases must be balanced against the dangers. In all other situations UV radiation contributes nothing to appearance and can cause irreparable damage. It should therefore be eliminated in the ways described above.

Summary

Daylight has the highest proportion of UV radiation and therefore must be filtered. Tungsten light does not need to be filtered. Fluorescent lamps have less UV than daylight but require to be filtered. There is no good glass UV filter. All are of plastic whether sheet or varnish.

MEASURING UV AND VISIBLE RADIATION

Since both UV and visible radiation are forms of energy, any scientist would point out that the most direct way of measuring either kind of radiation is to measure the rate at which the energy, whether UV or visible radiation, falls on a standard area (a square metre for example) where the exhibit is. In practice this is not done.

For visible radiation we use an instrument, the light meter or lux meter, which measures energy not directly but as the eye sees it. The eye does not see UV or IR so the meter does not respond to these ranges. The eye is more sensitive to green than to blue or red and so is the meter. This is because the instrument is most commonly used for relating light to visual tasks such as office work or learning at school. The best meters correspond very closely to the human eye in sensitivity (*Figure 63*, p. 177).

The scale of the instrument is graduated in *lux** and the measurement thus made is called the *illuminance* (formerly illumination value or level).

A light meter consists of a suitably protected sensitive surface, or photocell, connected electrically to a meter (*Figure 9*).

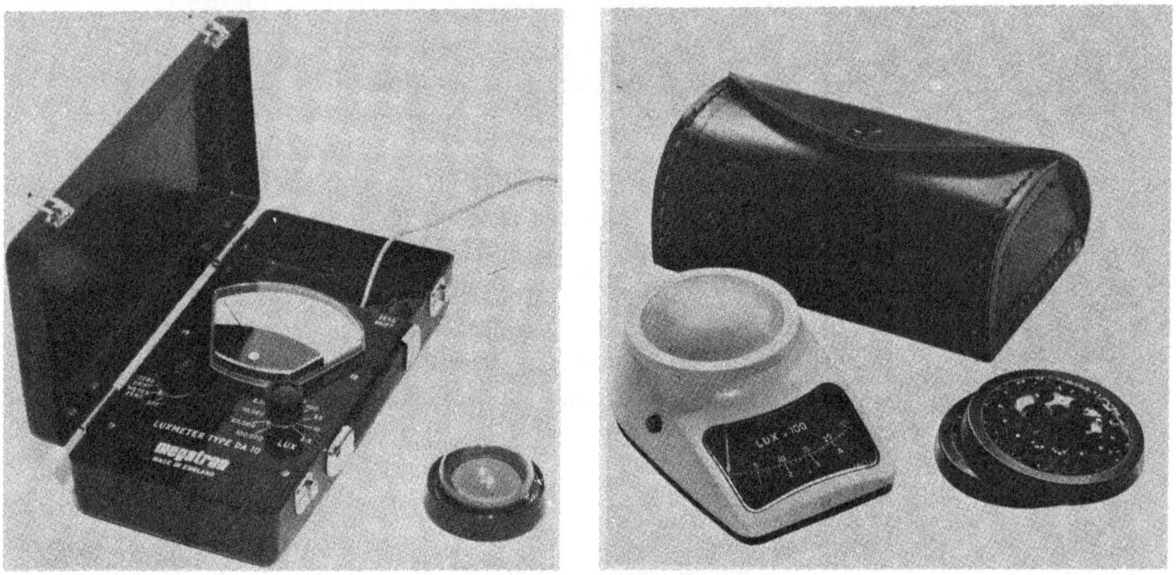

(a) *(b)*

Figure 9
(a) A battery operated selenium photocell light meter (Megatron DA 10), with highest range 0–100 000 lux and lowest 0–2.5 lux. Box 30 cm long. *(b)* A pocket light meter reading up to 5000 lux (Megatron S 511). The higher ranges are selected by putting masks over the photocell. Case 13 cm long

The selenium photocell, which is the commonest type, requires no batteries. In measurement the photocell is placed close to the surface of the exhibit or at an equivalent position, facing the light just as the exhibit does. For a painting, for example, the photocell surface should be parallel to the paint surface.

The UV monitor is designed to complement the light meter (*Figure 10*). We have already measured the illuminance and now we want to find out the *proportion* of UV in the light. The UV monitor will answer such questions as: 'Does this light source need a UV filter?', 'Does this filter need replacing?', 'How does the UV in this light source compare with the UV in tungsten light?'. Furthermore we want to get this measurement irrespective of the distance we are from the light source.

The measurement is actually in microwatts of UV radiation per lumen, but this need not concern most users. Readings for tungsten lights are around 60–80, and since we have said that tungsten lamps do not require UV filters, this is the highest tolerable level. Any source with a reading above about 75 therefore requires filtering. *Table 19* (p. 175) shows UV values.

*An *illuminance* of 1 *lux* equals 1 *lumen* per square metre. This unit of luminous flux, the lumen, could be described as the radiant energy flux as perceived by the human eye. Note that the old English unit, the foot-candle, equals one lumen per square foot. It makes no sense to talk about illuminance just in lumens because 1 lumen per square metre (1 lux) equals about 1/10 lumens per square foot (1/10 foot-candles).

Figure 10
A UV monitor which measures the proportion of UV radiation in the light as microwatts of UV radiation per lumen (of visible light). If a light source emits more than about 75 μW/lm it requires a UV-absorbing filter. There are two windows in the black panel on top of the instrument, one passing UV radiation and one visible to two photosensitive devices beneath. When the button on the right top is pressed one of two red light-emitting diodes on the centre of the side panel lights up. With the button depressed the knob is turned until the light flicks between one and the other of these diodes. The pointer on the knob then gives the proportion of UV in the light source. A lower-range model reading up to 300 is also available and would be the normal choice.

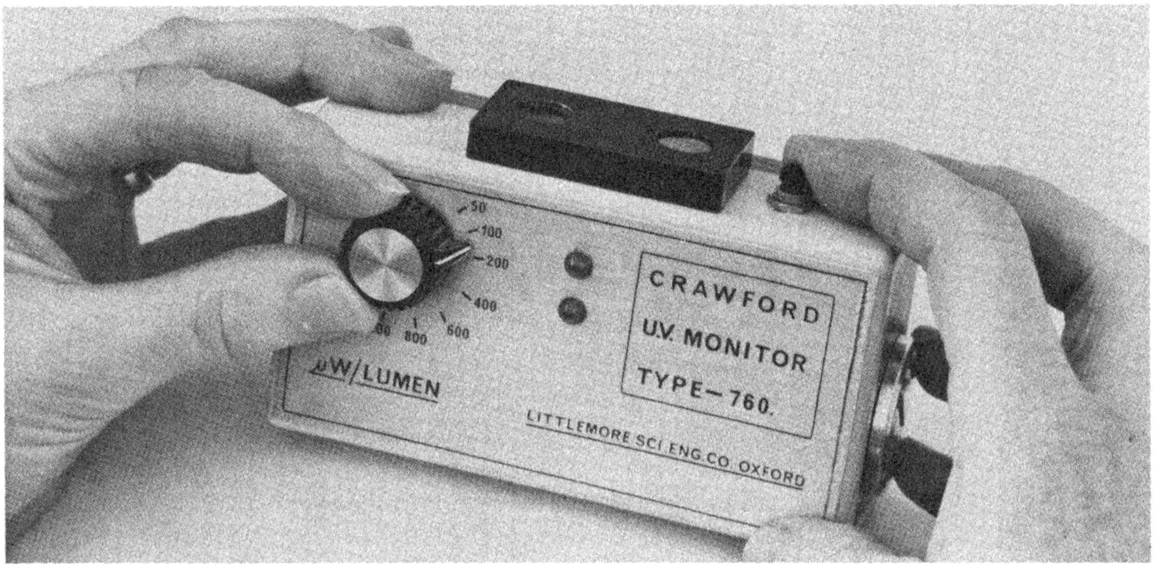

THE RECIPROCITY LAW

There is an acceptable rule-of-thumb which says that the same amount of damage will be produced whether by a strong light in a short time or a weak light in a long time: if we halve the illuminance we halve the rate of damage. This is a loose statement of the reciprocity law.

More precisely, the reciprocity law teaches us that light, like high-energy radiation, acts cumulatively, so that it is the total dose, the *exposure*, which matters. The *exposure* is the simple product of illuminance and time. For light to produce a given effect the exposure must be constant. If illuminance is measured in lux and time in hours, 100 lux on a painting for 5 hours gives it an exposure of 500 lux hours. 50 lux for 10 hours would give it the same exposure.

We use the reciprocity law every time we take a photograph, and in many experimental situations it holds exactly, though it may break down at very high or very low illuminances. In the museum, although there must be individual cases where the law does not hold, we are justified in using it for the general lighting situation.

Note that the reciprocity law does *not* state that twice the exposure will cause twice the amount of fading. Rates of fading commonly decrease with time, until there comes a point when no more fadeable material is left. The rate of fading is the obviously zero.

Summary

In reducing damage by light our most effective strategy is clearly to reduce both illuminance and time of exposure.

CONTROLLING VISIBLE RADIATION

Both UV and visible radiation can cause colour change and surface deterioration (*see* p. 15). It is not sufficient only to remove UV radiation.

We cannot eliminate the visible radiation as we have done the UV because we would then be left in darkness. We have to accept that a certain amount of damage is caused by the very act of display. And since we have to balance, by judgement rather than by scientific formula, two incommensurables – the amount of light needed for looking at exhibits against the damage which it causes – we are now in the realms of controversy.

To be prepared for this controversy we should have a good understanding of the visual significance of the exhibits, of the way the eye sees things, of the way exhibits can be lit, and of the rates of damage actually caused by light. Our knowledge in some of these areas is very poor indeed, so it would seem that we are ill-equipped to make any recommendation. But meanwhile precious objects deteriorate, often under conditions of unconsidered illumination.

Two courses of action must be pursued together. First, reduce illumination to no more than is necessary for proper viewing and reduce time of illumination where possible. Second, find out more about rates of fading in museums.

REDUCING ILLUMINANCE

The 200/50 lux illuminance levels shown in Table 2, or figures close to them, are now recommended, among others, by the U.K. Chartered Institute of Building Services[30], the French National Committee of ICOM[31], ICCROM[32], the U.S.S.R. Ministry of Culture[33], and the Canadian Conservation Institute[34,35]. *See also* Thomson[36], Brommelle and Harris[37], Brommelle[38,39], Harris[40-43], Brawne[44] and Allen[45].

TABLE 2. Recommended Maximum Illuminances

Exhibits	*Maximum illuminance*
Oil and tempera paintings, undyed leather, horn, bone and ivory, oriental lacquer	200 lux
Objects specially sensitive to light, such as textiles, costumes, watercolours, tapestries, prints and drawings, manuscripts, miniatures, paintings in distemper media, wallpapers, gouache, dyed leather. Most natural history exhibits, including botanical specimens, fur and feathers	50 lux

Notes

1 Although objects insensitive to light (e.g. metal, stone, glass, ceramics, jewellery, enamel) and objects in which colour change is not of high importance (e.g. wood) may be illuminated at higher levels, it is rarely necessary to exceed 300 lux. Large differences in illuminance between rooms give rise to adaptation difficulties.

2 Dust and dirt on lamp and reflectors reduce illuminance. Also the light output drops by about 25% during the life of the fluorescent lamp. Therefore in taking measurements under new installations up to +50% allowance can be made.

3 Lighting for restoration, technical examination and photography is not limited by the above Table. 2500 lux is a reasonable upper limit for those relatively brief periods of exposure (p. 44).

Since appreciation of museum objects is mediated mainly by our eyes any sacrifice in illuminance must be considered very seriously.

The 50 lux illuminance level is not at the bottom of what would in the past have been regarded as the normal scale of artificial lighting. In fact in the days before fluorescent lamps 50 lux was supposed a good artificial lighting level. More recently the taste for bright lights has been boosted by cheap energy and industrial interests, so that the task of control is difficult. Museums are expected to be lit as brightly as shops. Nevertheless it has been found that, provided glare is skillfully eliminated, 50 lux gives satisfactory lighting even of small objects with low contrast (*Figure 11*). The danger to fugitive materials in museums is so apparent that the 50 lux level has been fairly generally accepted as a necessary measure for conservation.

Criticism has been stronger against the maximum for oil and tempera paintings[46]. Whereas many curators have actually seen fugitive colours disappear in the course of years, we have to rely on less direct evidence for easel paintings, though we know that big changes have occurred (*Plates 6–10*). There is also the feeling that, for example, with a fully browned copper resinate on a painting otherwise healthy, we may be 'locking the stable door after the horse has gone'. In contrast to examples where the frame has protected colours on the edge of a picture, there exist countless examples where no such differences are to be found. Critics of lighting limitation may further point out that the attack by light of the paint medium has never been clearly demonstrated, so that a picture with fast pigments only, e.g. certain Rembrandts, may indeed be fast to light.

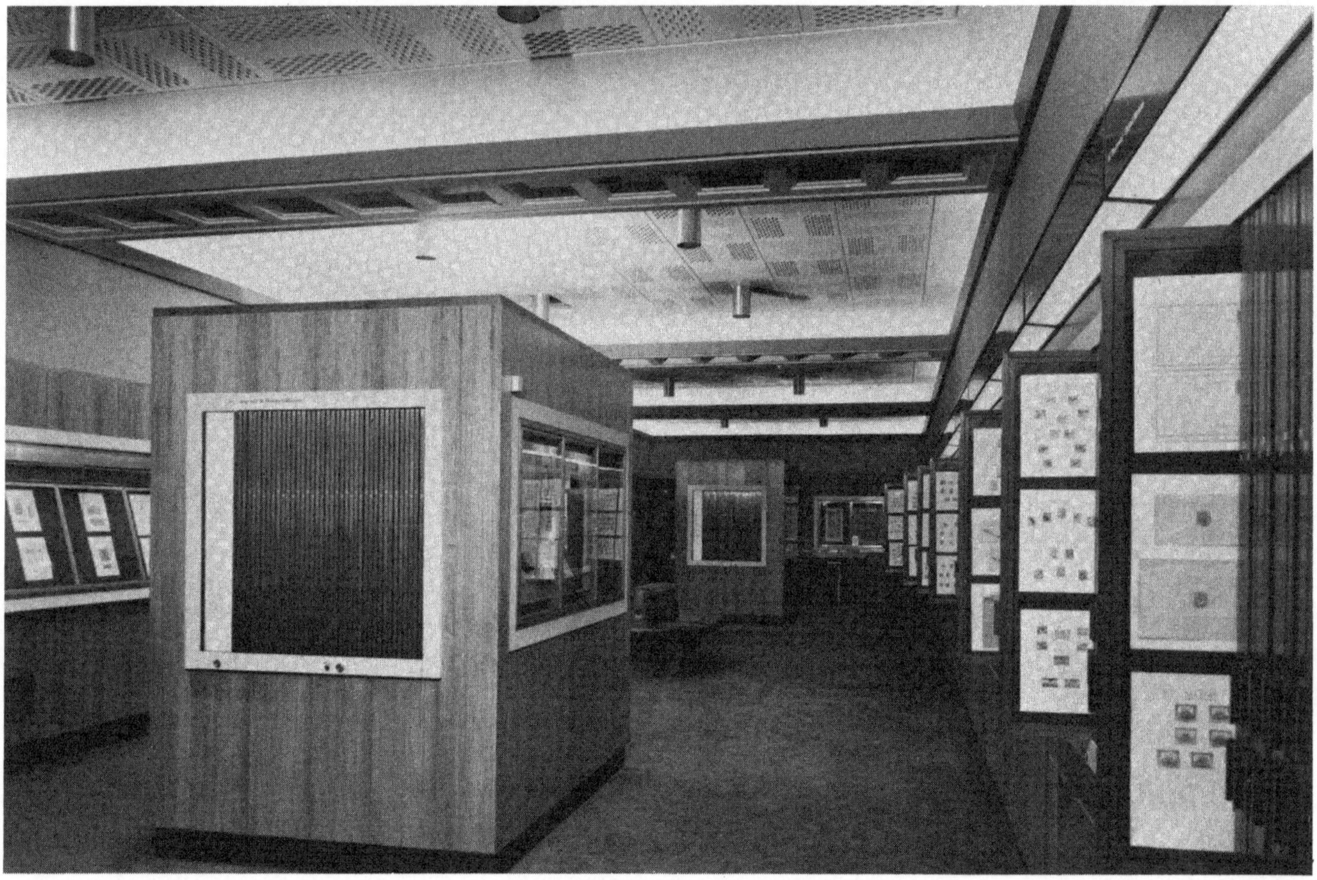

Figure 11
Part of the exhibition in the National Postal Museum, London, illuminated throughout at 50 lux. (By courtesy of the Post Office)

One attempt to investigate public response to illuminance in museums appeared in 1972: an American survey of the public's reaction to all aspects of display in 12 different museums and galleries in the San Francisco–Oakland area[47]. On lighting the survey concluded: 'the quantity of light in the exhibition spaces seemed to be less important than the balance of illumination levels between the works and the remaining space' . . . Users felt that 10–20 foot-candles (100–200 lux) were adequate for the perception of detail in the works of art.' More recently, and in response to the 150 lux limitation for easel paintings, Loe and others at the Bartlett School of Architecture in London, set out to re-investigate specifically the kind of light preferred by viewers. Their conclusion was that 'the preferred artificial lighting conditions for viewing works of art are a painting illuminance approaching 200 lux provided from lamps with good colour rendering characteristics'[48]. In the first edition of *The Museum Environment* the recommended figure was 150 lux. As a result of subsequent experience and particularly the work just quoted, this has been revised upwards to 200 lux.

The conservator who does not wish to accept the 50/200 lux maxima uncritically should at least do one thing: he should arm himself with a light meter and examine all kinds of lighting situations in his museum in order to judge for himself the various levels of illuminance which he finds, and how his viewing is affected by them, by his state of adaptation to the light in the room, by glare from light sources, and by other factors. He may conclude that these factors, glare in particular, are actually more important in determining viewing conditions than the level of illuminance itself, over a very wide range of illuminance.

50 LUX – ARTIFICIAL LIGHT

Let us consider the low level first, since the options are simpler. They are simpler because experience shows that, for 50 lux, artificial light rather than daylight is appropriate. Control of daylight to a closely pre-set level is a difficult and expensive business (p. 28). The 'coolness' (p. 49) of daylight when it has been reduced to 50 lux often gives the impression of gloom, especially when it is highly diffused. No one knows how deeply it has been built into our systems, but ever since our first ancestors sat around fires, and later used oil lamps and candles, the human race has been accustomed to 'warm' light in the home after dark. As a result the warm 50 lux from tungsten lamps appears to be brighter, and certainly more cheerful, than 50 lux of diffused daylight. For the same reason warm rather than cool fluorescent lamps should be chosen for 50 lux situations.

Glare and the adaptation of the eye to the light have been mentioned as the most important factors to control in order successfully to create an apparently bright situation. Glare has been defined as 'the discomfort or impairment of vision experienced when parts of the visual field are excessively bright in

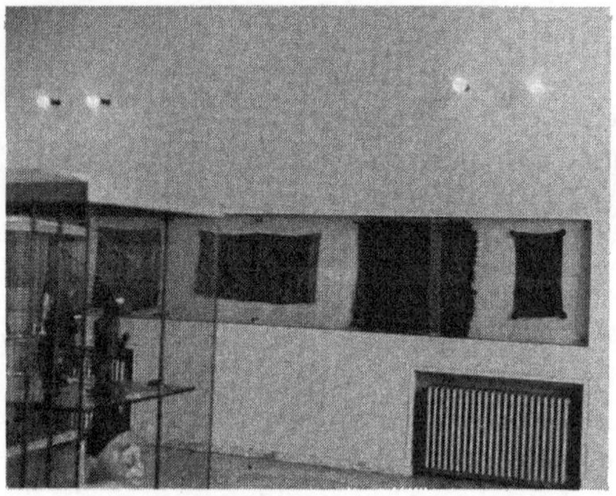

Figure 12
Glare from light sources. Where exhibits are placed all over the room and lit by small bright sources such as spot-lights it is very hard to avoid glare at all viewing positions, though this is a particularly bad example. Note also the radiator below an exhibit

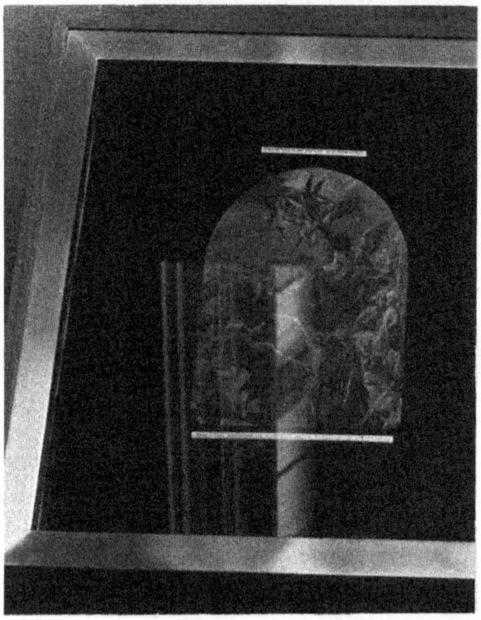

Figure 13
Reflections of roof lights in a desk case, from the normal viewing position. It is very difficult to avoid strong reflections in glass which is below the viewer and tilted upwards

relation to the general surroundings'[49]. In practical museum terms glare has a narrower meaning, since no part of the exhibit itself is likely to be the source of glare. Glare to us means bright spotlights in the corner of the eye (*Figure 12*), a bright window beside an exhibit, reflections in glass or glossy surfaces which obstruct viewing (*Figure 13*), etc. Glare can only be avoided by planning and adjusting the position of the lights, masking lamps from shining in unwanted directions and the use of dark surfaces in critical positions. Many museums consider it justified to remove glass from paintings when air-conditioning is installed, and this is of very great benefit to appearance.

We speak of adaptation of the eye in two senses: adaptation to the strength of the light and adaptation to its colour. Here we are concerned with the former. In bright light the pupil of the eye becomes smaller so that less light reaches the retina. But the eye can adapt itself to brightness over a range of more than 10 000 to 1, which is far more than can be coped with by merely varying the size of a hole. An important part of adaptation lies in the processing which signals from the retina receive on their way to the part of the brain which interprets their message. In just the same way a closed-circuit television camera has two ways of adjusting the brightness of the image on the screen: a diaphragm in the lens of the camera and electronic circuitry for gain control.

The time taken for the eyes to adapt to a new lighting situation is a matter of seconds in the usual indoor or outdoor situation. But, coming into a building from bright sunshine or vice versa, adaptation may take a minute or so. All this should occur before the visitor sees the first exhibit. This implies that adaptation to the lighting of the exhibition room should take place not in the room but in the entrance area of the museum. Furthermore, since adaptation can easily be upset by a view of brighter areas, whether through windows or doors, this should not be allowed to occur. But this does not imply that windows should be banned. There are ways of using windows which avoid both glare and loss of adaptation (p. 34). However the exhibiting of material insensitive to light, such as unpainted stone sculpture, at high illuminance can cause trouble, and the attractive idea of being able to walk straight from an inside gallery to a sculpture garden raises almost insuperable difficulties in light control, even at 200 lux, let alone 50 lux.

The last matter to consider here is the distribution of the light. Designers have sometimes supposed that, if the viewer is kept in darkness and all the light is thrown on the exhibits, these will look brighter by contrast. They may look brighter, and this is certainly an excellent way to eliminate reflections in glass, but it is not the best way to make detail most visible. Consider the two following examples.

To the doctor a very slight shadow on a radiograph of the thorax may indicate tuberculosis. He therefore needs to view the radiograph in conditions optimised for visual acuity. The commonsense arrangement is to put the radiograph transparency on an illuminated viewer and mask it round the edge so that it becomes the only luminous object in view. Experiment shows, however, that visual acuity is improved if the area surrounding the radiograph is made about as bright as the radiograph itself. This fact should be borne in mind when choosing a wall-covering in the exhibition room. If we want the best conditions for seeing detail it should not be very much lighter or darker than the exhibits. A dark painting is not seen to best advantage on a white wall.

The second example was an interesting revelation to me. In a European museum a new ethnographic display had been mounted, in which, for dramatic effect, walls, ceiling and floor were all in very dark colours. Carefully trained spotlights allowed the objects to stand out brightly in the darkness. Illuminance was here and there up to 1000 lux. The curator recounted how, in the original display, lighting had been less intense, but visitors had complained that the place seemed gloomy. It then became apparent that, however bright the lights, the lack of light from walls, ceiling and corners continued to give an overall impression that we were in an uncomfortably dark cavern.

Summary

Control of illuminance to 50 lux necessitates artificial lighting, 'warm' rather than 'cool'. Glare must be rigorously avoided. The eyes must be adapted to the light before the viewer enters the room. Lighting objects and leaving viewers in the dark is not usually the best solution.

DIFFUSION OF LIGHT

Objects lit by a point source of light throw sharp and dark shadows. Objects lit by completely diffuse light – light that comes from all directions equally – throw no shadows at all.

We depend a great deal on shadows for our comprehension of surfaces. On a large scale shadows delineate the contours of a sculpture, on a small scale, in textiles for example, they in-

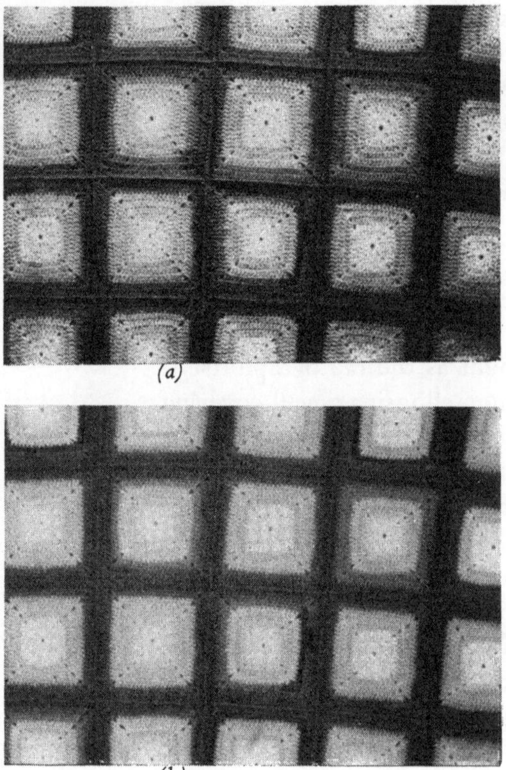

(a) (b)

Figure 14
A hanging rug lit *(a)* directly and *(b)* diffusely. In this and *Plates 11 and 12* the direct light is from a line of fluorescent lamps shining from above at 45°; the diffuse light is from a translucent ceiling aided by white walls and floor. Both are quite common museum situations, but most people would regard both as too extreme for this textured surface. Rug by K. Finch

form us of the surface texture. Our eyes have evolved in a world where light is partly directional (from the sun) and partly diffuse (from the sky). Only a small proportion of light comes from below. Thus in the museum our eyes receive visual information most comfortably and clearly when the lighting follows this pattern: from above and partly direct, partly diffuse (*Figure 14*).

There is a more particular effect that is dependent on the degree of diffusion of the light. An object with a shiny surface lit by a small source of light will reflect light in particular directions so that we get reflections and sparkle. A matt object reflects the light that falls on it in all directions. The eye notices all this in building up a conception of the nature of the surface. Completely diffuse light destroys this information. Standing in front of a painting lit by diffuse light you cannot tell whether its varnish is matt or glossy.

Highly diffuse light also makes glossy surfaces look paler for the following reason. When light from above strikes a glossy paint surface it is partly reflected down below the observer by the surface but partly enters the surface to be scattered, absorbed and reflected in all directions by the colours in the surface. These reflected colours reach the eye. Diffuse light comes from all directions, and so is scattered in all directions by the surface as well as by the colours. As a result we see not only the colours, but some white light superimposed on them: the colours look paler (*Plates 11 and 12*).

Summary

Light should be partly directional, partly diffuse. Light from sun plus sky is a good guide. Highly diffuse light can make glossy objects appear paler.

200 LUX – DAYLIGHT AND ARTIFICIAL LIGHT

Once we accept the need to limit the amount of light which is allowed to fall on an exhibit the difficulties in controlling daylight as opposed to artificial light become apparent.

In the open air the expected illumination at 50° North latitude at midday in July is ten times that at midday in December. When the sun is shining it contributes as much illumination as the whole of an average sky, and much more if the sky is very clear. Indoors these differences may be exaggerated, and the variations in the resulting illuminance will be further compounded by the inequalities of light distribution round the room[50].

In other words any system aimed to provide indoor illumination at a set level must be capable of being continually and unobtrusively adjusted. There are a number of ways, partial and complete, in which we can do this:

1 The simplest system is to block out so much light that illuminance at the brightest times of the year rises to not much more than 200 lux. At all other times daylight is supplemented by artificial light. If the object of the exercise is to light the room with daylight this is clearly unsatisfactory since outside the summer months the predominant contribution will be from artificial light. The scheme becomes more acceptable if a simpler summer regime is introduced, such as painting skylights in the spring with a paint designed to wear away by the end of the summer, or a paint which becomes transparent when wet.

2 We can go to the other extreme of complexity by using a translucent ceiling through which daylight is highly diffused so as to light all exhibits at the agreed level, the daylight being admitted through shutters or blinds which are automatically controlled by photocell. As the daylight fails, artificial light is introduced, also by photocells, under dimmer control. Unfortunately what once seemed a neat – though expensive – solution to the problem has not lived up to its expectations. The constancy of the illumination and its high degree of diffusion have been widely found to have a depressing effect. Technically the lighting situation obtained, although free from glare, is very like that under an overcast sky, being of the same colour temperature and directional quality. This may go a long way to explain the feeling. Certainly all that part of the character of daylight which depends on its variation is totally lost. In my opinion only an observer of rare colour discrimination would be able to tell whether the light coming through such a ceiling was from true daylight or from 'Northlight' fluorescent lamps.

3 Many existing daylight control systems, whether blinds or louvres, could be adapted, perhaps motorised, so that they were under the control of a room attendant. The output from a photocell on the wall could be led to a dial and the attendant would be required to reduce the light if the pointer on the dial moved onto a clearly marked part of the scale.

4 Instead of aiming to achieve a steady illuminance of 200 lux we can work out the total annual dose of illumination, or exposure, equivalent to this. Eight hours a day plus a low level of adventitious and security lighting (say 10 lux) amount to an annual exposure of about 650 kilolux-hours. Similarly an annual exposure for the 50 lux level would be about 200 kilolux-hrs*. Tables of illumination throughout the year at the locality of the museum can be obtained from meteorological organisations, and positions of blinds can be worked out to give 200 lux under average conditions (*Figure 61*). To carry out the scheme in detail requires the positions of blinds to be changed hourly (automatically) on a monthly schedule. This means a fair amount of computation, but simplifications will be evident. The result aimed at is a daylight illumination which retains some connection with the weather outside.

Naturally all these schemes require the daylight to be thrown fairly equally round the room, and daylight cannot be directed with the theatrical ease of spotlights. The simplest and almost inevitable answer is diffusion, but we have just seen that highly diffused light becomes unpleasant. If exhibits are placed all over the room, or, more difficult still, an adaptable display is demanded, there is no other method with daylight but diffusion. However, if the exhibits are limited to the four walls, the prospects of a light which does not illuminate the visitor more strongly than the exhibit are much improved. The light can be directed predominantly at the walls by admitting it round the edge of the ceiling but not – or not so much – through the centre (*Figure 15*).

Whatever has been done or planned it must be admitted that so far the full pleasures of daylight have proved incompatible with the requirements of conservation. But this is only a beginning. Let us at least attempt the difficult task of defining these pleasures insofar as they are relevant to looking at works of art.

First of all we can prune away the main false clue. Colour rendering, which is covered in other sections (pp. 48 and 199) often suffers under artificial light, but need not. Certain fluorescent lamps can be relied upon not to produce colour distortions except in rare circumstances[51] (p. 54). The feeling that we can only be sure of seeing a work of art in its true colours under daylight is not supported by investigation.

Secondly, some people dislike being hemmed in by a building in which there is no contact with the outside world, through

*Kilolux work on the same basis as kilowatts: lux × hours. For example, 50 lux for 24 hours amounts to an exposure of 1200 lux-hours or 1.2 kilolux-hours.

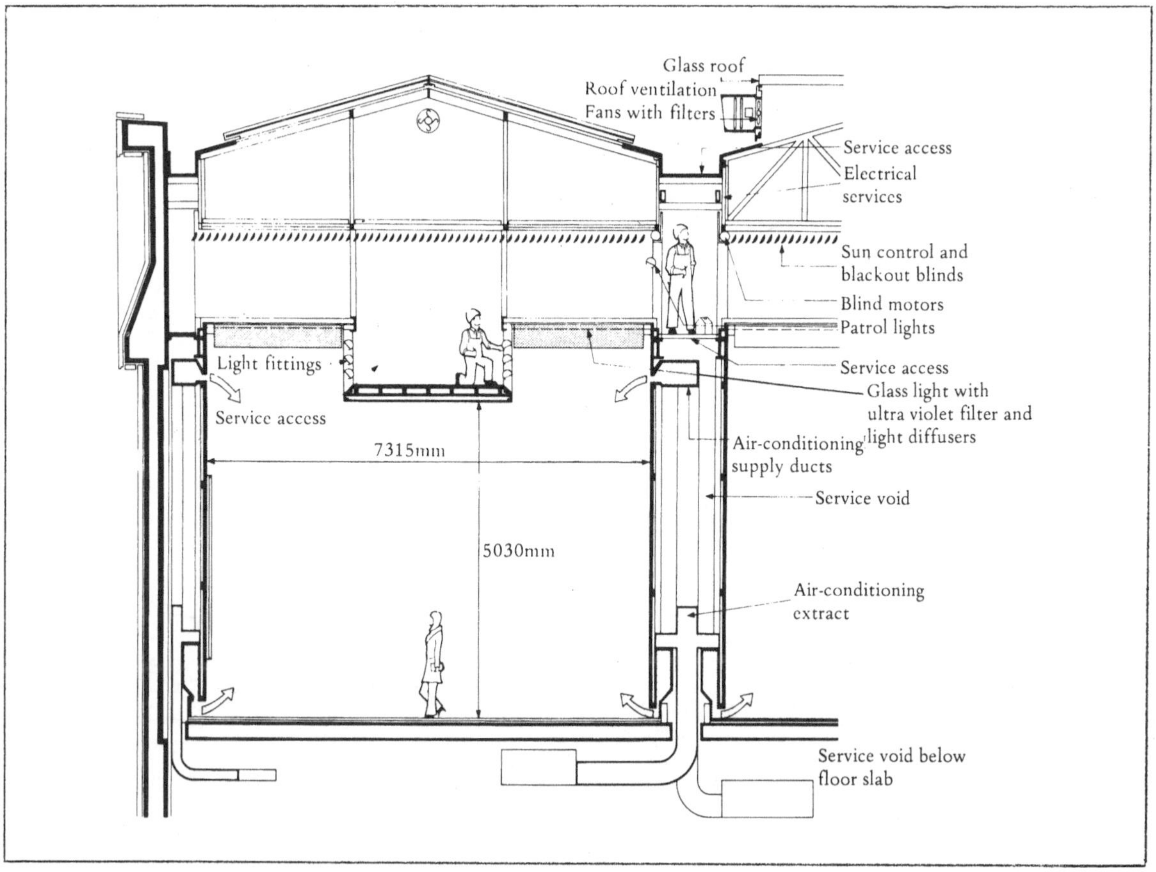

Figure 15
Lighting of the Northern Extension to the National Gallery. The daylight passes through skylights of lightly diffusing glass, through blinds which automatically control the amount of light, and finally through UV filters made of a plastic interlayer sandwiched between glass. The opaque lowered ceiling holds on its sides the fluorescent lamp fittings. Architect: D. Church, Department of the Environment, London[52]

windows or skylights. Windows can be provided, though they should be regarded as openings on the outside world rather than light sources (*Figure 16*), (*see* next section).

Thirdly there is what one might call the 'out of context syndrome'. Neither a tea ceremony bowl nor an African mask can ever be given more than a substitute setting in a museum. As a place of display for paintings the museum cannot rival 'real life' situations, be they church or palace. Are we sometimes unconsciously recognising this fundamental inadequacy by putting the blame on the lighting?

A love of sunlight is characteristic of the natives of those countries which do not have enough of it. In those that have too much a museum without daylight can seem like a cool haven of rest.

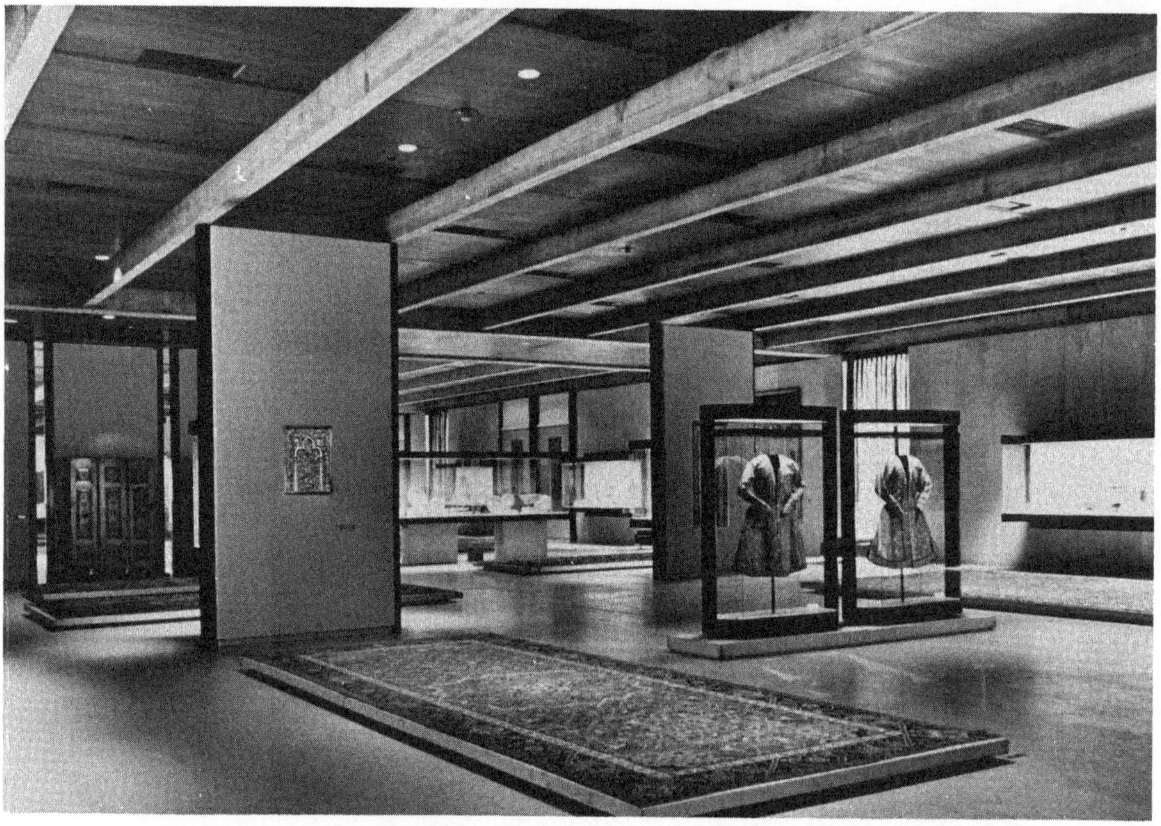

Figure 16
The Islamic room at the Calouste Gulbenkian Museum, Lisbon. The garden at the back of the museum can be seen through the floor-to-ceiling windows on the right, but their potential as glare sources is sufficiently reduced by a neutral grey glass plus a transparent net curtain. The exhibits are lit at 50 lux. Architects: Cid, Pessoa and d'Athouguia. Lighting policy consultant: W.A. Allen. Artificial lighting: M. de Amorin

Sometimes a work of art may be caught in a light that reveals rare and unexpected beauty. This is the nearest we may be able to get to identifying the peculiar quality of daylight – its infinite variability. People whose views one feels bound to respect have been deeply concerned to retain daylight in museums, and have remarked how new aspects were revealed to them by an accident of light. Out of doors everyone has felt exhilaration when the sun, emerging from cloud, exerts its pervasive influence, so very likely these accidents of light in museums have had a similar effect. Unfortunately sensitive material cannot stand up to direct sunlight, so such opportunities are lost both under artificial light and under controlled daylight.

Summary

The problem of daylight in museums is not yet solved. All we can do as conservators is to lay down a framework of rules within which the curator, architect and designer ought to work because of their common concern with the preservation of the collection.

CONSERVATION LIGHTING SPECIFICATIONS

A specification for all aspects of environmental control is suggested in the Appendix.

Concerning specifications for control of lighting, there is no problem for 50 lux material, which should be under artificial light in any case. No automatic control of the lighting is necessary since it is constant.

Under daylight the worst dangers can be avoided by rigorously ensuring that direct sunlight never strikes an exhibit, although even then illuminance may well rise to 500 or 600 lux. Therefore, if the decision is for manual control, vigilant operation of blinds and constant surveillance with a light meter is needed to keep illuminance within the low hundreds.

If we call the above a Class II specification, then Class I, the recommended specification for all very valuable light-sensitive material, would run as follows:

50 lux: as before.

200 lux:

(a) 200 ± 50 lux (150 to 250 lux) at the center of the exhibit.

(b) For large objects no more than a 2 to 1 ratio between the brightest and darkest parts. For picture galleries this would apply to the whole picture-hanging area.

The difficulty of keeping illuminance under automatic control should not be underrated. The equipment presently available requires regular maintenance and supervision by an electrical engineer capable of using a light meter.

UV control should be applied whether a Class I or II specification is decided on. The proportion of UV in the light should not exceed 75 μW/lm. This is easily attainable with a good commercial filter.

Summary

Class I and Class II specification for 50 lux material:
Illuminance should not exceed 50 lux on any part of the exhibit.
Class I daylight control:
200 ± 50 lux at center of exhibit.
Highest and lowest illuminance on large objects within 2:1 ratio.
Class II daylight control:
No direct sunlight.
Vigilance with manually controlled blinds and light meter.
In all cases:
UV elimination to less than 75 microwatts/lumen.

TREATMENT OF WINDOWS

Side windows are difficult, almost impossible, for the conservator to deal with to his satisfaction. Objects on the same wall as a window cannot be properly seen because of glare, while objects in cases close to the window may get 100 times as much light exposure as objects in a far corner (p. 39). In any old building converted to a museum, if you want to see severe fading you need look no further than the areas within a metre or so of side windows.

However, in spite of the dangers, side windows in old buildings cannot for architectural reasons simply be eliminated. Therefore we must find some defence against their depredations, especially where delicate materials such as textiles are on display.

The general principle is to reduce the light coming through the windows in some acceptable way, and at the same time supplement the light in the far parts of the room artificially.

It may seem obvious, but it is important to remember, especially on upper floors, that the light comes from above and the view from lower down. Venetian blinds can be lowered to obscure the sky yet retain the view. Some old houses have shutters divided, no doubt for this very purpose, in such a way that the upper parts can be closed separately. In other cases net curtains may be the preferred choice.

A very necessary step is to eliminate UV radiation, and for side windows a UV-absorbing varnish is the best answer. It is possible to eliminate UV and reduce illumination in one step by applying a neutral grey UV varnish or adhesive film. A more expensive alternative is to replace the window glass with neutral grey glass (which will still, however, need a UV-absorbing varnish). The effect of grey glass is not perceptible indoors, but from outside the windows look unusually dark. Aluminised film can make windows look, from the outside, like mirrors.

There is no reason why the architect of a new building should submit its exhibits to the risks described above. His side windows, however, must be sited and designed with care. Neutral grey glass has been mentioned. Other points to bear in mind are: keeping the height of the windows down, recessing them in thick walls and siting them preferably away from the main exhibition space, perhaps in connecting passages or ancillary areas. It must unfortunately be added that windows provide convenient access for robbers, and are not popular with security officers.

Summary

Side windows should be for views, not for lighting..

ANGLE AT WHICH LIGHT FALLS ON EXHIBITS

In exploring this problem we should keep outdoor illumination in mind. The eyes were evolved out of doors, and so interpret the scene best when directional light from the sun falls downwards on the scene and is supplemented by a hemisphere of scattered light from clouds and sky. Relatively little light is reflected from below, except when there is snow, desert sand or water. The angle of the sun is not critical, but it only rises to the vertical in the tropics.

With this lifelong experience in our minds, we could allow sculpture to be seen quite naturally under artificial light by using spotlights pointing down at around 45°, and keeping ceiling and walls light-toned, but the floor darker. A high ceiling makes the task easier. Spotlights need not all point in the same direction as sunlight does, but glare must be avoided. There is plenty of flexibility in lighting directions and wall coverings, but any excessive departure from this norm, for the sake of the dramatic, risks being detrimental to receptive viewing.

Hanging tapestries, paintings, showcases and other vertical surfaces require further thought. The difficulties are caricatured by photographs taken with camera-mounted flash units. If we light from too near the front there will be strong reflections from every glossy part of the surface. Let the light fall from too near

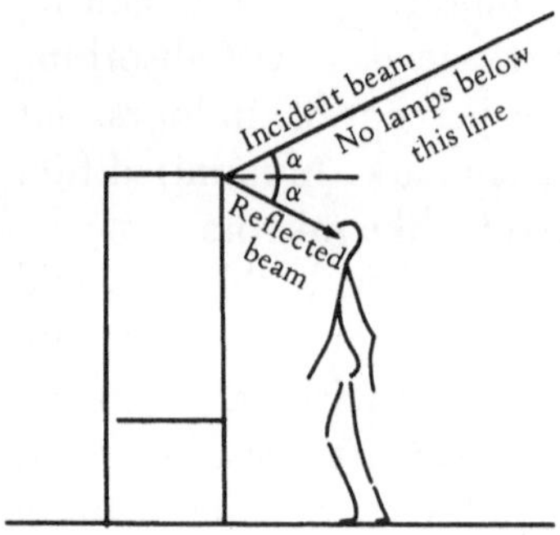

Figure 17
Will a lamp reflect into the vistor's eyes? Draw showcase and visitor to scale. Draw a line from his eye to top of showcase glass, and a line perpendicular to the glass at this point (broken line). Since angles of incidence and reflection must be equal, an incident-light line on the other side of the perpendicular will act as a barrier: no lamps should be below this line

the vertical and we make the shadows of any projections long and obtrusive, and may exaggerate the texture of a textile or the surface irregularities of a painting. It is easy to find out with a diagram whether particular lamps will reflect into the viewer's eyes (*Figure 17*). When the whole of a ceiling is a translucent light source the objects with the strongest reflections may be the tops of visitors' heads, and you cannot get the reflection of your head out of the way if you are looking straight at a showcase. Light as close to the vertical as 30° gives excellent glare-free viewing of vertical surfaces, but the light source should be extended (e.g. with a reflector) so as to soften the edges of shadows (*Figure 18*). Pictures with very deep frames may require a larger angle.

With a good reflector a single line of fluorescent lamps of good colour rendering (i.e. only moderate efficacy, see p.54) should give over 200 lux, with a ratio of less than 2:1 for the brightest and dimmest parts of the whole picture hanging area.

Desk-top showcases are impossible to light individually to high standards (*Figure 13*) and are usually best left to general room illumination. Lights inside showcases are bad because of the heat problems they cause, and in shallow desk cases the light falls too obliquely[30]. If we try to put the light behind the viewer he will shadow it and it will cause glare in some other viewing direction.

The use of point sources of light alone, such as tungsten lamps on a light track, shining in several directions in a low-ceilinged room gives almost inescapable glare problems, intensified if the general ambient illumination is kept low by narrow beams and dark walls. Unfortunately at the time of writing this is a popular exhibition technique.

As has been noted previously (p. 26) a more even lighting actually eases the task of keeping the illuminance on exhibits low but apparently well-lit. Extended sources, such as fluorescent lamps behind diffusers and light reflected from the ceiling, help to provide this ambient light, as does the avoidance of very dark-toned wall coverings. This is not meant in any way to discourage the use of spots for gentle accentuation.

Lighting sculptures and paintings from below is a gimmick which may very occasionally be justified.

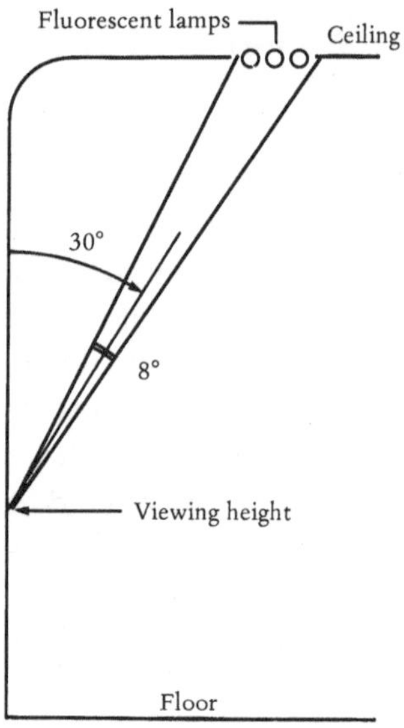

Figure 18
Simplified diagram of fluorescent lighting in a room at the National Gallery, London, which appears satisfactory. The light falls on the painting at about 30° to the vertical, and shadows are softened because the light source subtends 8° at painting distance. The same effect could be achieved with a single line of fluorescent lamps in wide reflectors

If the reader finds the general tenor of this section to be away from over-design and towards letting the exhibit show itself he will not be mistaken. Let us see the objects rather than reflections of the designer's ego.

REDUCING TIME OF EXPOSURE

A very important collection of Turner watercolours was bequeathed by Henry Vaughan to Cambridge, Dublin and Edinburgh museums on condition that they were to be exhibited to the public for one month only in the winter of each year. For the remainder of the year they were not to be illuminated except by special arrangement.

The custom in China and Japan for owners of classical hanging scrolls was to keep them rolled in boxes, to be brought out only for limited periods.

Current fashions do not favour this sort of tradition; the trend is for everything to be open to the public's view all the time, which is fine so long as the material is not endangered by the display. Curtains have even been banished from desk-type exhibition cases, thereby vastly increasing the exposure.

There are so many ways of reducing exposure in particular circumstances that the most that can be done here is to suggest a few general approaches:

1 *Limited exhibition of material brought out from store.*
This applies obviously to collections too large to be put completely on permanent display, but the principle can be extended for special material, as with the Turner watercolours.

2 *Illumination only during opening hours.*
Especially in daylit galleries during summer, exhibits are exposed for long hours with no one in the gallery. Concessions may have to be made for security after closing, but illuminance for this need not be more than 5–10 lux.

3 *Illumination only while on view.*
This is an extension of 2. In some of the less popular or more remote museums, rooms may often be empty of visitors for long periods. As in a private house there is no need to keep the lights on if there is nobody in the room. Also under this heading are the curtains over desk cases mentioned above, which cover the exhibit except when drawn aside by visitors. The electrical equivalent is a time switch, not necessarily to switch from abso-

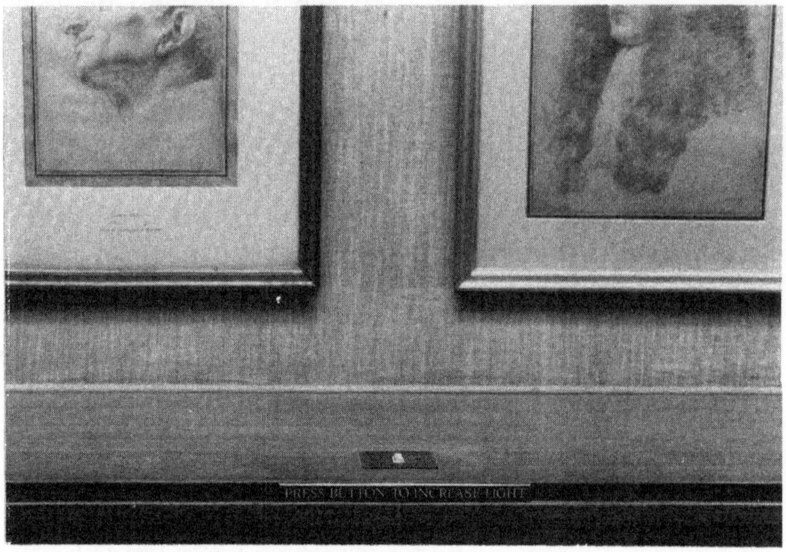

Figure 19
Drawings and watercolours on exhibition at the Barber Institute, Birmingham University. This is a side room normally illuminated by light scattered from the main exhibition room. To increase the light on the pictures the visitor presses the time switch shown

lute darkness to light but to provide extra illumination (*Figure 19*).

4 *Use of replicas.*
Whereas very satisfactory replicas can be made by casting and electrolytic methods of metalwork, sculpture and jewellery, the light-sensitive exhibits that we are concerned with must usually be photographed. In other words they must be two-dimensional. Even the texture and impasto relief on the surface of oil, tempera or gouache paintings is enough to prevent good replicas from being made without great trouble. However prints, drawings and watercolours reproduced photographically on the proper paper with individual control can form excellent replicas. In the interests of simple honesty all replicas should be so labelled.

There is a belief among some art gallery curators that the public go to great paintings purely for an aesthetic experience. If that were truly so, a replica exhibition of Turner watercolours would be as big an event as if the originals were to be shown, and a true lover of the art would be happy. Unfortunately, if all the tourists 'doing' the Louvre or the Uffizi, and all the seekers for the rare, the curious and the unusual, were to disappear, art galleries and museums would be empty places.

On the other hand, would not the curator of a public collection be exceeding his powers in this democratic age if, by the use of replicas, he were to keep his most sensitive material permanently in the dark, or available only in special circumstances?

For these sorts of reason the honest use of replicas is likely to remain rare.

A SUITE OF EXHIBITION ROOMS

Imagine a country house which has been turned into a museum and filled with a varied collection including paintings, furniture, tapestries and carpets. *Figure 20* is a plan of three of its exhibition rooms.

Room 1 is long, with two large windows at the south end.

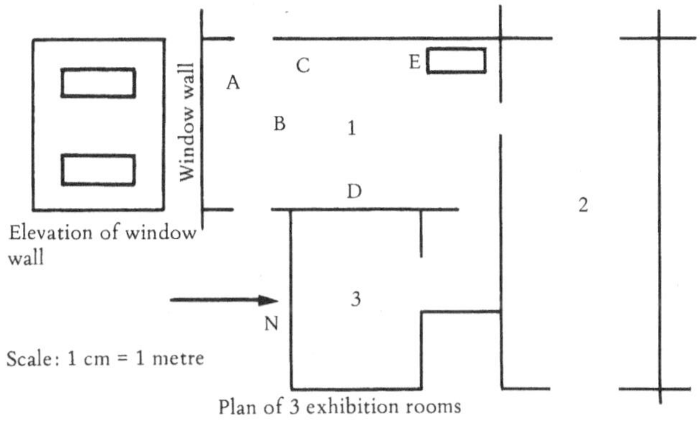

Figure 20
An imaginary suite of exhibition rooms. Lighting:
Room 1
side windows plus supplementary artificial light
Room 2
translucent ceiling, daylight and artificial light
Room 3
artificial light on demand

These windows afford views over a fine park and are important architecturally. Artificial light was until recently from chandeliers, but the ubiquitous light track has now begun to appear, and spotlights illuminate some of the exhibits at the darker end of the room, not only after dark but during most of the day as well. This is because the contrast between either end of the room is such that the end furthest from the windows looks dark even when plenty of light is available. Blinds are fitted to the windows, but are operated only spasmodically. Against the wall on either side are a set of 18th-century chairs. Some have their original covers, but these are now so faded and weakened as to be almost unrecognisable. There are valuable carpets on the floor, and a few desk-type exhibition cases, containing manuscripts, which have been allowed to retain their curtains. All paintings are covered with glass. A painting hanging at D has its importance emphasised by a special spotlight at 500 lux. The sun in its path across the sky causes direct sunlight to fall on all the exhibits near the windows.

Room 2 has recently been renovated. It is now lit by daylight through a translucent ceiling and by fluorescent lamps. Shutters operated by photocells ensure that the light remains always constant.

Room 3, entered via an anteroom, contains an important collection of watercolours. It is illuminated to 50 lux on a time switch set to 10 minutes. At all times a security light of 10 lux is maintained. Only a third of the watercolour collection is on display at any time.

Comments among the visitors include the following:

Everyone enjoys Room 1 the best, especially near the windows, which afford unobstructed views over the park. However both the original chandeliers and the new spotlights produce disturbing reflections in the glass over the paintings. There is a move afoot to take away the curtains over the desk cases because these seem out of keeping with fashionable design trends. When the sun is shining exhibits at the North end appear underlit even when the artificial light is on.

Room 2 has not found favour in its new form. It had been planned to set the automatic lighting at 150 lux, but in certain positions one can look through Room 1 to the bright windows so that ones surroundings look dull. So the illumination was raised to 250 lux. Even so the impression remains of dullness and over-diffusion.

Room 3, though lit only at 50 lux at the press of a button, is regarded as very satisfactory. The exhibits are recognised as delicate enough for special precautions to be taken. The room is approached through an anteroom lit only by light spilling over from Room 1. Thus the 50 lux seems bright enough.

It is possible to calculate in a very approximate way the exposure received every year by objects in various parts of this suite of rooms. The calculation for artificial light and for controlled daylight is straightforward because the light is constant. For the sun, however, some assumptions must be made of average sun-hours and the geometry of the sun's path at the location of the museum must be worked out. My calculations are for what I suppose to be a typical year in England, Northern France, Holland or Belgium for a house with no high buildings masking it on the south. They are summarised in *Table 3.*

The last fold on the face of a window curtain is exposed whether the curtain is open or shut. It receives almost as much light as the wall outside, but less UV, being behind glass. We shall take the exposure on this fold to be 55 Mlx h (million lux hours or megalux hours) per year (for lux hours *see* p. 22). At such an exposure even the toughest curtain will show signs of damage after ten years or so.

A carpet at A, in front of one of the windows and one metre from it, receives 13 Mlx h, more than three-quarters of it as sunbeams.

TABLE 3. Estimated Exposure in Million Lux hours per Year at Various Points in the Suite of Rooms, ***Figure 20***

	Sun	*Sky*	*Total*	*Exposure compared to Room 3*
Window curtain	40	15	55	5500
Carpet at A	10	3	13	1300
Case at B	4.5	2.5	7	700
Chair at C	4.75	1.75	6.5	650
Painting at C	3	0.75	3.75	375
Painting at D			1.75	175
Room 2			0.75	75
Desk case at E			0.03	3
Pictures in Room 3			0.01	1

Some early navigational instruments are displayed in a vertical case at B. Being of brass they are immune to light, but small items of leather and textiles are disintegrating. The case receives 7 Mlx h.

A chair at C beside one of the entrance doors also receives, on the side facing the window, about 7 Mlx h. It has been in this position for several years and the matching opposite side can be seen to be more strongly coloured. A painting on the wall behind the chair, being hung at right angles to the windows, receives rather less than the chair: about 4 Mlx h.

An exhibit illuminated during opening hours at 150 lux will receive about ½ Mlx h per year. Many of the objects well away from the window in Room 1 will be exposed at about this level, though the masterpiece hanging with its special lighting at D gets 1.75 Mlx h.

All objects in Room 2, now at 250 lux, receive about ¾ Mlx h per year. Strangely enough the attendants, while ignoring the blinds on the windows, are very quick to replace the curtains on the desk cases if visitors have not done so. Consequently exposure of the manuscripts in these cases, even though lit at 150 lux when the curtains are open, amounts, at half an hour per day to only 0.03 Mlx h per year.

Lastly, the press-button artificial light in Room 3 is busily used in the tourist season. Annually, however, it is estimated to be used for less than half an hour a day. Taking into account security lighting, annual exposure is 0.03 Mlx h. But only one in three of the pictures in this collection is on view, so one can say that each receives 0.01 Mlx h per year.

Ignoring the windows, which could never have retained their original curtains, one can see that the pictures in Room 3 would not receive in a thousand years the annual exposure of the carpet in front of the window. If this comparison is a little unrealistic, a 500 to 1 exposure ratio (the ratio of five centuries to a year) is in evidence between the objects near the window and those in Room 3.

Many ways of improving the situation in this suite of imaginary rooms will be evident to the reader, perhaps helped by the previous sections.

The light coming through the windows, particularly the direct sunlight, absolutely must be reduced. Venetian blinds, for example, could cut out direct sunlight and a lot of sky light while retaining the views of the park, but they would need constant supervision. UV-absorbing varnish would be required on the panes.

Within Room 1 the worst glare is on pictures with glass over them. This glass could be removed if funds were found to install

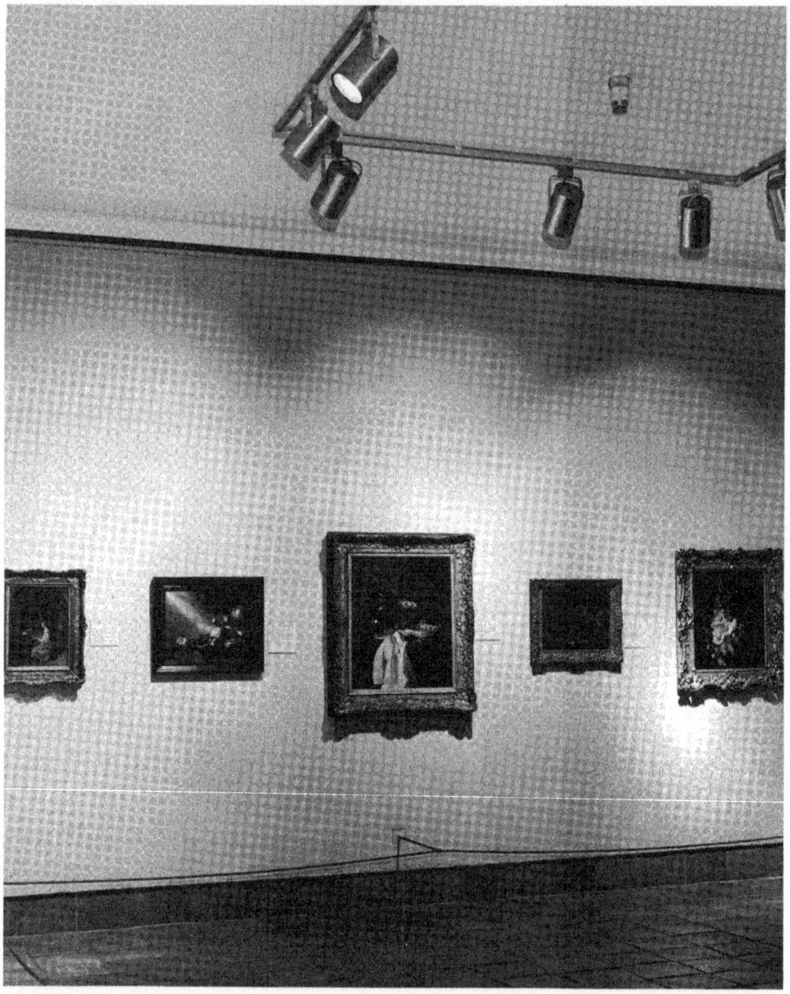

Figure 21
Avoiding glare from spotlights: the lamps are mouted in deep cylindrical fittings. (By courtesy of the Trustees of the National Gallery, London)

a ducted air-conditioning system to purify the air. Without glass, chandeliers fitted with low-wattage bulbs would cease to be a major glare problem. Spotlights are at their most distracting when shining right across the room, or otherwise at an angle near the horizontal. Placed so as to shine downwards, but not so acutely as to form deep shadows from picture frames etc., their light can usually be masked so as to be invisible from most viewing positions (*Figure 21*).

Curtains over desk cases are a splendid conservation measure which should be retained.

Complaints of the over-diffusion of light from a completely translucent ceiling, as in Room 2, are common, and I believe justified. An opaque central area would improve the situation greatly, but calculations or models would first have to be made to ensure that enough light would penetrate during most of the year. There is more about the control of daylight on p. 28.

HEAT

Those who are new to the study of conservation usually suppose that the most important thing to do to the museum climate is to control the temperature. It is not. Humidity, light and air pollution are all more important. But that is not to say that we must ignore temperature.

Let us try to enumerate the various ways in which a rise in temperature can affect exhibits. First of all there are four straightforward results of a rise in temperature:

1 The rates of those chemical deterioration processes which do not need light increase. For example if we raise the temperature (at constant relative humidity (RH)) of cellulose from 15 to 20 °C or from 20 to 25 °C we will increase its rate of deterioration in the dark by about two and a half times in each case (*see* p. 192).

2 The rates of physical processes such as movement of water and air through solids also increase. Typically the same rise in temperature of 5 °C would speed up these processes by about one and a third times. Roughly this will also apply to the evaporation of traces of volatile material, perhaps formed by deterioration, since such evaporation is primarily limited by diffusion. Age-embrittlement is related to these processes.

3 Materials expand by a very small amount. For moisture-containing materials such as wood, paper, bone, ivory, leather,

textiles and paint this effect can be ignored, since swelling due to water is far greater. For brittle, non-moisture-absorbent composite objects whose parts expand at different rates – for example, enamel in metal cloisons – an unusual change in temperature might be dangerous.

4 Biological activity as a rule increases in warm weather.

But temperature can also have an indirect effect, often much more important, through its influence on RH and moisture content:

5 Unless the RH is kept independently constant, which it should be, a rise in temperature causes drying, which results in embrittlement, for example of wood, paper, animal glue and leather.

6 A rise in temperature by causing drying, will actually reduce the rate of fading of most dyeings[5,6,53]. In these cases – contrast with 1 above – rate of fading increases with increasing moisture content. This effect is not usually considered enough justification for keeping textiles in specially dry conditions, since this might also affect their flexibility (*see also* p. 86).

7 Perhaps the most important effect of all: a temperature rise due to radiant heat, as from direct sunlight or a strong spotlight will cause drying even when the RH of the room or showcase is kept constant[54,55] (*see* p. 45).

It should be noted carefully that changes which are caused directly by light are almost unaffected by temperature in the museum range. Any deterioration process caused by light, a colour change for example, is not just one chemical reaction but a whole series. The first change, without which none of the others can occur, is caused by light alone (or even more likely by UV) and it is truly unaffected by temperature. But the changes that follow may not need light, and as such are temperature dependent. The nett result for the overall photochemical change is a slight temperature dependence such that, in a typical case, one might have to raise the temperature from 20 to 90 °C to double the rate of deterioration.

Summary

A small change in temperature can have several effects, some good, some bad, but temperature change is not as important as humidity change except when it in effect causes humidity change by drying (e.g. direct sunlight or a strong spotlight).

CONTROL OF TEMPERATURE

Here are some practical results from the foregoing section:

1 It does not seem worthwhile to put all our most precious exhibits in refrigerated showcases though simple cooled display cases have been made[56]. Temperatures in exhibition rooms can be on the economical side. But warmth in winter is of no benefit to museum material, which is happier at low temperatures, though above freezing.

2 For storage, as opposed to display, a low temperature (say between 5 and 10 °C) can be of real benefit to archival material and textiles (i.e. cellulosics). This will also reduce biodeterioration problems. But there is a danger: water vapour may condense on objects brought directly from cold store to office or exhibition room. The dew-point in a room at 20 °C, 55% RH is 11 °C, so that condensation will occur on any object whose temperature is below 11 °C (*see* p. 90). Passage through a preconditioning zone is necessary. Colour film, being very colour-fugitive, also benefits from low temperature. The safest storage environment for all photographic material, whether colour or black-and-white, is a temperature as low as possible, preferably below freezing, and an RH of around 30%[57–59].

3 We must be on our guard against excessive radiant heat. The most likely causes of trouble are direct sunlight, over-bright spotlights and lamps in showcases.

4 Sometimes detrimental heating other than by radiation occurs, as with objects above or too near radiators (a rising current of air may also deposit dust). Lamps in showcases can heat by conduction as well as by radiation.

Substantially all the radiation absorbed by the exhibits, both visible and invisible, whether daylight or artificial light, is converted to heat. Strong light directed at an object will therefore heat it above the general room temperature with consequent dangers of cracking and warping. Light sources emitting a high proportion of IR radiation (e.g. tungsten lamps) heat more strongly than those emitting little (e.g. fluorescent lamps), but all heat to some extent.

If lighting is controlled to 200 or 50 lux it will be found that heating is also controlled to reasonable limits. For example 500 lux of tungsten light will heat a surface by 2 or 3 °C depending on its colour. The temperature rise on a black surface will be about 50% more than on a white[54]. The heating effect of fluorescent light is less than half this, and of daylight less still. At 200 lux the heating effect of tungsten will be reduced to about 1 or 1½ °C, and of daylight and fluorescent proportionally less (*see* p. 186).

In addition to radiant heat, light sources themselves get hot and pass the heat to the air circulating past them. Therefore lamps should be kept outside exhibition cases and ventilated with air which will not directly travel past the exhibits.

Lights are usually switched off at night, so that strong lighting can set up a daily fatigue cycle of expansion and contraction, either by direct effect of heat (e.g. enamel on metal), or more usually by drying (e.g. wood, ivory).

Where tungsten spotlights are preferred but heating must be minimised, dichroic reflector spotlights, sometimes called 'Coolbeam' lamps, can be recommended. These direct only visible radiation at the object and transmit the IR through the back of the lamp, which must be well ventilated. The lighting is a little cooler, but colour-rendering is not disturbed.

Summary

Storage at low temperature can be of benefit to archival material and textiles. Excessive radiant heat must be avoided, but there should be no problem at 50 or 200 lux.

LIGHTING FOR PROFESSIONAL PHOTOGRAPHY, TELEVISION AND RESTORATION

Heating is the main trouble here. Photographers often use more light than they need, and they should therefore be warned of the dangers and supervised. With present-day colour film and television cameras lighting need never exceed 1000 lux. Nevertheless, more is sometimes required for various reasons. Metal halide lamps such as the Wotan/Osram HMI are popular and these are relatively cool sources. The recommendation is, therefore, that tungsten illuminance should be limited to 1000 lux, but that under metal halide lamps this figure can go up to 2500 lux. Though these lamps are high in UV, exposure to them is so short that a UV filter need not be insisted on. As the sensitivity of television cameras increases less will be needed and less should be used. Lighting should be switched off or reduced except when photography or adjustment of lighting is actually taking place. Fans and heat filters[31,54,55] can be valuable for reducing the surface temperature of exhibits under strong lighting.

Electronic flash (*see* next section) has the great advantage that heating is for all practical purposes eliminated.

Within reason restorers should use what light they need, but here 2000 lux seems a reasonable upper limit.

ELECTRONIC FLASH

Today the usual photographic flash system for both professional and serious amateur consists of a discharge tube filled with xenon gas. The tube itself is usually made of borosilicate glass. The spectral composition of the light from such a flash is shown in *Figure 59(f)*. It is rich in both UV and IR.

We have just seen that the radiant heat from tungsten incandescent lamps used at high intensity must be carefully controlled. The advantage of electronic flash, even though considerable IR radiation is emitted, lies in its negligible heating effect. This is because no more light falls on the object than is required for the actual exposure. By contrast, tungsten illumination, even if switched off until everything is ready, is for most of the time illuminating the object while the camera shutter is closed.

Since flash is also very satisfactory for colour rendering, using daylight film, looked at in this straightforward manner it seems that we could not do better than use flash for all still photography, and indeed this is probably the case.

There has been a certain theoretical worry, however, that the very strong pulse of light in the flash may have some special effect. With this in mind Hanlan[60] carried out some tests, exposing colours to equivalent amounts of light from flash and from steady illumination. He concluded: 'There seems to be no reason to regard normal use of electronic photoflash lamps as unduly hazardous.' And indeed from his results we can fairly safely assume that in real conditions and prudently used, repeated electronic flash is likely to be less damaging than repeated exposure under incandescent lamps.

The French National Committee of ICOM have also studied this question and come to similar conclusions. Bearing in mind that it is the most important objects that get the most photography they laid down certain safety rules for flash[61].

Rather than try to limit the power, number and distances of several flash units, the most straightforward approach is to give an upper limit to exposure. Just as we have used 'kilolux hours' to measure exposure in museums over years, we conveniently use the much smaller unit 'lux-minutes' for measuring a single flash exposure. 20 lux-minutes is equivalent to a flash exposure at f22 on 100 ASA film, and this can be taken as a reasonable upper limit. Indeed the recommendation could be read as 'no smaller aperture than f22 at 100 ASA or equivalent'. 20 lux-minutes might seem a small exposure compared to the million or so lux-hours per year we can expect in a museum. But it is as well to limit flash even though no special effect has been demonstrated.

Care should always be taken that neighbouring objects do not receive a bigger exposure than the one being photographed. This applies of course to all photographic lights and dreadful mistakes of this kind have occurred.

Where one, perhaps very famous, object is being repeatedly flashed, a UV filter over each flash should be specified.

It may not be the museum's policy to allow amateurs to use flash. Nuisance value to other visitors should be taken into account.

COLOUR RENDERING

The key difficulty in the comprehension of colour rendering lies in the eyes' remarkable powers of colour adaptation, so that, whatever the light source, as long as it can roughly be called white, the eyes will see white objects as white under it. This adjustment, which is not under conscious control, will have its influence on the perception of all colours.

Before the reader comes to think that there may be no such thing as an ideal or perfect white, it should be said that this is in fact an easy thing to define and even to make to a close approximation. An ideal diffusing white or grey surface reflects all colours equally in all directions. If it reflects, say, 37% of light at 400 nm, it will also reflect 37% in the green at 550 nm and in the red at 760 nm. Its reflectance spectrum is thus a straight horizontal line. Coming quite close to this in practice at about 97% reflectance is a block of magnesium carbonate, a white chalky material which is soft enough to scrape clean, revealing a new surface for measurement purposes.

If we wanted to compare a set of colours under two lamps our first thought might be to make two little booths with identical objects in each but each lit by only one of the lamps. We could then stand back so that our eyes could wander from one booth to the other. Unfortunately the process of colour adaptation makes this ploy worse than useless. In such a situation the eyes will adapt to one of the light sources, making colours under the other appear distorted or 'wrong'.

To emphasise how flexible our visual mechanism is, consider photographic colour film, which is completely inflexible. Two types are widely available, for 'artificial light' (tungsten incandescent) and for 'daylight'.

If a 'daylight' film is used to photograph a scene under tungsten illumination everything will have a red--brown cast, since tungsten illumination is much stronger in red and weaker in blue than all forms of daylight.

If we try things the other way round and go out of doors with an 'artificial light' film then there will be a blue cast over the final print (*Plate 13*), whereas the 'daylight' film should have things about right (*Plate 14*). But daylight is so variable that our film can only look after the more common conditions. Under a very blue sky the 'daylight' film, having no powers of adaptation, shows things as too blue, and snow, for example, becomes blue-tinged (*Plate 15*). On the ski slopes under a clear sky, are the shadows 'really' blue? The snow which is lit by both sun and sky (but much more strongly by the sun) looks white to our eyes. The snow in the shadows is lit only by the sky and so looks bluish. But here our eyes are trying to do yet another correction to deny the blue colour. This is the 'colour constancy' operation. Another example of colour constancy: the different sides of a room get different amounts of light, and sometimes different kinds. Curl one hand and look through the hole so formed at a corner of the ceiling. Looked at out of context in this way it is sometimes quite hard to believe that the three sides all have the same colour. Normally one of the mind's colossal simplifications takes place: we know the colour is the same on all walls and therefore it appears to be the same. In contrast to colour adaptation, the colour constancy operation can be suppressed by conscious control, as every artist knows. So on the ski slopes the snow in shadow will be perceptibly blue if we turn our attention to its colour. But if we then move to a pine wood, mostly under shadow, colour adaptation will occur and the blue tinge of the snow will disappear.

We now need some means of classifying light sources in respect of being either bluish and cool at one end of the scale or reddish and warm at the other. The numerical scale for this classification is known as 'colour temperature'.

Any object heated to incandescence will glow with a colour characteristic only of its temperature and independent of its composition. This loose and perhaps surprising statement bears the important message that colour can be connected to temperature, but it must straight away be qualified to render it more exact. First of all many things burn with a coloured flame when they are heated to incandescence, so we must use a non-combustible material. Secondly objects at room temperature are coloured by virtue of the light they reflect, but this self-colour becomes negligible when they glow under strong heat. The filament of a tungsten lamp might be green or purple for all we know from the light it emits when it is switched on, because its light depends on its temperature and not on its self-colour. However, in order to avoid the complications of flames and self-colours, we invent a mythical 'black body' (sometimes called a Planckian radiator) which is non-combustible and reflects no light at all. The colour of such a body, if it existed, when it is heated up could be used absolutely accurately as a measure of its temperature.

Something quite close to this ideal black body is a small hole in the side of a furnace. With the furnace cold and shut the hole looks black. As the temperature rises the hole changes colour from red to yellow to white to bluish. An instrument called a radiation pyrometer is in fact regularly used at furnaces for measuring their temperature from the colour of the light which comes through a small inspection hole.

The colour temperature of the surface of the sun is about 6000 K. The light is a little changed as it passes through the sun's outer atmosphere, but this is close to its true surface temperature. Its interior is very much hotter. Colour temperature is conventionally measured on the Kelvin scale. 0 K is known as absolute zero, and is equal to −273.15 °C. To convert from a value in K to °C subtract 273.

It is a little unfortunate that the age-old convention of calling reddish lights warm and bluish lights cool goes in the opposite direction to the colour temperature scale. The red glow of a log fire is warm but the blue of an iceberg is cool. When you're hot you go red but cold turns you bluish. And so we get our convention. Yet the driver of a steam locomotive knew that if his fire glowed red it was going out: the whiter the glow the higher the temperature, even to the extent of a really high temperature being bluish. A tungsten lamp has a colour temperature of about 2800 K and is regarded as 'warm', while the colour temperature of an overcast sky is about 6500 K and this gives a 'cool' light. A 'cool' fluorescent lamp, therefore, has a higher colour temperature than a 'warm' one.

Now that we have connected the colour of the sun to its

TABLE 4. Some Correlation Colour Temperatures (CCT)

	CCT
Clear blue sky	15 000–30 000 K
Overcast sky	6500
Zenith sun (in clear sky)	6200
Sun 10° above horizon (in clear sky)	4100
Tungsten halogen lamp	up to 3500
Normal gas-filled tungsten lamp	2700–3100

Notes
1 The term Colour Temperature should strictly be confined to black body radiators, and the term Correlated Colour Temperature used for all other sources.
2 For fluorescent lamps *see Table 5*

temperature the importance of the concept in lighting becomes apparent. It turns out that not only the sun but all the light sources which we have been using up to the present, from oil lamp to tungsten lamp, correspond in colour quite closely to one or other colour temperature. Hence, our eyes have learned to travel up and down the colour temperature scale adapting in such a way as to eliminate colour casts, and to recognise objects by their colour whatever the light. Some colour temperatures are given in *Table 4*.

The system breaks down when there is very little light around. Then the cones in the retina, which are used for normal vision, become too insensitive, and the rods come into action. The rods are 1000 times as sensitive to light as cones, but they are not sensitive to colour. The rods start operating together with the cones at illuminances below about 30 lux in surroundings of normal reflectances (*see* p. 60). Below this level colours lose their strength until at about 1/10 lux the rods dominate and different colours are no longer discernible. But the eyes are capable of seeing things down to about a hundred-thousandth of a lux.

The characteristic of both theoretical black body radiation and of the light sources which closely approximate it, whether hot or not so hot, such as the sun, the furnace and the incandescent lamp, is that their spectra are all fairly smooth – there are no missing colours and no very large peaks (*Figure 59(a) and (d)*, even though the balance between the blue and the red varies considerably. It has been said above that our eyes can adapt to all these sources. Therefore the convention has been adopted that all black body radiators and those real light sources close to them give good colour rendering. This includes daylight and tungsten lamps. Indeed the whole problem of measuring colour rendering did not arise until the introduction of fluorescent lamps, some of which could be seen to distort colours in ways to which the eyes could not adapt. For example you could go into a butcher's shop lit by one of these lamps and you would not be able to judge the quality of the meat from its colour, nor would a doctor be able to make a sound judgement on his patient's health from his appearance. It is as well to remember that it is in such situations as these that the eye is really keyed up in colour rendering. In the art gallery the visual mechanism is, as it were, in a different and more relaxed posture. Matters of survival are no longer in question. Faces of curious colour are so common in paintings as to pass without question.

In spite of all this many artists have painted colours with extreme delicacy and it is most important that their colours are seen without distortion.

The problem of colour rendering then becomes the problem of finding out the extent to which a fluorescent lamp or other illuminant not belonging to the black body family departs from this black body standard. First of all a Correlated Colour Temperature (CCT) is assigned to the lamp under test, being the temperature of a black body of closest match to the test lamp (for colour temperatures in the daylight range we turn to standard figures to find a mixture of sun and sky of closest match). Then a mathematical process is carried out for comparing the appearance of colours to the *adapted* eye under the test lamp and under the standard reference illuminant at the same CCT. The mathematics is based on experiments carried out with observers (*see* Part II p. 199). However in essence we could make our judgement without mathematics if only we could carry out the following procedure:

1 Adapt our eyes to the illuminant under test.

2 Look at a set of representative objects under it and accurately memorise their colours.

3 Adapt to the reference illuminant.

4 Look at the same objects under this second illuminant and compare the colours to the colours in our memory.

The intervening periods of adaptation are essential, and of course our memories are not up to this exercise.

What has been said above implies that, for example, tungsten illumination is just as much a standard for colour rendering as the light from an average sky. But some colours do *not* look the same under these two illuminants. There are two reasons that might be invoked for this. The first is unquestionable, the second open to argument.

1 The first is called *metamerism*. It is a long-established custom to distrust any light source but daylight for matching textiles in a shop. In the past this was often a simple matter of insufficient light, but mismatches can also occur due to metamerism. Metamerism occurs when colours match under one light source but not under others. Colours which have the same spectral reflectivity characteristics, that is to say reflect the same amount of light as each other at each wavelength, will match under any light. But two colours which match under one light may not have the same spectral reflectivity, especially if one is made of a mixture. They may then not match under another light and are called metameric.

Examples of metamerism which occur in paint usually involve the red end of the spectrum. The reflectance spectra of Azurite, Prussian Blue and Manganese Blue peak at the blue end of the spectrum and fall at the red end, as one might expect. But other blue pigments have a surprisingly large reflectance at the end of the red, where the eye is not very sensitive: Cobalt Blue and Ultramarine for example. They tend slightly in the direction of the purple of *Figure 2*. A paint of the pure blue group, say Azurite, may match with one from the 'tendency to purple' group, say Cobalt Blue, in a low red light such as north sky light. But bring the two under tungsten, rich in red, and the Cobalt Blue will look more purple: a metameric mismatch will appear[62,63].

Metameric trouble with the pigments used in paintings and the natural dyes used in old textiles is not uncommon, particularly between one fluorescent lamp and another. There is scope for mismatches among the modern dyestuffs with their strong colours and sharp spectral absorption bands, especially when they are mixed to produce dull colours.

2 The second matter relates to the abundance of red and shortage of blue in tungsten illumination compared to average daylight. Our first expectation would be that blues tend to darken and reds stand out in objects viewed under tungsten. But the properly adapted eye compensates efficiently for this, as has been explained above. Even allowing for adaptation, might not an area painted in dark blue tones lose some of its significant detail because of the shortage of blue light under tungsten, especially at 50 lux? Colin Thompson has written of 'circumstances where the distinction between one colour and another is critical, as anyone knows who has darned navy blue socks with black wool and looked at the results next morning.[46]. Navy blue socks are more popular than green or red socks of the same

Munsell Value (p. 59), and one might in fact find these just as difficult to darn. I have searched, so far without success, for a painting in which colour passages loose significance because of the colour temperature of the illuminant[51]. Experimental work mentioned in Part II (p. 206) appears to confirm this negative finding, but the question has not yet been precisely resolved[64].

In the light of the foregoing we might now ask ourselves a number of questions before choosing our illuminant:

1 Is good colour rendering important? It usually is in museums. If so, our choice is at present limited to daylight, tungsten lamps (including tungsten iodine, sometimes called quartz halogen), or one of the fluorescent lamps of good colour rendering (*see Table 5*).

2 If the light is to be artificial, do we want warm or cool light? The main point here is that at low illuminance a warm light more easily creates an apparently bright situation. At 50 lux, artificial light, for this reason and because it is more easily controlled, is to be strongly preferred.

3 Has the UV radiation been dealt with? UV filters are essential where sensitive material is on display under daylight or under most kinds of fluorescent lamp (p. 16). They need not be used for tungsten lamps (but glass is needed for tungsten iodine).

TABLE 5. Some Fluorescent Lamps or Good Colour Rendering

	Colour temperature category	R_a	R_W	*Crawford class*	*UV content*	*Best efficiency at 2000 h*
Matsushita FL 40S-N-EDL	Cool	98	96	B	61	?
Sylvania Design White	Cool	89	87	B	111	?
Wotan Maxilux 19 DL	Cool	93	91	B	95	?
Philips 94	Int.	95	86	C	43	65
Philips 38	Int.	90	83	C	81	45
Sylvania Cool White DL	Int.	89	83	C	127	?
Thorn Kolor-rite	Int.	89	82	C	85	45
Wotan Maxilux 22 DL	Int.	94	87	C	?	63
Philips 93	Warm	93	81	C	49	65

Notes

'Cool' colour temperature category is about 5000 K, 'Int.' (Intermediate) about 4000 K, and 'Warm' about 3000 K.

There are lamps of good colour rendering at about 6000 K, such as the Northlight/ Colour Matching, but these are too cold for normal exhibition use.

Recommendation for good colour rendering.

R_a about 90 or better,
R_W (worst R) about 80 or better,
and Crawford Class A, B or C

(For an explanation of these colour-rendering indices see p. 199.)

UV Content is in microwatts/lumen. If less than 75 no UV filter is required.
Efficacy (see p. 186) varies a little by lamp size. The best value after 2000 h running is quoted.
Countries of origin: Matsushita – Japan. Philips – Netherlands. Sylvania – USA. Thorn – UK.
Colour rendering and UV content computed in the National Gallery Laboratory from spectral data kindly supplied by the manufacturers.
This a selection, not an exhaustive list.

4 Will radiant heat be a problem? Radiant heat is discussed on p. 45. Briefly, tungsten light, whether as a narrow spotlight or widely spread, causes appreciable heating at over 200 lux, especially if directed at closed cases. If the light source is in the case, which it should not be, the problem is even greater. Where there is a particular sensitivity to radiant heat, use either fluorescent lamps or tungsten lamps with dichroic reflectors ('Coolbeam' lamps, p. 46).

Summary

The problem of colour rendering began with fluorescent lamps. It cannot be solved by looking at identical groups of objects lit by different lamps in adjacent booths. The colour rendering quality of a lamp can, however, be calculated in a way which takes into account the colour sensitivity and the colour adaptation capability of the normal human eye. The convention is to regard all phases of daylight and tungsten light as standards. Spectral measurements on the lamp to be assessed are then compared to the same measurements on a standard of the same 'correlated colour temperature'. Fluorescent lamps of good colour rendering have been selected in Table 5. Some colours which match under one light fail to do so under another. This is called metamerism.

THE MEASUREMENT OF COLOUR

An adequate explanation of this subject requires a book[65–67]. However the reader may even be surprised that colour can be measured, so that a very brief summary of the principles involved may be helpful.

A restorer often needs to colour a small patch so as to match it to its surroundings in a way that makes its presence disappear from view. To do this he has to match not only the colour but other visual properties such as gloss, surface texture and relief. The scientist investigating colour change has a simpler problem because he can ignore the other properties and concentrate on the colour, that is, on the wavelength characteristics of the light reflected from the colour as they affect the eye. We shall take this second approach.

We can use for our measurements either a reflectance spectrophotometer or a colorimeter. Both instruments are designed to illuminate the test area with light of a known colour and to measure the strength of the reflected light. The arrangement must be fixed so that the light always falls on the test area in the same way and the reflected light is always measured from the same direction. The difference between the two instruments is that, whereas the colorimeter uses only a few colours, commonly three, the reflectance spectrophotometer measures reflectance throughout the visible spectrum at small enough intervals for a continuous reflectance curve, as in *Figure 2*, to be built up. This reflectance curve gives a colour specification which is completely independent of the illumination and which can in some cases be used for chemical identification. Since our main purpose in measuring colour is to detect changes due to deterioration, the reflectance spectrum is the best form of data presentation.

However we may now wish to represent the colour as the eye sees it by taking type of illumination into account. In the ocean at depths greater than about 30 m, red light from the sun and sky is so completely absorbed by the water that red sponges and fishes look black. More alarming still to the amateur diver, as the author has confirmed, ones blood becomes undoubtedly green[68]. The reflectance spectrum remains unchanged, and if we take a torch down with us red objects are seen as red, yet in the ambient light colours are dramatically changed because red is absent.

Of course such exotic effects do not occur in the museum because only 'white' light sources are used. Nevertheless the colour which the adapted eye sees is always a combination of the reflectance properties of the object (its reflectance curve), the spectral power distribution of the illuminant, and the response of the human eye. This leads us directly to the Chromaticity Chart (*Figures 22 and 23*) and to the system of colour measurement of the C.I.E. (Commission International de l'Eclairage). To use this system we have to have figures for (a) the reflectance spectrum of the colour patch, as measured by spectrophotometer, (b) the spectral power distribution of the illuminant, usually from standard tables[66,69], and (c) the response of the average human eye, also from standard tables.

The result of a rather lengthy computation gives us (x,y) values which allow us to place our colour on the Chromaticity

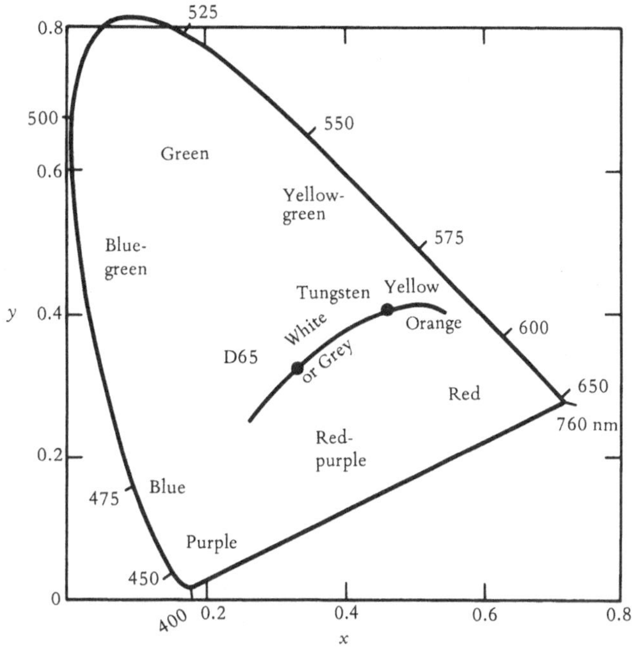

Figure 22
The CIE Chromaticity Chart of the colours. All the pure spectral colours are to be found round the curved edge of the enclosed area, from blue at bottom left to red at bottom right. As we move in from this curved edge the colours become desaturated, that is to say, mixed with grey or white, and the whites are all found in the lower central area. Cutting through this area is a tinted curve, the black body curve, on or near which are to be found all the museum light sources. Average daylight (D65 Standard Daylight) and tungsten illumination are marked on this curve. All colours are contained on the diagram, though their appearance may be modified by surface texture and reflection (1964 10° CIE diagram)

Chart, just as on any other graph. A colour change in our sample due to damage by light or other factors will now result in a new (x,y) point, and this shift in colour can thus be measured on the Chart. It is best to be frank and realise that the only way to understand the colour measurement system is to master the calculation, and this requires some elementary mathematics (*see* Part II, p. 195). However no mathematics are required to appreciate the usefulness of the diagram itself, and here is an example.

If we draw a line between any two colours on the diagram all their mixtures will be found on a line between them. Two complementary colours if mixed in the right proportion produce white or grey. Therefore to find the complementary of a colour join it to the light source by a straight line and continue the line through the light source and beyond. The complementary colour, in varying degrees of saturation will be found on this line. Further properties are given in the captions to the *Figures 22 and 23.*

It should be noted that, although the (x,y) position gives all the necessary information on the composition of the colour, one item of information, called the *luminance*, has been dropped out on the way from the reflectance curve. Luminance, conventionally represented by Y, can be calculated from the reflectance curve together with the amount of light falling on the colour, but not from the Chromaticity Chart. Luminance is a measure of the amount of light coming to us from the colour. The illuminant, the white light source used for illumination, must also be put on the chart, since its colour can also

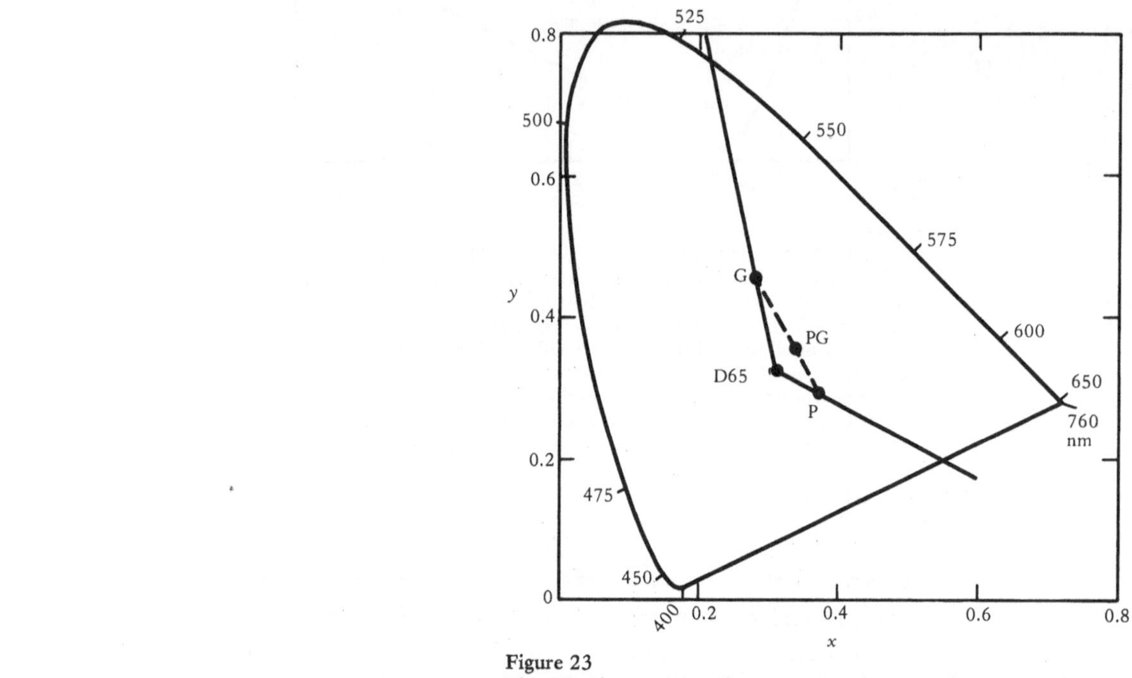

Figure 23
The CIE Chromaticity Chart (continued). A non-luminous colour means nothing on this diagram unless the light illuminating it is specified by placing it also on the diagram. Here G is the green of *Plate 2*, P is the purple, and PG is the grey formed by mixing them. The illuminant is Standard Daylight D65, and this also is marked. All mixtures of two colours lie on a line joining them, so PG is on the line joining G to P. To find the 'dominant wavelength' of a colour, which is its nearest pure spectral colour, draw a line from the light source through the colour to cut the boundary curve. The dominant wavelength of G is about 530 nm, which could have been roughly guessed from *Figure 2(a)*. But P is interesting because the purples, being mixtures of blues and reds, lie outside the spectral range. The straight line joining 400 and 760 nm on the diagram bounds all these purples which have no spectral equivalent, and indeed the line from D65 through P cuts this purple line rather towards the red end. Finally even the grey PG has a dominant wavelength, though since it is so near D65 we can only say that it is in the yellow-red direction. The original PG, if not its reproduction in *Plate 2* is certainly a warm grey

be measured. We would find that a neutral grey colour patch would come very close to the illuminant on the chart, in spite of one being dull and the other bright. The luminance Y provides the missing information by putting the colour on a scale from light to dark, as measured by the amount of light it reflects.

We now have two ways of measuring colour, the first and most informative by obtaining a reflectance spectrum, the second by finding the position of the colour, as it appears under a certain light source, on the Chromaticity Chart. With certain forms of colorimeter this second method can be carried out directly and quickly.

A third method, quick and relatively cheap, should now be mentioned: the colour atlas. The best-known colour atlas is the Munsell System, issued in two volumes containing arrays of standard colour patches, the *Munsell Book of Color*[70].

Imagine a globe like the earth containing all the colours

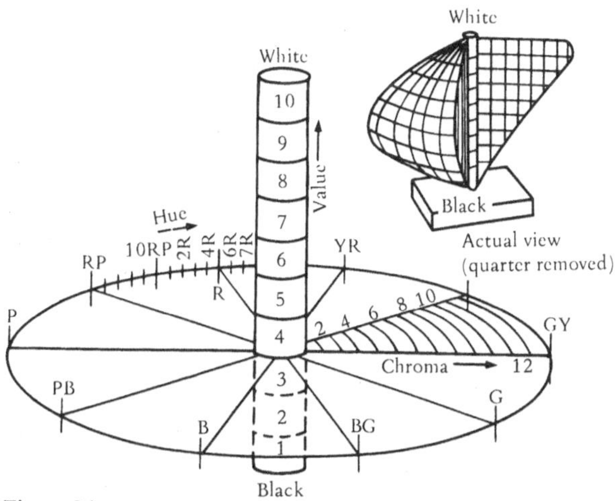

Figure 24
Diagram of the Munsell Colour System, showing the three Munsell dimensions of HUE, measured as longitude (round the equator), VALUE along the axis from south pole (black) to north pole (white), and CHROMA, measured from the axis (where there is no colour) outwards. Each Hue (P, RP, etc.) is divided into 10, which is about as much as the eye can discern (P = purple, R = red, Y = yellow, G = green, B = blue). These three dimensions are used to specify any point within the globe, and every point represents a colour. There is, however, a most irregular surface to this globe. Some colours, like blue, cannot take a high Value without being mixed with white. This reduction in Chroma pulls the blue Hue surface towards the axis. Others, such as yellow can have a high Value and remain spectrally pure (of high Chroma). The effect of this is to push in the globe near the north pole in the blue direction and to stretch it out in the yellow direction. The Munsell figures of the green and purple of *Plate 2* are 2.5 G 6.5/9 and 5 RP 4/6.5. This means that the Hue of the green brush-stroke is 2.5 divisions along the scale from the end of GY, while its Value is 6.5 and its Chroma 9. Similarly the purple is 5 divisions towards RP from P, its Value is 4 and Chroma 6.5. (Diagram adapted from IES[71], pp. 5–10)

(*Figure 24*). The South Pole is black and the North white. All the greys from dark to light are placed along the axis between them. The position on this axis is called the *Value*. Round the equator are placed the pure colours. They are called pure or fully 'saturated' because they are unmixed with white or grey. The position along the equator is called the *Hue*. But the majority of colours are neither fully saturated nor neutral grey, therefore we need a third dimension defining the distance out from the grey central axis. This is called the *Chroma*. Thus each colour is defined by three dimensions: Hue, Value and Chroma.

Munsell readings can be converted to (x,y) values on the Chromaticity Chart. Thus it can be seen that, with some loss of accuracy, a valuable record of colour change as change in Munsell Hue, Value and Chroma can be kept while completely avoiding the cost of instrumentation and specialised staff.

Sometimes the Munsell Value alone is a useful scale on which to designate lightness, as of a wall-covering (*Table 6*). In practice reflectance percentage can be obtained from Munsell Value V by the formula $V(V-1)$, though this formula breaks down at the extremes.

TABLE 6. Munsell Values[62]

Value	10	9	8	7	6	5	4	3	2	1
Approx. percentage of light reflected (luminous reflectance)	100	80	60	45	30	20	12	7	3.1	1.2

Summary

Colour can be measured using a spectrophotometer or a colorimeter. Using the results from either of these instruments colour can be defined by its coordinates (two numbers) on a chromaticity chart (Figure 22) plus a value for its luminance, the amount of light coming to us from the colour. Another way of defining colour by three numbers (for Hue, Value and Chroma) is the Munsell system, using a colour atlas of small colour patches which are compared with the sample.

THE LIGHTING SITUATION AND THE PROCESS OF SEEING

Human vision was not evolved for the purpose of wandering round a museum under carefully chosen illumination, but for survival. There was survival value in a system which could detect sources of food or of danger by their colour in all situations. It is as well to remember that a keynote of the visual system is its adaptability. The eye is not like a photographic camera[65,72]. It is more like a closed-circuit television system linked to a computer. The computer comes between the TV camera and what we 'see'.

The previous section on colour rendering has shown that the visual system says 'white' for many different surfaces and lights because of this unconscious computing. There is no absolute standard of white in the brain against which we can compare the signals from our eyes.

Similarly there is no absolute standard of brightness. The visual system is very good at detecting that one thing is brighter than another when both are near at hand (*Figure 25*). But we can easily be misled into thinking that bright lights do not give sufficient illumination or that a low illuminance is apparently bright.

Recent experiments have shown that we can even be made to see curved lines as straight. For this purpose the subject wore distorting spectacles continuously for a period of time[74]. Gradually the distortions produced by the spectacles disappeared and the world looked as it had without them. If the subject had then been asked to draw a straight line on a canvas with the spectacles still in place, what would he have drawn? A straight line, of course, since drawing involves checking what you draw against what you see, and both are affected by the same distortions. (Yet one still comes across theories that El Greco distorted figures because of defective vision.)

In limiting the brightness of museum lighting we have to gather what knowledge we can of the effects of illumination level on our ability to see small but significant differences in form and colour. Put a little more precisely we need to know the effects of illuminance on (a) visual acuity and (b) colour discrimination. Bearing in mind what has been said about the relative nature of the eye's judgements it can be readily seen that this is no easy task.

Neither visual acuity nor colour discrimination are likely to be at their maximum however the museum is lit, whether or not there has been an attempt at control. The situation is different in the restoration studio. The restorer like the surgeon needs

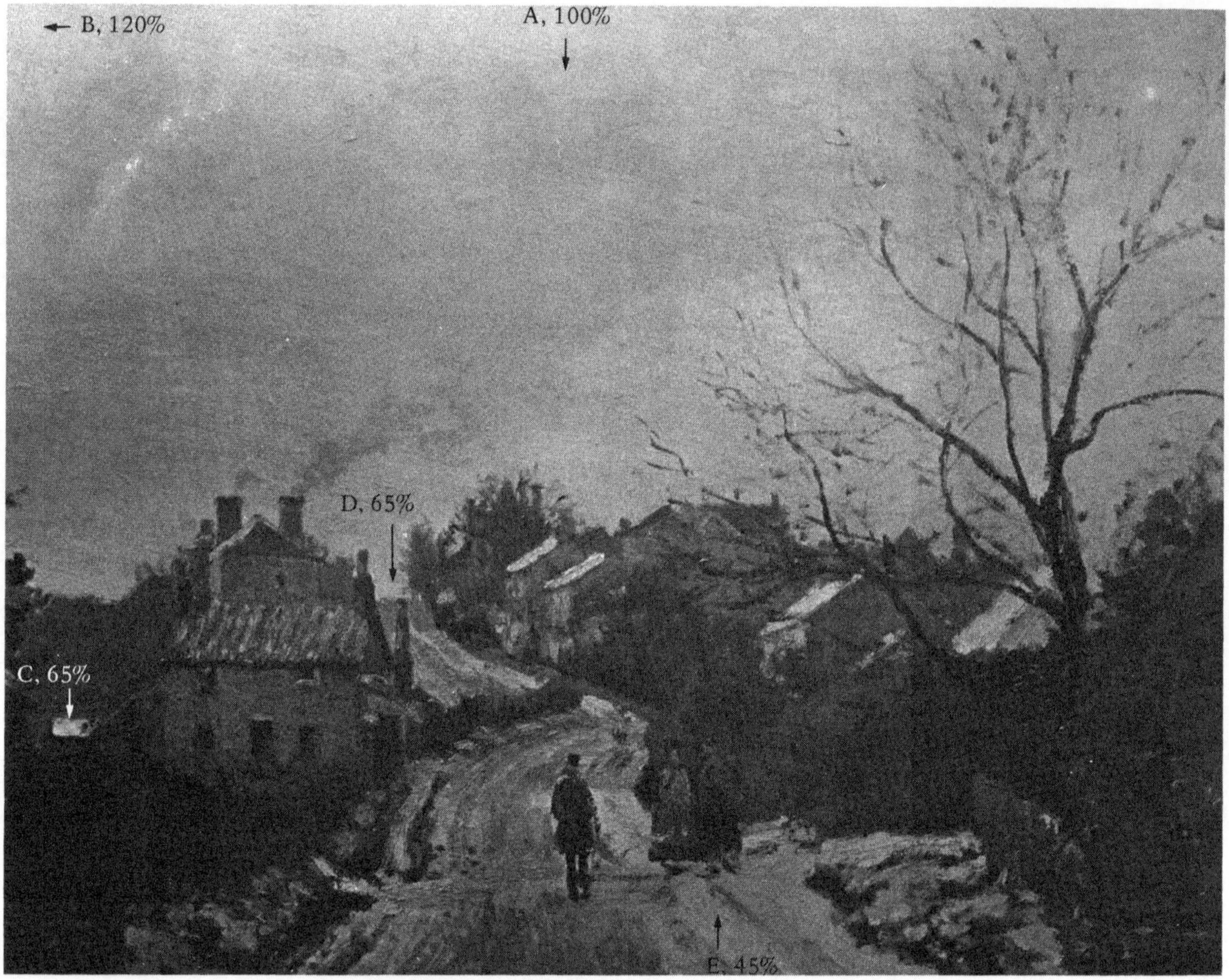

Figure 25
Pissarro, *Lower Norwood under snow* (N.G.No.6351). Looking at either the original painting or this monochrome reproduction it is impossible to ascertain whether or not the picture is evenly illuminated or to compare relative brightnesses in different parts of the scene. This is because the eye has no inbuilt scale of brightness and relies mainly on noticing brightness differences between adjacent areas[73]. In fact for this photograph the picture was much more strongly lit at the top. The relative brightnesses (on the print from which this block was made) of some light-toned areas marked are as follows: A 100%; B 120%; C and D 65%; E 45%. However it may appear that the small roof (C) is the brightest part in the picture, and that the light areas on the road (E) are as bright as the sky

to see the object under his hands in all possible detail, and he needs strong illumination in order to do this (p. 42). But many of the things a restorer sees are not relevant to the message which the artist wished to convey to his audience. I think one would be correct to suppose that artists have always assumed that their creations would be viewed in a variety of situations, not all lit ideally, and have designed their work accordingly. Even the Impressionists and others who made a point of completing their canvasses in the open air did not expect them to be so viewed. When De La Tour painted a candle-lit scene he painted it in such a way that the scene would look candle-lit under any reasonable lighting (for a contrary view see Weale[75]).

But it could also be said that, however robust the work of art, it will look better in some lighting situations than in others, and so we should bend our efforts to finding the best possible situation. Within the limitations of the museum one cannot but agree, provided there is indeed a consensus of perceptive opinion on what is best, and provided that the damage caused by light is kept under control.

There is certainly no mathematical treatment whereby we can equate the viewer's gain against the exhibit's loss. However there has been considerable research on the visual process as it is affected by the lighting, and Brommelle[76] has carefully related the experimental work to the museum problem.

Brommelle considered what happens to deterioration and what happens to our visual sensitivity when we double the amount of light falling on an exhibit, for example, by doubling the recommended 50 lux. There is some increase in sensitivity, but it is nowhere near doubled, whereas we can assume that damage is indeed doubled from the reciprocity law (p. 21). For further details *see* p. 180.

On the question of colour discrimination Crawford[77] has written on the museum problem. Crawford set up a situation in which the observer looked at a rectangle which could be brightly or dimly lit with any colour or white. Another colour, the test colour, could be added progressively to the lower half of the rectangle. The observer was required to note the point at which the test colour being added became just visible to him. The surrounding area was lit at a slightly lower but similar level to ensure adaptation. Crawford concluded that discrimination of colours begins to fall off below 30 lux: 'Much higher levels have been proposed for the attainment of a full appreciation of the saturation of the colours of a picture, but it is uncertain whether such higher levels are necessary if the observer is fully adapted to the level of 30 lux and is not distracted by other higher levels which are in view. Full adaptation and absence of distraction obviously call for careful and appropriate design of the whole of a picture gallery.'

As we get older our eyes become less effective. Are light

levels which are satisfactory for the young also adequate for the old?

The most obvious change with age is the reduction of focusing ability. We can no longer adjust our eyes to see both near and far objects in sharp focus. The effect is more pronounced in dim light since the pupil is then at its largest and the depth of field consequently at its smallest, just as in a camera. For this reason older people find their focusing sharper in strong light. However I think we must assume that visitors come to museums equipped with suitable spectacles. Other diseases of the eye more prevalent in old than young, such as incipient cataract, of course lower discrimination.

The old eye is also yellower. Both the lens and the macula (the area on the retina corresponding to our centre of vision) become more yellow with age. This does not mean that things look yellower: the computer in the brain sees to it that the necessary adjustment is made. But it could mean that, for example, a dark blue on a black ground becomes less visible with age. Even if there were such cases of depressed blue sensitivity, the detection of light blue tones appears unaffected[78].

Lastly, one would not be surprised if the eye, becoming older, also becomes in a general way a little less effective in picking up messages as happens to the ear, when it becomes not deaf but less capable of sorting out conversation from background noise. In electronics terminology we would say that the communications channel may become more 'noisy'. Surprisingly no experimental work has been done on this that I can find, so we cannot say whether it happens or not.

Many experiments are not easy to interpret in real-life terms. At the beginning of this section our visual system was compared to a TV camera plus computer. This system presents a view of the world to the higher centres of the brain, which attempt to interpret it. But the computer has done a lot of processing of the messages from the retina before our consciousness comes into play. There is discussion of some further experiments on the processes of seeing in Part II p. 180.

Summary

The eye is a camera plus a computer. In reducing damage by light to a minimum we must ensure that exhibits are lit for comfortable and discriminating view. Maximum visual acuity is only attained at very high illuminance and is appropriate to restoration rather than normal viewing. The 50/200 lux illuminance levels are safely above the levels at which colour and contrast discrimination are seriously affected.

Humidity
Part One

Humidity

THE IMPORTANCE OF HUMIDITY

Even where it is not visible, water is to be found almost everywhere. Our bodies are composed of water to the extent of 65% of our weight. Since plants and animals contain a great deal of water it is not surprising that products made from them also retain moisture. If the moisture is taken away from wood, ivory or bone they contract and very likely split and warp. Laminar organic products such as paper, parchment, leather and natural textiles become less flexible so that their fibres become easier to break.

By contrast, in very damp conditions direct physical damage is less likely to occur, but these are the conditions that suit the growth of moulds and fungi.

The absorption of moisture makes objects swell and *vice versa*. In changing size they may also change shape, or warp. Many museum objects are made of different materials joined together. These rarely respond to moisture changes in the same way so that such changes may cause breakage. Even objects composed of pieces of the same material joined together may crack, either because of the glue, or because the material swells differently in different directions. Wood, bone and ivory swell much more across the grain than along it. In the most sensitive category are panel paintings, veneered furniture, musical instruments and wooden objects from the tropics.

All of this makes moisture change potentially disastrous in the museum. We must therefore study how to control it. And it is as well to realise at the outset that control of humidity is a great deal more important than control of temperature.

The amount of water dispersed in the air as vapour (that is to say as a gas, not droplets) is normally quite small but always very significant. If all the water vapour were to be condensed out of a cubic metre of air on a spring day in a temperate climate we would get about 10 ml of liquid water. But this amount of water is contained in a mere 6 cm cube of wood in the same conditions. Nevertheless in a normal ventilated situation it is the water vapour in the air which controls the amount of water in everything else. Thus to control the moisture in the exhibits we control the humidity in the air.

To control something to a certain level we must be able to measure it, which means in turn that we must have a scale to put this measurement on.

MEASURING THE HUMIDITY IN THE AIR

The most obvious suggestion for a scale of measurement would be the amount of water contained in a given volume of air, say 10 grams per cubic metre (10 g/m^3). The trouble with this situation is that hot air with 10 g/m^3 of water vapour will cause dangerous drying, whereas cold air with 10 g/m^3 can be so damp that moisture condenses from it. This is because hot air has a greater capacity for water than cold air. Air which dries clothes on a washing-line in the shade during the day will re-moisten them as it cools in the evening. Thus this scale, which we will return to and which is called the *absolute humidity* of the air, will not do for our general scale of measurement. We need something that is more or less independent of temperature.

Let us forget about the physics of the situation and look around for the most practical measurement. We wish to prevent objects from changing size and shape because of moisture changes. Therefore we take a sample of moisture-absorbent material such as a block of wood, and measure its length across the grain in different situations. We say that if the wood swells that means that the air has got damper and vice versa. If we manage to control the air so that the block of wood stays the

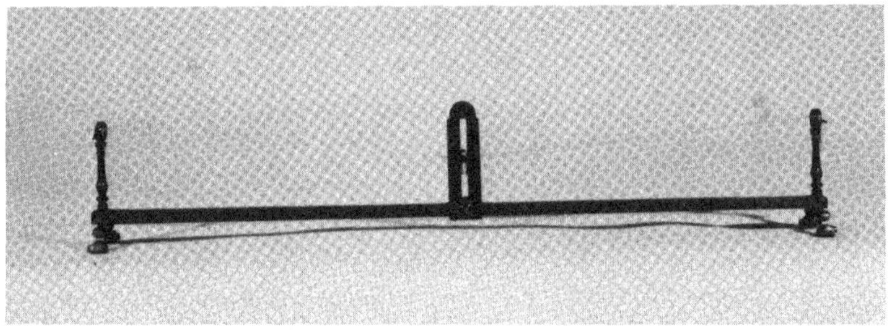

Figure 26
An elegant early hygrometer which consists simply of a long set of human hairs attached to the frame at both ends and supporting a weight against a scale in the centre. As the relative humidity (RH) rises the hairs lengthen and the weight with its pointer falls down the scale

same size we have solved our humidity control problem. We have in fact used the scale of *relative humidity*, which is what we need. An instrument could indeed be constructed using a block of wood, but it would be far too slow in reacting. We can use instead a bunch of human hair, which reacts in the same way but much more quickly. A hair hygrometer was constructed in 1783 and such instruments are still in common use today (*Figure 26*).

This RH scale is usually expressed as a percentage and it can be defined as follows:

$$\text{RH} = \frac{\text{amount of water in a given quantity of air}}{\text{maximum amount of water which the air can hold at that temperature}} \times 100\%$$

Air at 100% RH is holding all the water it can, and is said to be saturated. Saturated air at 10 °C holds about 10 g/m^3 moisture, at 20 °C about 17 g/m^3 and at 30 °C over 30 g/m^3. The RH is a measure of the percentage saturation of the air. Air at 50% RH, whatever the temperature, is therefore holding half the water it can.*

For any level of RH each material will have a characteristic water content, provided that the temperature is not allowed to run to extremes. For wood at 55% RH this is about 12%. But we do not have to measure the moisture in materials – all we have to do is to measure the RH of the air surrounding them and keep that constant.

It is common knowledge that materials expand when the temperature rises. But for moisture-absorbent materials temperature expansion is small compared with that caused by humidity change. For example the same amount of expansion across the grain follows either from a rise of 4% RH or a rise of about 10 °C at constant RH. But whereas the RH change is the kind of fluctuation one gets even with the best air conditioning, 10 °C represents a very large change in temperature (*see* Part II, p. 222).

The hair hygrometer in its eight-day recording form, often combined with a temperature recorder (*Figures 27 and 28*), has become a common sight in museums all over the world, and testifies to a growing awareness of the importance of climate control. *But it is only half an instrument.* Because it can so

*To be very strict, Relative Humidity and Percentage Saturation are two different scales, RH being the ratio of partial pressures of water vapour and PS being the ratio of masses. But for our purposes the two can be taken as indistinguishable.

easily slip out of calibration, either through a jolt or by slow drift, so that its readings are no longer true, there must be hundreds of humidity records stored away in museums which are in fact worthless. Ideally the hair hygrometer should have its calibration checked monthly. This is a simple matter, but requires a second form of hygrometer whose readings can be relied on. This becomes, in a manner of speaking, the other half of the instrument.

Until recently the hygrometer used for this recalibration has always been one of the wet-and-dry-bulb instruments such as

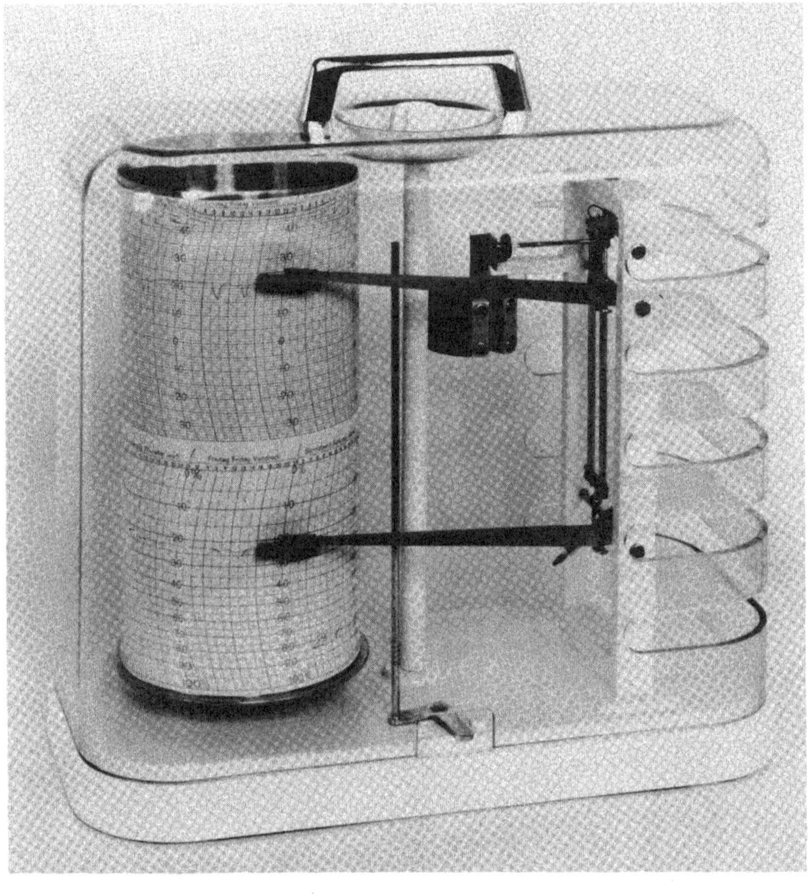

Figure 27
Recording thermohydrograph (Procema). The charts hold a one-week record, temperature above and RH below. The RH sensor is a bunch of human hairs or of a plastics substitute hanging vertically on the right. Its upper end is attached to the frame, its lower end to a lever carrying a pen at its tip. The pen bears on the drum. A rise in RH causes lengthening of the hairs, making the pen drop. The temperature sensor is a bimetallic strip bent into a cylinder. On a rise in temperature one of the metals expands more than the other, causing the cylinder to uncurl. This moves the temperature pen upwards. Before the instrument is moved or the charts changed, the pens must be withdrawn from the paper by actuating a lever in the centre of the base

the sling hygrometer (*Figure 29*). These devices may still be the only form available in some parts of the world. In the right hands they are highly reliable, but in the museum one cannot be sure they will be operated correctly. Therefore the description of the wet-and-dry-bulb hygrometer will be followed by some words on choice of a good electronic instrument.

Summary

The primary humidity scale in the museum is the scale of Relative Humidity (RH). We can say that in practice, moisture-containing materials, such as wood, paper, ivory, leather, parchment and natural textiles, will neither appreciably expand nor contract so long as the RH remains constant.

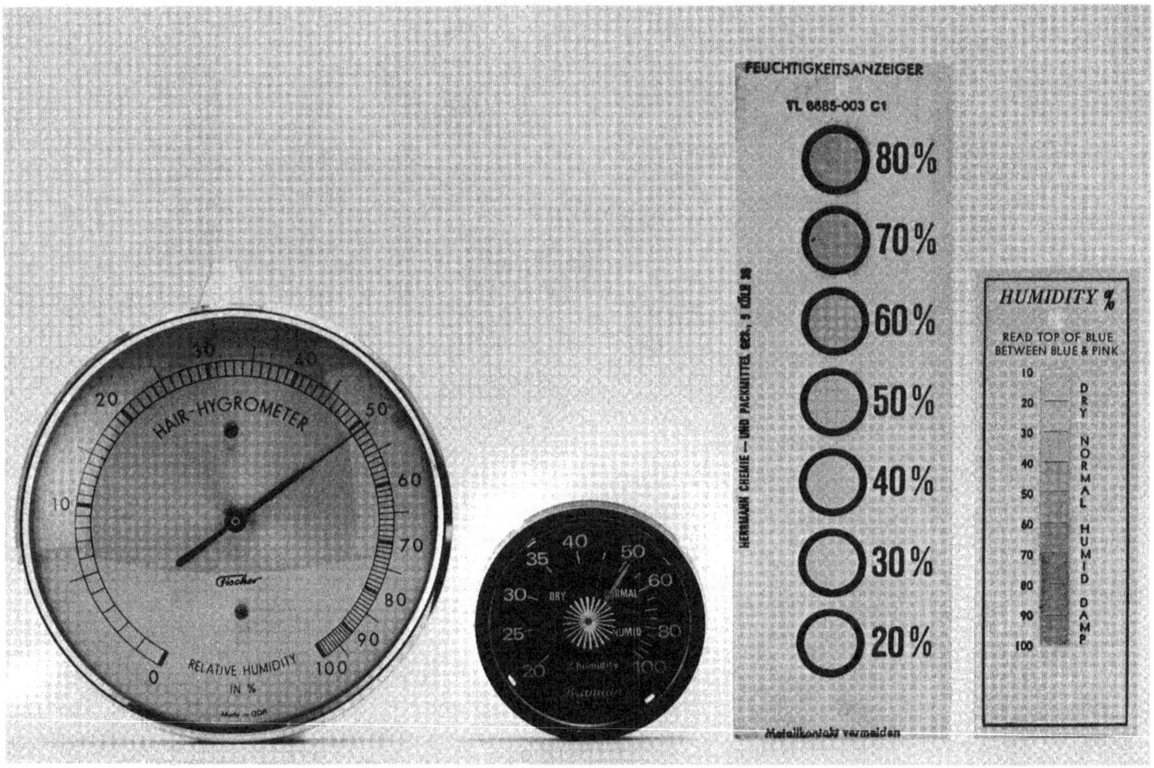

Figure 28
Dial hygrometers – usable if regularly calibrated. Humidity indicating cards

THE WET-AND-DRY-BULB HYGROMETER

The simplest, most inexpensive type of wet-and-dry-bulb instrument is the *sling* or *whirling hygrometer*, sometimes called the sling psychrometer.

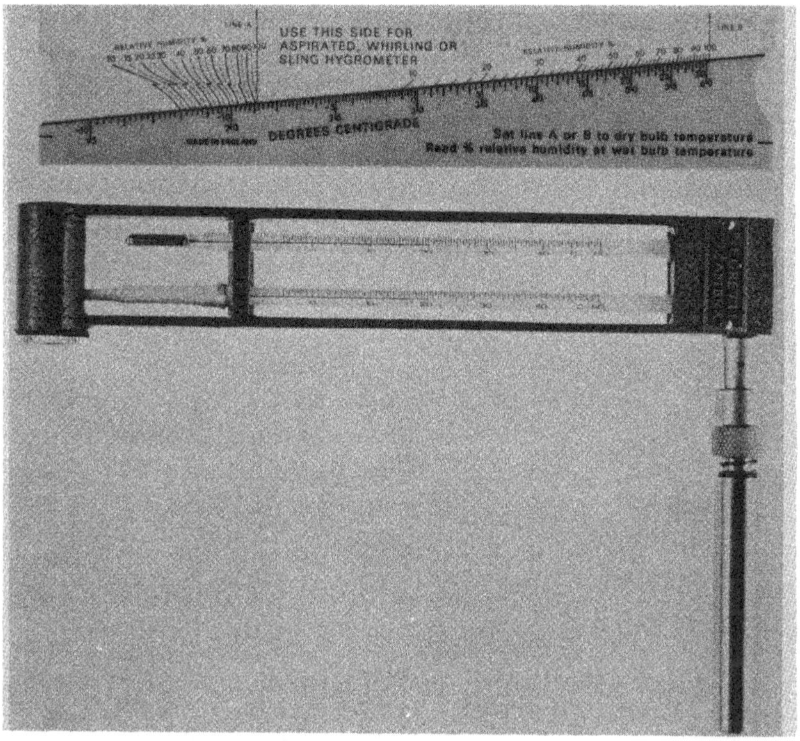

Figure 29
Sling hygrometer of standard pattern with slide rule for obtaining RH from wet-and-dry-bulb readings. The upper of the two thermometers is the dry bulb. The lower, enclosed in a sleeve, is the wet bulb. The sleeve is kept moist by filling a reservoir on the extreme left of the instrument with distilled water. Of the two temperature scales on the slide rule, the upper more expanded scale will usually be appropriate, using line B

The sling hygrometer, like all wet-and-dry-bulb instruments, consists of two ordinary thermometers (*Figure 29*), now called the wet bulb and the dry bulb. The dry bulb is unmodified, and therefore simply reads the air temperature. The only modification to the wet-bulb thermometer is a fabric sleeve or wick fitted over the mercury bulb. This must be kept wet with distilled water.

It is easy to see that moving air past the wet bulb will cool it by evaporation of the water, just as standing a porous pot on the window-sill cools its liquid contents. This effect has been known since antiquity. Thus in a current of air the wet bulb will always show a lower temperature than the dry bulb, provided that the air is not fully saturated with moisture, and the drier the air – the lower its RH – the greater this difference will be. The difference also depends on how fast the air moves past, but surprisingly, only to a limited extent. Therefore, if we mount the two thermometers, as in *Figure 29*, so that they can be swung round and round, after a short time the wet bulb will settle down to a definite temperature. Wet and dry bulb temperatures are then both recorded. To find the RH from these two readings we need to refer to a table, *Table* 7 or a chart, *Figure 31*, or a slide rule provided with the instrument (*Figure 29*). To use these we need the wet bulb depression, which is the difference between wet and dry bulb readings.

Sling hygrometers are simple and reliable, yet can be wrongly used. The most common error is too high a wet bulb reading, leading to too high an RH estimate. This can be because of: (a) insufficient swinging, (b) too long a pause before finding wet bulb reading (which starts to rise as soon as swinging ceases), (c) hands or breath too near thermometer bulbs, or (d) dirty wick.

In using the sling hygrometer, these points should be kept in mind:

1 Use distilled water only for wick. Keep it clean and change when necessary.

2 Sling well away from body and walk slowly round room.

3 Keep hands and breath away from thermometer bulbs. We need room temperatures, not body temperatures.

4 Keep strong light away from thermometer bulbs.

5 Ensure that the wet bulb does not dry out.

6 The wet bulb temperature falls during slinging. Its true value is not reached until successive readings agree.

7 Read wet bulb quickly, before it starts to rise. Dry bulb can then be read at leisure.

8 Read to ¼ °C. It is not practical to read to much less than this, but in the middle of the range, ¼ °C error in either wet or dry bulb reading amounts to an error of 2% RH. Therefore we cannot expect better than ±2% RH accuracy.

We now have a reliable RH-measuring instrument which will also enable us to perform a monthly calibration of our eight-day recording hair hygrometers (thermohygrographs): the wet-and-dry-bulb hygrometer. *Always mark on the chart where a re-calibration has been done and write the correct reading.*

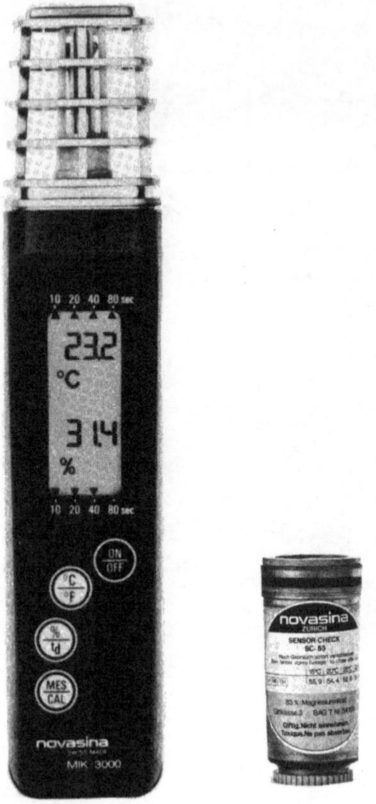

Figure 30
An electronic hygrometer which can be self-calibrated: the Nova Sina Mik 3000. To recalibrate, the guard on the sensor at the head of the instrument is removed and replaced by the 'Sensor-check'. This carries a salt solution behind a semi-permeable membrane and, in effect, surrounds the sensor with a known RH. When the reading with the Sensor-check in place settles down to a steady value this reading can be checked and if necessary adjusted to the RH provided

Figure 31
Simplified hygrometer chart. An alternative to the hygrometric table for finding RH from wet- and dry-bulb readings. Select the vertical line corresponding to the dry-bulb reading and the sloping line corresponding to the wet bulb. Estimate the RH from the curve where they cross. A fuller explanation of the reading of this type of chart is given on p. 210. For practice in estimating positions between the marked lines and curves it is helpful to make markings on a photostat of this diagram

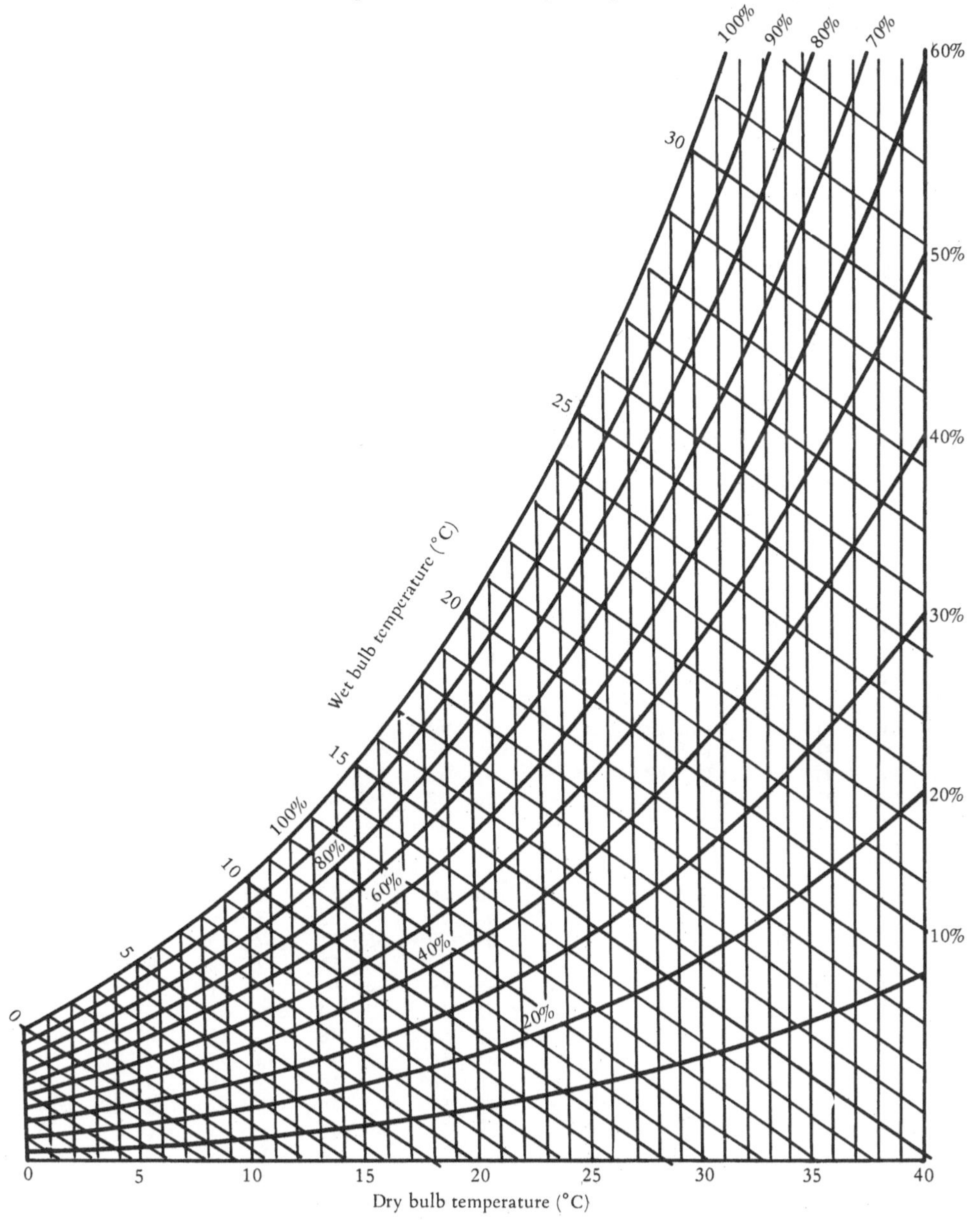

TABLE 7. Hygrometric Table.
For calculating wet- and dry-bulb readings on a sling hygrometer or other aspirated wet-and-dry-bulb instrument. Record the dry-bulb temperature and the wet-bulb depression (difference between wet- and dry-bulb). Locate the row corresponding to the dry-bulb temperature and the column corresponding to the wet-bulb depression. The RH is where they cross. For example, a dry bulb of 20 °C and a wet bulb of 14½ °C gives a depression of 5½ °C. Where the row corresponding to a dry bulb of 20 ° crosses the column for 5½ ° C depression we read RH = 55%

Dry Bulb (°C)	*Depression of the wet bulb (°C)*																			
	0	**½**	**1**	**1½**	**2**	**2½**	**3**	**3½**	**4**	**4½**	**5**	**5½**	**6**	**6½**	**7**	**7½**	**8**	**8½**	**9**	**9½**
40	100	97	94	91	88	85	82	80	77	74	72	69	67	64	62	60	57	55	53	51
39	100	97	94	91	88	85	82	79	77	74	71	69	66	64	61	59	57	54	52	50
38	100	97	94	91	88	85	82	79	76	74	71	68	66	63	61	58	56	54	51	49
37	100	97	94	91	87	85	82	79	76	73	70	68	65	63	60	58	55	53	51	48
36	100	97	94	90	87	84	81	78	76	73	70	67	65	62	60	57	55	52	50	48
35	100	97	93	90	87	84	81	78	75	72	70	67	64	61	59	56	54	51	49	47
34	100	97	93	90	87	84	81	78	75	72	69	66	64	61	58	56	53	51	48	46
33	100	97	93	90	87	83	80	77	74	71	69	66	63	60	58	55	52	50	47	45
32	100	97	93	90	86	83	80	77	74	71	68	65	62	60	57	54	52	49	46	44
31	100	96	93	90	86	83	80	77	73	70	67	64	62	59	56	53	51	48	45	43
30	100	96	93	89	86	83	79	76	73	70	67	64	61	58	55	52	50	47	44	42
29	100	96	93	89	86	82	79	76	72	69	66	63	60	57	54	52	49	46	43	41
28	100	96	93	89	86	82	79	75	72	69	65	62	59	56	53	51	48	45	42	40
27	100	96	92	89	85	82	78	75	71	68	65	62	59	55	52	50	47	44	41	38
26	100	96	92	88	85	81	78	74	71	67	64	61	58	55	51	48	46	43	40	37
25	100	96	92	88	84	81	77	74	70	67	63	60	57	54	50	47	44	41	38	36
24½	100	96	92	88	84	81	77	74	70	66	63	60	57	53	50	47	43	40	37	34
24	100	96	92	88	84	80	77	73	69	66	62	59	56	52	49	46	43	40	37	34
23½	100	96	92	88	84	80	77	73	69	65	62	59	56	52	49	46	42	39	36	33
23	100	96	92	88	84	80	76	72	69	65	62	58	55	51	48	45	42	39	36	33
22½	100	96	92	87	83	80	76	72	68	64	61	58	55	51	47	44	41	38	35	32
22	100	96	92	87	83	79	76	72	68	64	61	57	54	50	47	44	40	37	34	31
21½	100	96	91	87	83	79	76	71	67	63	60	57	53	50	46	43	39	36	33	30
21	100	96	91	87	83	79	75	71	67	63	60	56	52	49	46	42	39	35	32	29
20½	100	96	91	87	83	79	75	71	67	62	59	56	51	49	45	42	38	35	31	28
20	100	96	91	87	83	78	74	70	66	62	59	55	51	48	44	41	37	34	30	27
19½	100	96	91	87	82	78	74	70	66	61	58	55	51	47	44	40	36	33	30	26
19	100	95	91	86	82	78	74	70	65	61	58	54	50	46	43	39	35	32	29	25
18½	100	95	91	86	82	77	74	69	65	60	57	53	50	46	42	38	34	31	28	24
18	100	95	91	86	82	77	73	69	65	60	56	52	49	45	41	37	34	30	27	23
17½	100	95	90	86	81	77	72	68	64	59	56	52	48	44	40	36	33	29	26	22
17	100	95	90	86	81	77	72	68	64	59	55	51	47	43	39	35	32	28	24	21
16½	100	95	90	86	81	77	72	68	63	59	55	51	47	42	38	35	31	27	23	20
16	100	95	90	85	81	76	71	67	62	58	54	50	46	41	37	34	30	26	22	18
15½	100	95	90	85	81	76	71	67	62	58	53	49	45	41	37	33	29	25	21	17
15	100	95	90	85	80	75	71	66	61	57	52	48	44	40	36	31	27	24	20	16
14½	100	95	90	85	80	75	70	65	61	56	52	47	43	39	35	30	26	22	18	14
14	100	95	90	84	79	74	70	65	60	56	51	46	42	38	33	29	25	21	17	
13½	100	95	89	84	79	74	69	64	60	55	50	46	41	37	32	38	24	20	16	
13	100	95	89	84	79	74	69	64	59	54	49	45	40	36	31	27	23	18	14	
12½	100	94	89	84	78	73	68	63	58	53	49	44	39	35	30	26	21	17		
12	100	94	89	83	78	73	68	63	57	53	48	43	38	34	29	24	20	16		
11½	100	94	89	83	78	72	67	62	57	52	47	42	37	32	28	23	19	14		
11	100	94	88	83	77	72	66	61	56	51	46	41	36	31	26	22	17			
10½	100	94	88	83	77	71	66	61	55	50	45	40	35	30	25	20	16			
10	100	94	88	82	77	71	65	60	54	49	44	39	34	29	24	19	14			
	0	**½**	**1**	**1½**	**2**	**2½**	**3**	**3½**	**4**	**4½**	**5**	**5½**	**6**	**6½**	**7**	**7½**	**8**	**8½**	**9**	**9½**

Depression of the wet bulb (° C)

Notes

1 The steps are in ½ °C, but we can read to ¼ °C or better, and an examination of the table will show that in some cases there are quite large jumps from one depression column to the next. It is therefore particularly important to record the depression as accurately as possible, and to interpolate between the figures on the table if necessary. For example, a dry bulb of 11¼ °C and a wet bulb of 4 °C gives a depression of 7¼ °C. A depression of 7 °C at 11¼ °C reads ½(26 + 28) = 27% RH. A depression of 7½ °C at 11¼ °C reads ½(23 + 22) = 22½% RH. The RH corresponding to 11¼ °C dry bulb and 7¼ °C depression is therefore ½(27 + 22½), which rounds off to 25% RH.

2 Some tables read the *wet* bulb on the left, rather than the dry bulb, but otherwise follow the same procedure. Note the column headings with care.

3 The range of this Table is limited to between 40 and 10 °C and a minimum RH of about 15%. For figures outside these values a more comprehensive set of tables[79], should be consulted.

Dry Bulb (°C)	*Depression of the wet bulb (°C)*																		
	10	**10½**	**11**	**11½**	**12**	**12½**	**13**	**13½**	**14**	**14½**	**15**	**15½**	**16**	**16½**	**17**	**17½**	**18**	**18½**	**19**
40	48	46	44	42	40	38	37	35	33	31	29	28	26	24	23	21	20	18	17
39	48	46	44	42	40	38	36	34	32	30	28	27	25	23	22	20	18	17	15
38	47	45	43	41	39	37	35	33	31	29	27	26	24	22	20	19	17	16	
37	46	44	42	40	38	36	34	32	30	28	26	24	23	21	19	18	16		
36	45	43	41	39	37	35	33	31	29	27	25	23	21	20	18	16	15		
35	44	42	40	38	36	34	32	30	28	26	24	22	20	18	17	15			
34	44	41	39	37	35	33	30	28	26	24	23	21	19	17					
33	43	40	38	36	34	31	29	27	25	23	21	19	17	15					
32	42	39	37	35	32	30	28	26	24	22	20	18	16						
31	41	38	36	33	31	29	27	24	22	20	18	16							
30	39	37	34	32	30	27	25	23	21	19	17	15							
29	38	36	33	31	28	26	24	21	19	17	15								
28	37	34	32	29	27	25	22	20	18	15									
27	36	33	30	28	25	23	21	18	16										
26	34	32	29	26	24	21	19	16											
25	33	30	27	25	22	19	17												
24½	32	29	26	24	21	19	16												
24	31	28	26	23	20	18	15												
23½	30	28	25	22	19	17													
23	30	27	24	21	18	16													
22½	29	26	23	20	17														
22	28	25	22	19	16														
21½	27	24	21	18	15														
21	26	23	20	17															
20½	25	22	19	16															
20	24	21	18	15															
19½	23	20	17																
19	22	19	15																
18½	21	18																	
18	20	16																	
17½	19	15																	
17	17																		
16½	16																		
16	15																		
15½																			
15																			
	10	**10½**	**11**	**11½**	**12**	**12½**	**13**	**13½**	**14**	**14½**									

ELECTRONIC HYGROMETERS

Beware of cheap electronic hygrometers! I was once shown by the proud owner how his new instrument was so sensitive that you could find an RH difference between floor and table. True, there was a difference, but both readings were over 10% wrong. It is safest to regard all electronic instruments as wrong unless recently checked. This looks as if we were back where we started, because we are meant to be finding an instrument which will check our thermohygrograph. But the answer is at hand: surround our electronic RH sensor with a miniature climate of known RH and adjust it until it registers this RH.

At the time of writing there is one instrument which does this is a simple and handy way: the Nova Sina Mik 3000, marketed in the U.S.A. under the name Beckman (*Figure 30*). This will therefore be used for explanation.

The Mik 3000 is a hand-held instrument which measures both RH and temperature. In a manner similar to that used for the Crawford UV Monitor a reading is obtained by turning a knurled wheel back and forth under two small red light sources. The correct position is found when the light just jumps from one to the other and the RH is then read off a dial.

But the essential item which makes this instrument what we require is the 'Sensor-check': a cap which, when fitted over the sensor, provides it with a standard humidity, say 55%. If the instrument then reads incorrectly an adjustment can be made to a recessed screw.

It must not be forgotten that this instrument, like its predecessors, requires to be aspirated before it is read: it must be swung gently so that room air is always moving past the sensor until its readings settle down. The readings should then be reliable to at least ± 2%.

One more cause of inaccuracy in such instruments must be noted: the instrument should be at the temperature of its surroundings. For example if brought into a room from outdoors it should be allowed ten minutes to settle down.

NON-MECHANICAL HYGROMETERS

Anything that operates without moving parts is worth investigating since failures are usually mechanical. Humidity indicating cards consist of a series of patches impregnated with cobalt chloride (*Figure 28*). Each patch is labelled with an RH in 10% steps. Their colour ranges from pink at the dry end to blue at the 100% end. The RH is read at the point of change between

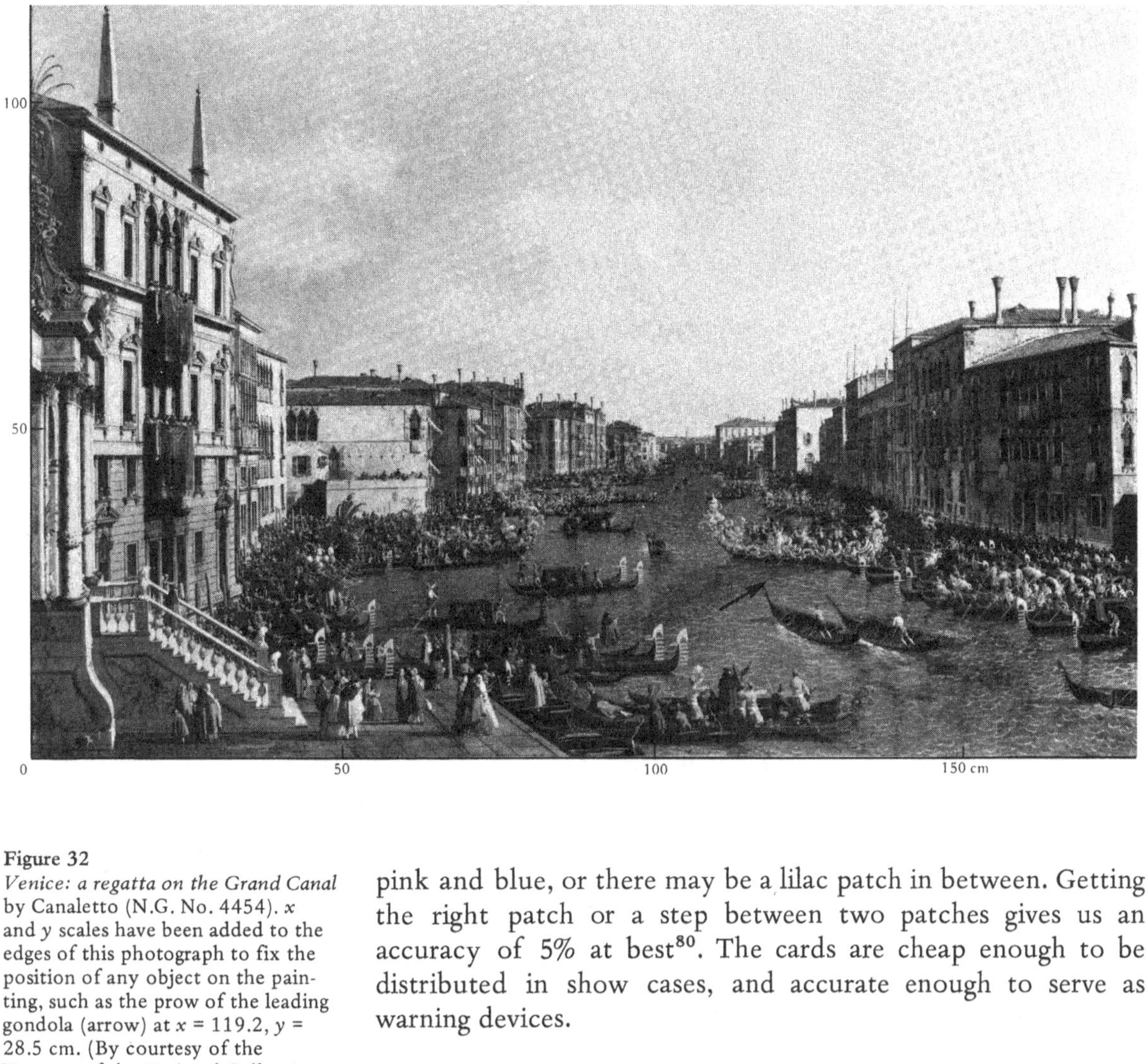

Figure 32
Venice: a regatta on the Grand Canal by Canaletto (N.G. No. 4454). x and y scales have been added to the edges of this photograph to fix the position of any object on the painting, such as the prow of the leading gondola (arrow) at x = 119.2, y = 28.5 cm. (By courtesy of the Trustees of the National Gallery)

pink and blue, or there may be a lilac patch in between. Getting the right patch or a step between two patches gives us an accuracy of 5% at best[80]. The cards are cheap enough to be distributed in show cases, and accurate enough to serve as warning devices.

Summary

Use of a reliable electronic hygrometer with self-checking device, or of a wet-and-dry-bulb instrument is essential for calibrating other RH-recording instruments.

UNDERSTANDING THE HYGROMETRIC CHART

The best way to understand the relation between temperature, RH and the amount of moisture in the air is to study the Hygrometric or Psychrometric Chart (*Figures 33–36*). To those unfamiliar with graphs the chart looks formidable. Therefore its use will be described in simple stages from first principles.

If we want to point out a certain feature in a picture, we can of course describe it – for example, the tip of the prow of the leading gondola of the pair racing in the foreground (*Figure 32*). But for record purposes it might be safer to take measurements.

We would say that our feature was 119.2 cm from the left edge of the painting and 28.5 cm from the bottom. Numerically minded people would say $x = 119.2$, $y = 28.5$ or even record the feature by the abbreviated expression (119.2, 28.5).

The next step is to say that the *x* and *y* scales or axes can be used for anything that can be measured -- time and temperature for instance. (A familiar example is the chart of a hospital patient's temperature.)

Beginners looking at more difficult curves denoting relations between things sometimes forget the simple fact that the *x*-axis is always just a scale for measuring from left to right (so that the *x*-axis scale could just as well be at the top of the diagram) whereas the *y*-axis is for measuring upwards (the *y*-axis scale can be on the left or right).

Thus the ordinary graph relates two things, one measured from left to right by the *x*-axis scale, and one measured upwards by the *y*-axis scale. Unfortunately our immediate task is to relate, not two things but three: RH, Temperature and Absolute Humidity. We do this in effect by putting several graphs or curves on top of each other (*Figure 33*).

Consider first the room temperature scale in this diagram. Since we are dealing with hygrometry we must specify it as the 'dry bulb temperature', and it is marked along the *x*-axis at the bottom. We measure the room temperature and find it to be 20 °C, so we draw a vertical line upwards from the 20 °C point on the scale (*Figure 33*). Every point on this line is at 20 °C.

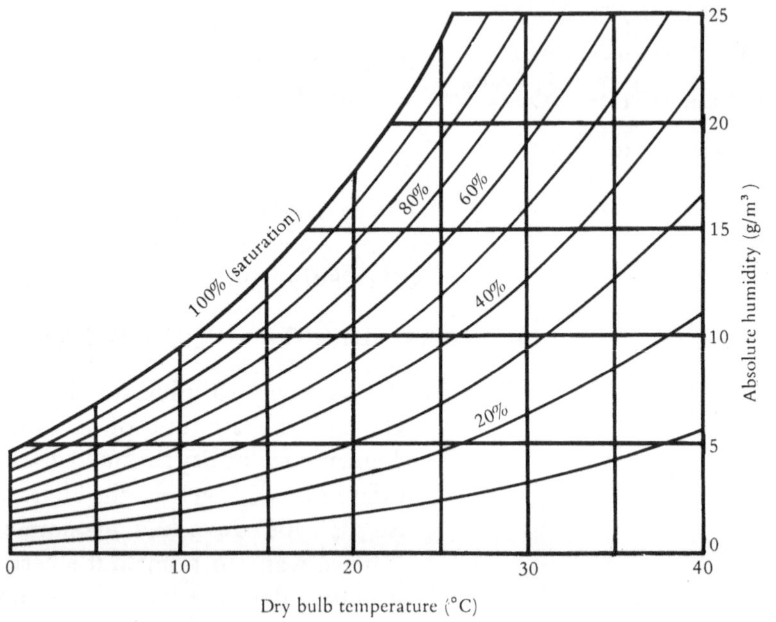

Figure 33
Simplified Hygrometric chart. The colour line is at 20 °C dry bulb

Nothing else on the diagram is at 20 °C. Going left from this line the temperature falls, going right it rises. Therefore the climatic conditions which we are measuring in the room must be somewhere along this line.

We now measure the room RH and find it to be 55%. Now RH in our diagram is not marked along a scale but by another method. Each curve in *Figure 33* represents a fixed RH. The line for each RH is not straight, as with temperature, but curved. This is a matter of convenience. The reader may come across other hygrometric charts where each curve represents a fixed temperature. They can be tackled in just the same way. Which variable of the three (RH, temperature and absolute humidity) goes where is entirely a matter of choice, so that there is no need to worry if the hygrometric charts in some books look quite different.

The RH curves go up in 10% steps, so that there is none at 55%. We therefore take the nearest on either side, 50% and 60% and draw or imagine a curve midway between them (*Figure 34*).

Every point on this dotted curve that we have drawn is at 55% RH, and nowhere else is. Since the room temperature of 20 °C means that we are somewhere on the green vertical line, the only place we can be is where this vertical line crosses the dotted curve (*Figure 34*).

But there is a third variable, the Absolute Humidity, marked up the right hand side as the *y*-axis*. Now that we have fixed the temperature and RH there is only one possible value for the Absolute Humidity, found by measuring the height of our point by means of the Absolute Humidity scale (*Figure 35*). The height on the Absolute Humidity scale corresponds to about 10 g/m^3 : every cubic metre of air in the room contains 10 g of water.

Now it can be seen that the three variables (RH, temperature and Absolute Humidity) are so related in the chart that if we know any two of them we can find the third.

Measure the volume of the room where you are sitting. A small exhibition room might have a floor area of 6 × 10 m and a height of 4 m, which equals a volume of 240 m^3. We have just found that, at 20 °C and 55% RH each cubic metre of air contains 10 g of water, therefore the whole room contains 240 × 10 = 2400 g of water, or about 2½ litres.

*The explanatory chart used here and in the following diagrams carries certain simplifications, the main one being that Absolute Humidity is used scaled in g/m^3 whereas on the standard chart the scale is on a weight/weight basis, for example kg moisture/kg dry air, and is termed Moisture Content. This results in a small loss of accuracy which is further discussed on p. 210.

Figure 34
Colour curve is at 55% RH

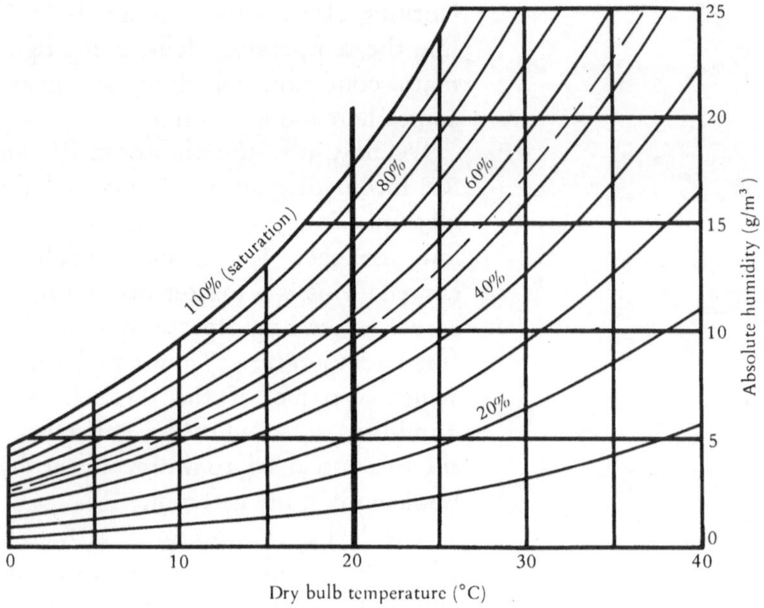

Figure 35
Horizontal colour line is at 10 g/m³ Absolute Humidity

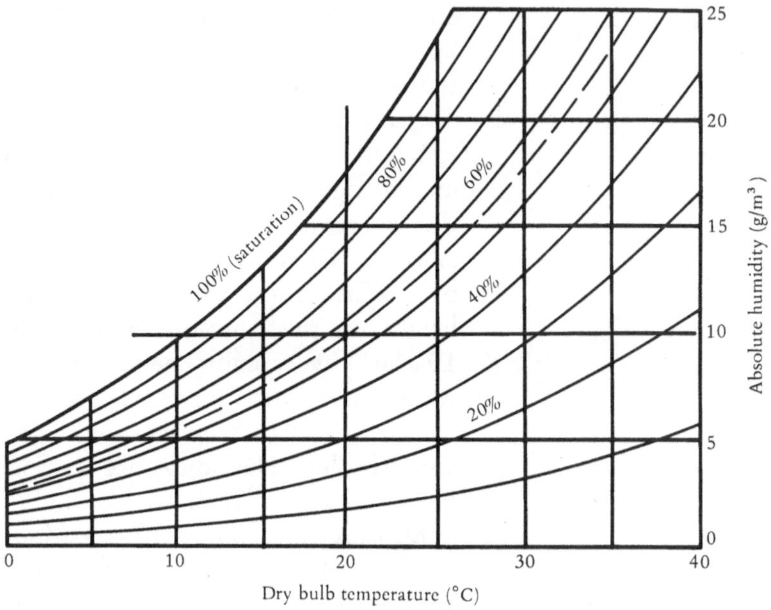

Now suppose that your museum director comes into this room on a winter morning, finds it unpleasantly cold, checks that all windows and doors are closed, and orders the electric heating to be switched on. Before the heating commences your quick check gives 12 °C and 50% RH. What will the RH be when the heating is switched off at 20 °C?

To solve this problem with the aid of the Hygrometric Chart all one needs to note is that an electric heater neither adds nor takes away water from the air. The chart shows that when the Director enters the room the Absolute Humidity is about 5.5 g/m^3, and so it must remain at this value (*Figure 36*). This limits us to travelling horizontally right or left. Right is the direction of rising temperature, therefore right is our direction of travel. When we stop our horizontal line touches the vertical line at 20 °C, since this is the final room temperature. We now find that the humidity is 30% – an undesirably low level, but a common enough situation in European and North American museums.

With this basic understanding of the relation between humidity and temperature we must now look into the best RH to be aimed for and the levels to be avoided.

Figure 36
Colour arrow: temperature rise at constant Absolute Humidity

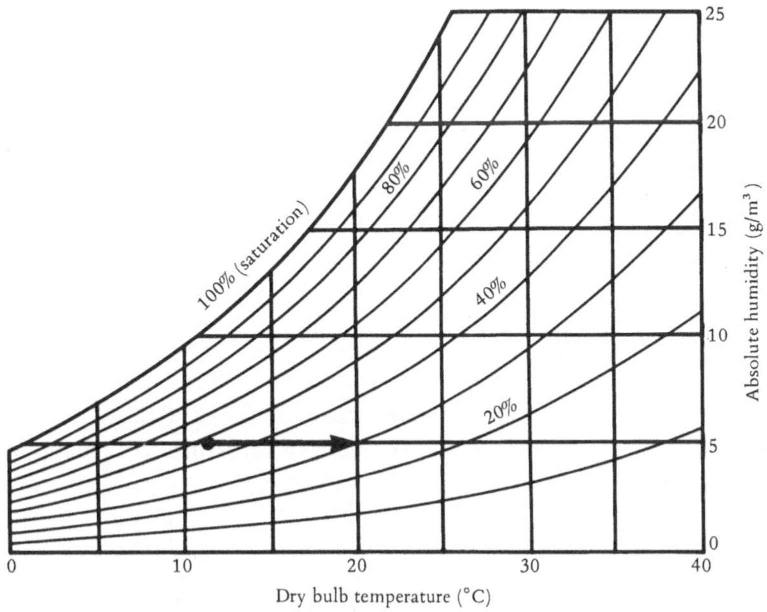

Summary

The Hygrometric Chart in its simplest form relates the temperature of the air in the room (the dry bulb temperature) to its RH and Absolute Humidity.

RESPONSE OF MUSEUM MATERIAL TO RH

We need only an overall picture of how the various classes of museum material respond to RH. For greater detail the standard textbooks should be consulted[81-83].

There are three different modes of deterioration that are influenced by RH:

1 Change in size and shape
2 Chemical reaction
3 Biodeterioration

One could say these are the physical, chemical and biological modes.

1 CHANGE IN SIZE AND SHAPE

All moisture-absorbent materials, such as wood, bone, ivory, parchment, leather, textiles, basketry and matting, and adhesives (important), swell when the RH rises and shrink when it falls (*Table 34*), causing warping, dislocation between parts, splitting, breaking of fibres, etc, especially at low RH.

A bulky object will respond to a new RH very slowly from the outside inwards, so that a man-sized wooden sculpture might take two or three months to complete its response to a large change, while a sheet of paper will respond in minutes.

One of the most sensitive examples is a panel painting. The paint on the picture side acts to some extent as a barrier to moisture. If the RH falls, this causes the air to start drying the panel, but moisture will leave it more easily through the back than through the front. Consequently the back will contract more than the front and the painting will take on a convex warp. To accomodate itself to this warp the paint layer may have to crack. Indeed all old panel paintings now have a characteristic pattern of cracks, longer along the grain since the change is mainly across the grain (*Figure 37*). If there is no further change in RH, after a period the moisture content of the panel will even out, causing the warp to disappear[84]. But picture restorers will be quick to point out that panel paintings commonly have a permanent convex warp. This most probably occurs as follows. During a period of high RH moisture enters freely through the back of the panel which tries to expand. If it is prevented from doing so, either by the drier internal parts of the panel or by a constricting frame, the wood at the back will be forced into an internal adjustment so that, though swelled by moisture, it remains of the same volume as before. This is, in fact, a plastic compression at the cellular level which causes no change in appearance but is permanent. Thereafter, when the RH reverts to its usual level, the back layers will shrink in the normal way and will therefore be smaller than they were before the process started. But the front layer under the paint will be unchanged. This forces the picture to take up a convex warp, and a number of similar cycles will augment the effect, which is known as 'compression shrinkage'[85,86].

But the RH falls as often as it rises. Would not the reverse – a concave warp – occur as the back dries and compresses the front? The answer is that damp wood is both weaker and more plastic than dry wood, so that the greatest compression shrinkage occurs in the dampest conditions.

The effect is of wide occurrence in wood constructions. Straightforward examples are the loosening of axe-heads, the shrinkage of wood in barrels and the widening of gaps between

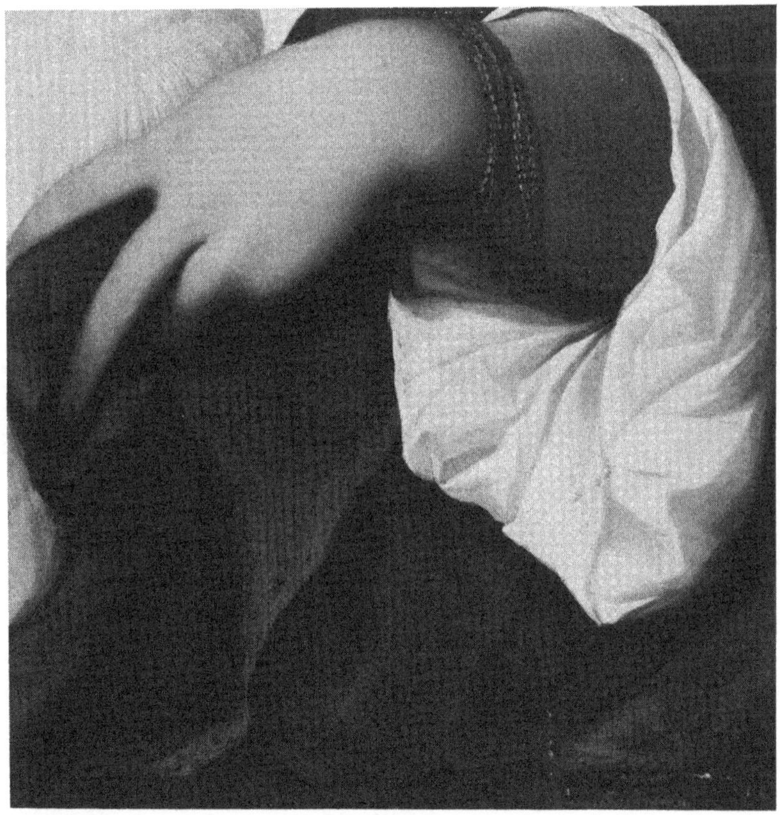

Figure 37
Detail from a portrait by Palma Vecchio (died 1528) (N.G. No. 3939) showing the vertical grain of the panel by the craquelure in the dark paint of the costume. Note that this craquelure is missing in the areas of flesh and white paint, which indicates that in this case such paint is less brittle. (By courtesy of the Trustees of the National Gallery)

floor boards. In each of these three cases cross-grain expansion following a rise in moisture level is prevented: by the steel axe-head, by the barrel hoops and by the walls of the room. Internal compression occurs while the wood is damp and in a dry period the wood shrinks to leave gaps.

It has been said above that all moisture-absorbent materials swell in a rising RH and vice versa. An apparent contradiction is manila rope or certain multiple-ply threads which tighten on wetting. But this contraction along the rope is a result of swelling across the fibres which has the effect of tightening the twist. A canvas painting is usually a painting on a glue-sized canvas. In damp conditions animal glue will always expand but the canvas thread may contract. In high RH there is often a once for all contraction in textiles, due to release of tensions caused during manufacture. But thereafter we can expect the usual situation: expansion in high RH and contraction in low[87,88].

2 CHEMICAL REACTION

Two quite different classes of chemical reaction are favoured by high humidity: (a) the corrosion of metals, and (b) the fading of dyes and the weakening of paper and textiles. Certain less common reactions of museum material are also known to be affected by moisture (*see* (2c)).

2a The corrosion of metals Whereas the moisture-absorbent materials discussed above must be neither too damp nor too dry, all metals benefit from dryness, particularly iron and its alloys and copper and its alloys.

Since most museums form mixed collections it is indeed unfortunate that the very level chosen as the *lower* safe limit for moisture-absorbent materials (40–45% RH) should also be the upper limit, not to be exceeded, for unstable iron and bronzes with traces of chloride[89–91] and in addition, as we shall see below, for unstable glass.

Fortunately good quality iron (e.g. low sulphur), especially if clean and suitably coated, and bronzes with stable patinas will exist in permanent safety at 55%. This implies that special vitrines may have to be built for valuable but unstable metal objects. If dehumidifiers are used for vitrines containing metals, lithium chloride should not be used as a desiccant. Lithium chloride is an electrolyte and, as such, could initiate corrosion if inadvertently transferred in the air stream from the dehumidifier to the metal. Dehumidifiers using silica gel are available[107].

Protection against the corrosion of lead, tin, pewter and silver, though probably aided by very dry conditions, is not a problem which needs low RH, though the tarnishing of silver is accelerated by high RH[92]. Gold, of course, is unaffected.

A pillar of iron 1600 years old stands in the precincts of the Qutb Minar in old Delhi. This is often sentimentally cited as an example of a lost art of ancient technology, for the pillar has not corroded away in the way that any unpainted mass of iron would do in a modern city in a very short time. However samples of the pillar have been shown to rust in a humid atmosphere in the normal way, and indeed below ground the pillar is coated with rust about 1 cm thick[93]. Certainly the pillar is of a stable composition (low in sulphur and high in phosphorus) and the weather of Delhi is drier for most of the year than that of a European town, but there is a wet season when humidities are continuously high. There are two important reasons for the stability of the pillar. Firstly, the absence of acid in the atmosphere at the time of its erection, and indeed the probable presence of alkali from human habitation. This would allow a firm thick protective oxide film to form. The second reason is that the pillar is massive enough to act as a heat store in the warm climate of Delhi, so that rain is quickly evaporated from the surface. However with the spread of industrialisation the acidity of the air is already rising, and this ancient monument may yet suffer attack (*see also* p. 146).

2b The fading of dyes It can be simply stated that the great majority of cottons, linens, wools and silk fade more rapidly at high RH than low. The textile itself is affected by light also, and all those which contain moisture (meaning all natural textiles) weaken more quickly by light at high RH than low. But probably the ratio of rates of fading or damage at 30 and 60% RH, for example, is less than two in most cases, and so the effect is not great enough to justify altering conditions in a mixed museum just for the sake of the textiles. This also applies to storage, since in any case the absence of light should eliminate fading in storage.

However the pure textile museum (note: no valuable furniture or paintings) in a dry or temperate winter climate could benefit both in conservation and economy by opting for a lower RH than 55% – say 45% (*see also* p. 44). Brommelle[94] has suggested that we have here a point in favour of using tungsten light rather than fluorescent, for illuminating textiles, since the radiant heat plentiful in tungsten would lower the effective RH on the textile below that in the room.

2c Weeping glass and other materials Since glass vessels have always been used for water it may come as a surprise that some glass objects in museums are sensitive to moisture. However it appears that in certain glass compositions sodium and potassium ions retain a slight solubility. These can then be leached out by moisture as sodium and potassium hydroxides, which are quite quickly converted to carbonates by the carbon dioxide in the air. But sodium and potassium carbonates are deliquescent – they attract moisture – and hence droplets are formed on the glass. During the progress of this leaching tiny cracks are first formed, but in later stages the glass may become opaque, greyish or milky and tiny scales may flake off. If we can keep the glass dry there will clearly be no further leaching. An RH not exceeding 40% has been recommended[95]. However specialist advice should be sought, since it appears that in certain cases an incipient crizzling (cracking) of unstable glass objects can be accelerated by dryness[96].

Finally geological museums have troubles with iron pyrites minerals and pyritised fossils. During the course of 'pyrites disease' pyrites, FeS_2, is converted into ferrous sulphate, $FeSO_4$, and sulphuric acid[97]. Though bacterial action was once postulated, this now seems unlikely in the museum[98]. Most pyritic material is safe at RH 55% and below, but ideally 50% should be the upper limit.

3 BIODETERIORATION

Although biodeterioration is outside the range of this book, we have already noted that mould growth can be prevented by keeping the RH below 70%. Bacteria require even higher humidities.

Most insect pests flourish at the higher humidities. However even though the optimum RH for *Tineola* (Clothes moth) is about 65–75% and for *Hofmannophila* (Brown house moth) is about 90%, to take two examples, an RH as low as 20% will not eradicate them[99–101]. Very low humidities would be preventative but impracticable. We must therefore conclude that insect damage may be discouraged but cannot be prevented by humidity control. Nevertheless one ought to remember that the entry of insects into an air-conditioned building, with dust filters and closed windows, is made very much more difficult.

BEST RH FOR MOISTURE-ABSORBENT MATERIALS

We first find upper and lower danger limits and then decide the most suitable RH within this range.

It is widely agreed that there is danger of mould growth above 70% RH. Some cautiously lower this upper danger limit to 65% RH. Therefore RH should certainly never exceed 70% where circumstances are particularly favourable to mould growth. The mould-growth limit depends on other factors, two in particular: air movement and temperature. To flourish, moulds require warm and stagnant air. Japanese antiquities are commonly kept in closed cases at 65%, 20°C. This is probably the safe upper limit. The temperature in historic houses and churches can be allowed to fall to 10°C during winter even though the corresponding humidity might be 80%. But there must be no damp surfaces preparing conditions for mould growth in warmer weather (*see* p. 119).

The lower limit is set by cracking, breaking of fibres, and damage by embrittlement. This becomes a danger for wood, bone, ivory, basketry, parchment and heavy leather book bindings at about 40% RH. Paper in use in libraries also becomes more liable to break at these levels. It might be objected that paper and textiles seem perfectly usable in the driest weather, but this can no longer be guaranteed if the material has become weak with age*. Therefore RH should preferably be above 45% and should not be allowed to fall below 40%.

Between these limits the choice once made should be maintained day and night, winter and summer, wet and dry season.

For Europe the choice is commonly *55%*, midway between the danger limits, but variations on this will be discussed under 'Climate'.

Summary

A change of RH causes changes in size and shape of moisture-containing materials. High RH causes mould to flourish (above 65–70% RH) and metals to corrode. Low RH (below 40% RH) causes embrittlement of moisture-containing materials. Dyes fade more rapidly at high RH than low. Biodeterioration is greater at high RH but cannot be prevented through RH control alone.

CLIMATE INSIDE AND OUTSIDE THE MUSEUM

The natural life of an exhibit may be taken to be its expected life in its climate of origin under shelter, that is to say protected

*Banks recommends that books and documents, where they are not handled, are best at 30–40% RH. Where they are handled: 50–60%[102].

from rain, direct sun, strong wind and freezing temperatures. In this sense a museum without any form of climate control constitutes neither more nor less than a shelter, and the lifetime of a native exhibit should be unaffected.

In practice one must unhappily say that the lifetime of an exhibit in a museum may be severely reduced below its natural lifetime in any of the following ways:

1 Not using the shelter of the museum adequately.
Examples: Allowing the ingress of damp or direct sun.

2 Imposing through modern design new and deleterious conditions.
Examples: Dryness due to winter heating. Excessive illumination.

3 Transferring an exhibit to a climate foreign to it.
Examples: Wooden sculpture from equatorial Africa brought to Europe.
Books packed into bookshelves in the wet tropics.
Steel from a dry clean climate brought to the acidic pollution of industrial Europe.

There are sufficient examples under (3) to warrant a simple examination of climatic zones.

We will have to take a rather different approach to that of the orthodox climatologists who divide the world into hot–dry, hot–humid, temperate coastal and inland zones, etc.

Our primary measuring rod is average indoor relative humidity, and our chief secondary concern is whether the climate is cool enough for winter heating to be widely practised. On this basis we define two extremes: a Humid Zone and an Arid Zone.

TABLE 8. Choice of RH Level according to Climate

65%	Acceptable for mixed collections in the humid tropics. Too high, however, to ensure stability of iron and chloride-containing bronzes. Air circulation very important.
55%	Widely recommended for paintings, furniture and wooden sculpture in Europe, and satisfactory for mixed collections. May cause condensation and frosting difficulties in old buildings, especially in inland areas of Europe and the northern parts of N. America.
45–50%	A compromise for mixed collections and where condensation may be a problem. May well be the best level for textiles and paper exposed to light.
40–45%	Ideal for metal-only collections. Acceptable for museums in arid zones exhibiting local material.

Note
International exhibits and loans require international agreement on RH levels, and introduce a bias towards the median levels 50–55% RH.

Figure 38 shows that a large part of the Far East, including southern India, Sri Lanka, the Calcutta area, Indo-China, Malaysia, Indonesia, the Philippines and part of the coast of China remains above 65% RH out of doors for the whole of the year, and for long periods in the 80s. Although fluctuations will be smaller, we can assume that indoor climate also will be above 65% since heating is never necessary in this part of the world. This is our Humid Zone, and its primary conservation characteristic is a continual mould problem. Where air conditioning is installed humidity should be maintained, not at the temperate European level of 55%, but as high above this as is compatible with the health of the other exhibits – say 65% (*Table 8*).

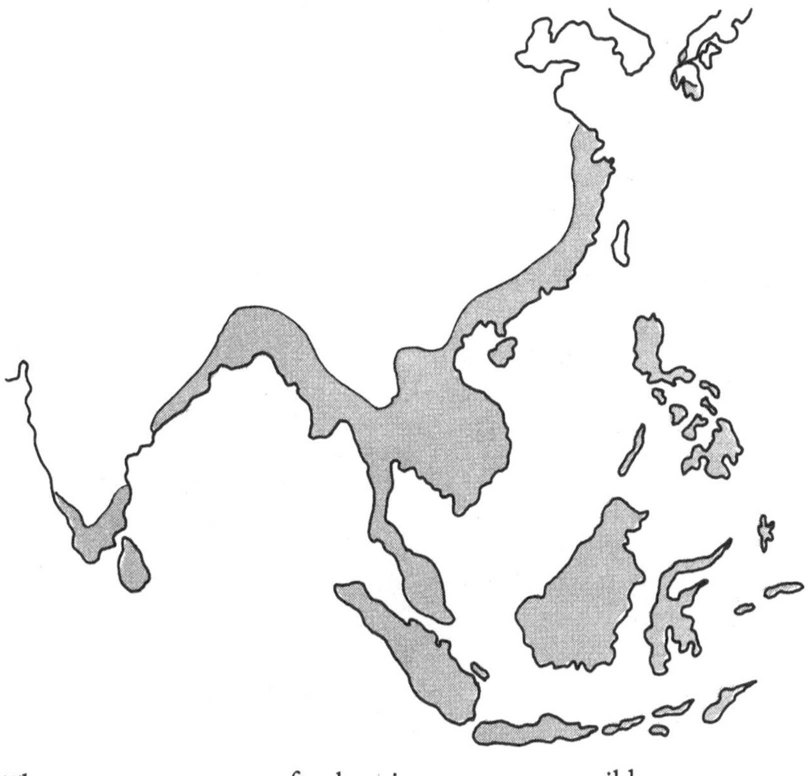

Figure 38
The Asiatic humid tropics. The shaded zones have daily average RH over 65% throughout the year, and over 80% in the wettest month

There are two reasons for keeping as near as possible to average outside conditions, and these apply wherever the museum is situated, first because timber old and new will be seasoned to the average prevailing humidity, and secondly because the nearer we are to outside conditions the smaller will be the cost of running air conditioning.

In contrast to the Humid Zone, our Arid Zone will have a different extent to, that of the orthodox climatologists. *Table 9* shows that there are naturally arid zones, but we have also to consider the artificial arid zones produced by winter heating. Tamen Rasset (*Figure 39*), a town-sized oasis in the centre of the Algerian Sahara, is so dry that plumbing, though attempted, is impracticable. London (*Figure 39*) is generally considered a rather damp city. Yet if we allow for winter heating to 20 °C,

TABLE 9. RH and Climate in the Warmer Parts of the World (no Winter Heating)

Climate type	*Examples*	*Average RH*
Equatorial rain forest (rain at all seasons)	Malaysia, Indonesia	75–80%
Hot coastal	Bombay, Madras, Durban	70–75%
Tropical rain with dry season	Inland India	50–70%
Steppe with little rain	W. Pakistan, Isfahan	40–50%
Desert	Luxor	40% and below

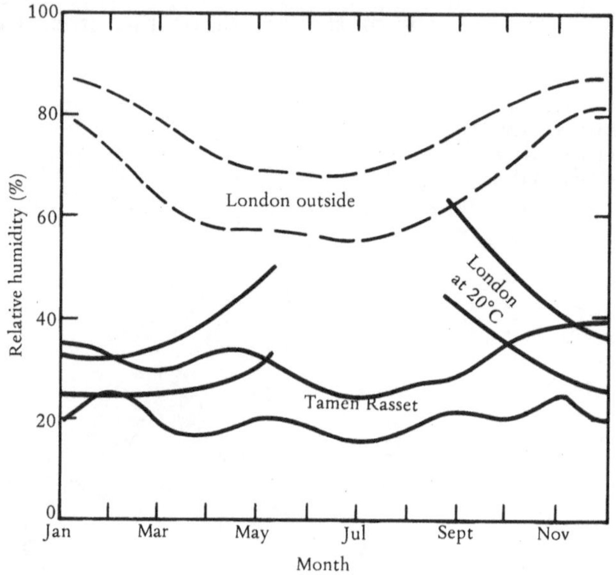

Figure 39
Annual RH variation for London and the Sahara (Tamen Rasset). Black broken curves: London outdoor RH. Black full curves: London indoor RH after heating to 20 °C. Colour lines: Tamen Rasset RH. At each site the average monthly RH both early in the morning and in mid-afternoon are recorded. The RH in the early morning will usually be the highest for the day, and that for mid-afternoon the lowest. Therefore for each month there are two points, forming for the year an upper and a lower curve. e.g. London outside morning RH in Jan = 87%, afternoon = 80%. The curves show that RH conditions indoors in mid-winter correspond to Sahara conditions[103].

the humidity climate in indoor London throughout the winter is very close indeed to that of the Sahara.

One of the worst possible situations – but also distressingly common – makes the comparison with the Sahara even closer. If the heating is switched off at night, including strong spotlights on or in showcases, a daily 'fatigue' cycle, perhaps worse than that occurring in the Sahara, is set up, of hot–dry days followed by cool–damp nights. The inland and northern areas of Europe and North America fare even worse than London and the maritime areas of Western Europe, as *Figure 40* indicates. There is further discussion of the problems of heating and

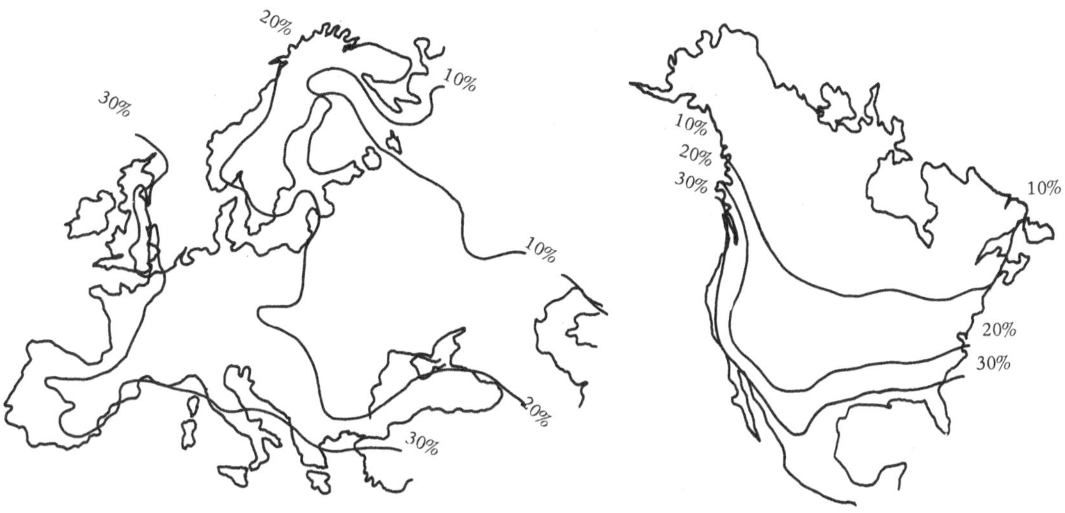

(a) *(b)*

Figure 40
(a) and *(b)*. January indoor RH climates at 20 °C. The contours show the RH which results from heating outdoor air on an average January morning to 20 °C, as occurs when a museum is heated without humidification. They illustrate how low the RH falls in winter in the middle of a continental land mass, and in particular emphasise the acute problems of Canada and central U.S.S.R. 5% RH (approx. 0.8 g/kg) has been added to the calculated figures to allow for adventitious moisture gain. (Based on figures from the Met. Office, U.K.[103])

humidifying museums in these areas of low winter temperature on p. 116.

It can now be seen that museum climates range between two extremes: damp all the year round with dehumidification required in the Museum Humid Zone; tolerable summers but extremely dry winters in the Museum Arid Zone (most of Europe and North America) (*Figure 40*).

Is there such a climate as a museum Temperate Zone, where throughout the year average daily RH remains within the moderately safe limits of 40 to 70% and heating is rarely required? There are a few such favoured places scattered along the Mediterranean littoral, where curators can almost get away with being lazy about climate control, but not enough of them to form a zone. It seems that wherever we are in the world we are fated to have to interfere with the interior climate in some way so as to ensure the reasonable safety of the exhibits.

Later on (p. 116) we must consider further factors affecting choice of RH, but first it is necessary to return to the Hygrometric Chart.

Summary

Large areas of the humid tropics have an RH all the year round high enough for mould growth. By contrast all areas which require winter heating will get too dry during winter. Lists have been published of most suitable RH for different classes of object[90,104]*, but these will also be affected by climate, loans policy, etc. A classification by climate is given in Table 9.*

CONDENSATION AND THE DEW POINT

The Hygrometric Chart will show us what happens when the air gets saturated with moisture. This term, saturated, means that the air is holding all the moisture it can. As air cools it becomes able to hold less moisture, and there comes a point where it is holding all the moisture it can, and is thus said to be saturated. If it is cooled still further it must give up some of its moisture as condensation. The temperature at which condensation first occurs (the temperature at which the air reaches saturation) is called the *dew point*. If we know the temperature and RH of the air we can find its dew point from the Hygrometric Chart.

Out of doors on a winter's day each breath of air we inhale is warmed and humidified by our lungs. On exhalation this air is cooled below its dew point, thus giving rise to visible condensation. Indoors in winter the room air circulates to window-

Figure 41
Arrow shows the effect of cooling an air mass to its dew point and beyond

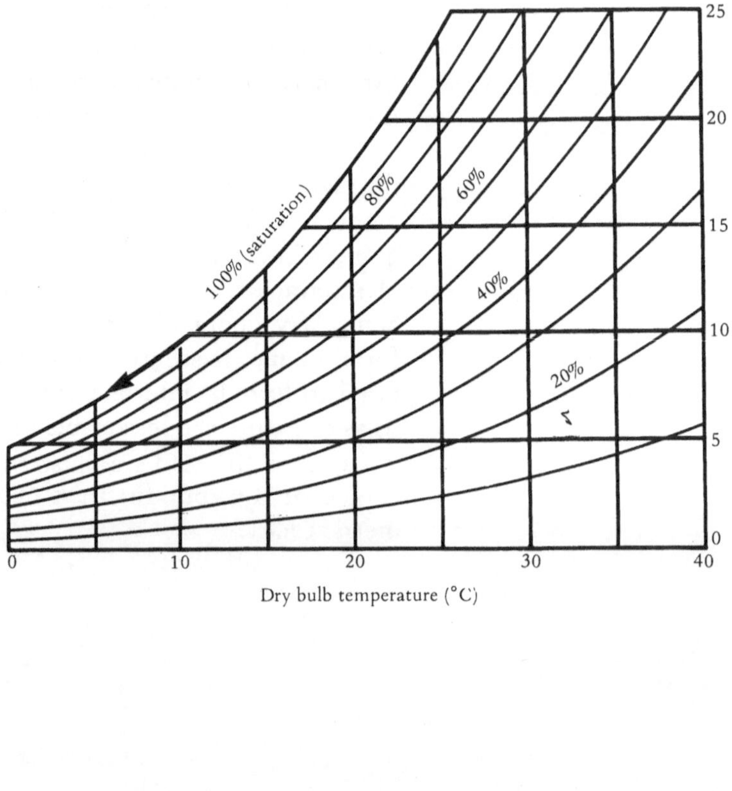

panes which are often cold enough to cool the air below its dew point. Drops of water then appear on the windows.

Suppose that some of the air in our exhibition room at 20 °C, 55% RH, and therefore 10 g/m^3 Absolute Humidity, circulates to the window and consequently starts to cool. Before condensation begins there is only one direction we can move on the diagram: because no moisture is given out or taken in we cannot move away from the 10 g/m^3 line of Absolute Humidity (*Figure 41*). The temperature of the air is falling, and therefore we move horizontally left on the diagram, and we can see that as this happens its RH rises rapidly. When the air reaches 100% RH we can go no further horizontally. The air has become saturated and has reached its dew point temperature, which can be read off the bottom scale at this point: about 11 °C. The temperature of the air is still falling, but now condensation is occurring so that the Absolute Humidity also falls – below 10 g/m^3 – though it remains saturated. It only gives up the moisture it cannot hold. The only route for the air on the Chart is now down the curve of 100% RH, so we travel down this curve until either the air reaches the temperature of the window-panes or returns to the room by circulation. As it returns to the room it is re-heated to 20 °C but, having lost some moisture, it is now at a lower RH. We will see later that this principle is made use of for drying by refrigeration.

HUMIDITY CONTROL

The only complete answer to humidity control is air-conditioning of both exhibition and store rooms, and this is important because air-conditioning will also remove dirt and gaseous pollution from the air.

But there are many situations where air-conditioning cannot be the answer, at least for the present, most likely because of cost, but also in many cases because something has to be done quickly. For these situations we now investigate room humidity controllers. These are most peoples' introduction to air control and so they will be described first.

RH CONTROL IN A ROOM

There can be said to be two standards of climate control in a museum. The lower standard involves no more than eliminating the major dangers. Where humidity is concerned the major dangers are mould growth due to high RH and cracking due to low RH. The higher standard of humidity control means constant RH day and night throughout the year.

Control of room RH by means of free-standing units can in favourable cases maintain the higher standard, but more often gives only the lower for the following reason.

One piece of equipment will either humidify or dehumidify, not both; but in most climates the danger lies either in excessive humidity or excessive dryness, not both. In such cases one will be able to eliminate dangerous conditions without being able to guarantee that for short periods humidity will be kept within the best limits.

Thus in Europe, even maritime Europe, the danger to avoid is dryness due to winter heating. If cool rainy periods during the summer cause interior conditions to become too damp this can be corrected by a little heating. In the monsoon tropics the problem is humidity during the monsoon, and, though in the dry season humidity may fall below the ideal, the dryness danger level is not likely to be reached. Even in excessive dryness due to *natural* conditions the objects are likely to remain free of gross damage, especially those of local origin which have lived all their lives in such conditions.

THE HUMIDISTAT

Neither humidifiers nor dehumidifiers can operate effectively without a humidistat. It is obviously impracticable to have someone continually measuring the RH of the air and switching on the machine where necessary. The humidistat does this automatically just as a thermostat switches a heating system on and off.

A common type of humidistat works on the same principle as the recording hygrometer described on p. 68. A bunch of hairs operates an electrical relay by expansion and contraction. Some humidity-controlling devices have built-in humidistats, which can be satisfactory provided that one does not rely on their calibration. This goes for all humidistats. In other words a humidistat set at 55% RH should certainly keep the RH constant, but this may be at a higher or lower level than the marked figure of 55. One then has to use the sling hygrometer or equivalent (p. 70) to find the setting on the humidistat which corresponds to a true RH of 55% in the room.

Considering winter dryness in the temperate parts of the

world, since this dryness is caused by heating, one could obviously improve the situation by reducing the heating. But for the comfort of both staff and visitors this drop in temperature must be moderate. A simple and ingenious way of doing this automatically is to control room heaters by connecting a thermostat and a humidistat in series[105,130]. The humidistat would be set to the lowest allowable RH, say 45%, and would switch on the humidifier in the normal way to maintain this RH. But current can only pass to the heaters if, firstly, the thermostat requests it and, secondly, the humidistat allows it (i.e. RH is above 45%).

HUMIDIFYING EQUIPMENT

There are two kinds, both electrically operated:

1 Atomising

2 Exporative

The problem is to transform a sufficiently large and controlled quantity of liquid water into the vapour state reasonably fast. Putting water containers on radiators, for example, is not a sufficient solution to the problem because evaporation is slow and uncontrolled.

1 ATOMISING HUMIDIFIERS

The atomising humidifier uses a swiftly revolving spindle to draw a fine sheet of water up from a reservoir to strike fixed blades. The tiny droplets formed are fanned upwards and evaporate within a metre or so of the device.

The first snag with atomising humidifiers is that the whole of the water goes into the air, and that includes the minerals which all natural water contains: 30–1000 parts per million. This may not sound much, but could amount to a kilogram of minerals per winter period from each humidifier, much of which ends up somewhere in the room, including the surfaces of the exhibits. And indeed a grey deposit will become visible on dark surfaces after some months of use. Distilled or deionised water can obviate this problem, at extra expense, though there is no ready way to check that such water is indeed being used.

The second snag with atomising humidifiers is that they do not 'fail safe'. In other words a failure of the humidistat to switch off the humidifier when the required humidity has been reached can cause water to be added to the air until it starts running down the walls.

This type of humidifier has also been called 'centrifugal' since the water is thrown out by centrifugal force from the central spindle.

For the reasons given above atomising humidifiers have been supplanted in museum exhibition rooms by the unheated evaporative types.

2 EVAPORATIVE HUMIDIFIERS

A boiling kettle or any container of water heated by an electrical immersion heater is the simplest heated evaporative humidifier, and can still have uses in an emergency. Otherwise this type of humidifier is more common in ducted systems. One successful room humidifier of this type actually passes the mains alternating current directly through the water to heat it. The forced evaporative humidifier is at present the preferred type for exhibition rooms. The principle is to soak up water in an absorbent material and blow the room air through it. One effective type consists of a band of foamed plastic carried on

Figure 42
Defensor 4000-V evaporative humidifier, shown on the right with the drum raised for cleaning

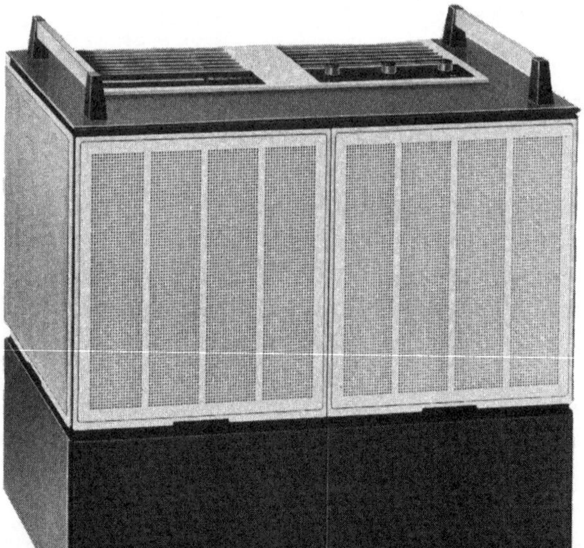

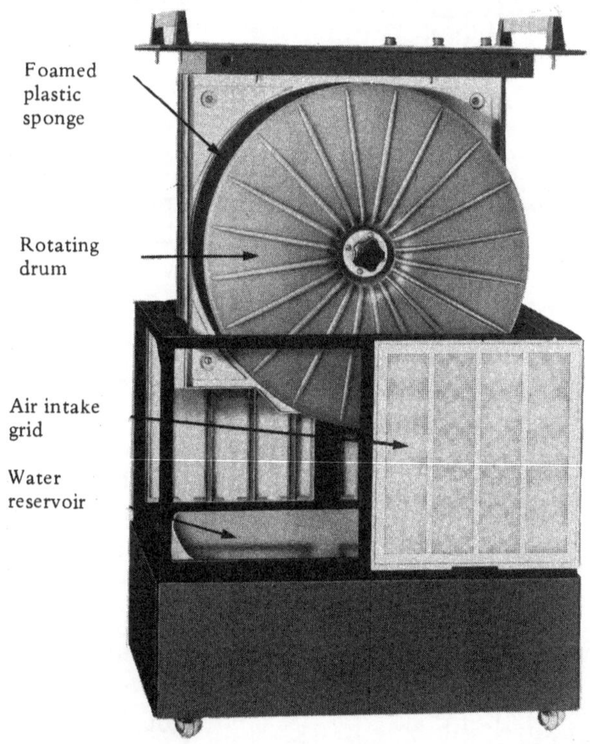

a drum framework which dips into a reservoir of water. The drum slowly revolves while a fan blows the room air through the wet plastic sponge (*Figure 42*). Since water as it evaporates draws heat from the surroundings this operation cools the air, so that some models are equipped with a heater. Every litre of water evaporated per hour requires 700 W of heating to offset this evaporative cooling.

Imagining now that the humidistat on one of these unheated evaporative humidifiers gets stuck in the 'on' position, the RH will indeed rise above the pre-set level, but the more it rises the less effective will the humidifier become. Damp air cannot take water from a damp sponge. In fact such humidifiers are not expected to be capable of attaining an RH much greater than about 70%. In this sense the unheated evaporative humidifier fails safe.

When water evaporates it leaves its minerals behind so that the moisture added to the air is pure. But the minerals leave a scale on the sponge which could eventually become like that on the inside of an electric kettle. Maintenance involves gently cleaning the sponge in water at intervals to break up the scale and disperse it. If tap water is used for humidification, cleansing is needed once a month. This period can be extended to three months if demineralised water is used[106]. The growth of algae and other organisms on the sponge can be avoided by adding a small amount of algicide to the water, after cleaning.

Naturally all humidifiers depend on a regular supply of water. This requires the organisation of a daily filling routine. Alternatively most humidifiers can be supplied with a ball valve and plumbed into a mains supply of water, though they then cease to be transportable.

DEHUMIDIFYING EQUIPMENT

There are two kinds, both electrically operated, and the choice depends on climate:

1 Desiccant dehumidifiers

2 Refrigerant dehumidifiers.

1 DESICCANT DEHUMIDIFIERS

We call this kind *desiccant* since moisture is removed from the air by blowing it through a drying or desiccating material, which can be lithium chloride (*Figure 43*) or silica gel.

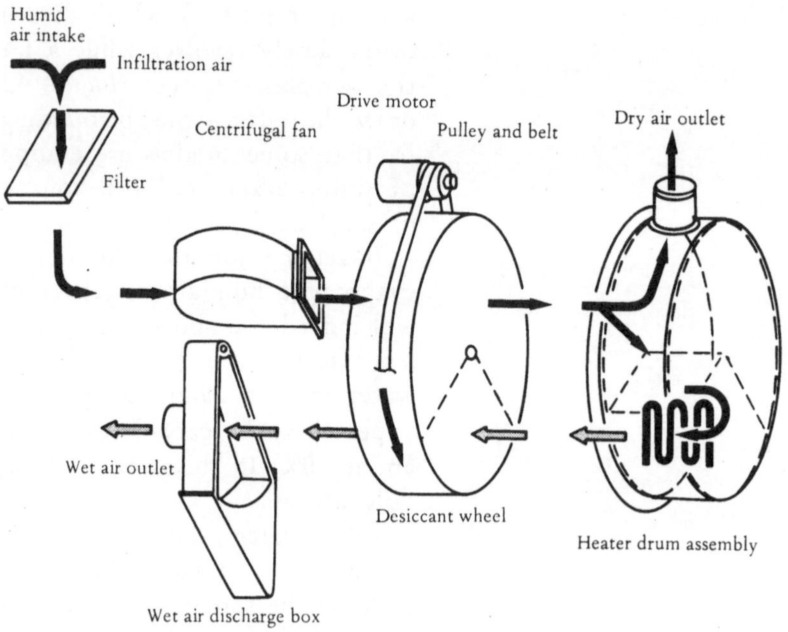

Figure 43
Rotaire desiccant dehumidifier in diagrammatic form. In this model the room air passes through filter, fan and desiccant wheel, after which it returns dry to the room. But part of this dried air is diverted through a heater to drive off water from a zone of the desiccant and discharge this as wet air outside the room

This is carried in a slowly revolving drum. The room air is blown through one sector of the drum, moisture being extracted by the desiccant in the process. But since the desiccant can only absorb a limited amount of moisture it must be periodically regenerated. This is done as the drum revolves, causing the sector which was used to dehumidify to pass to a position where heated air is blown through it to carry away its absorbed moisture.

The apparatus has three ducts: an entry and an exit duct for the room air, and an exit duct for the damp air. With the apparatus placed in a room to be dehumidified, say a damp basement, the entry and exit ducts for the room air open straight into the room from the apparatus. But the exit duct for the damp air must have a pipe on it leading out of the room. Since the regenerating air must be reasonably dry, a fourth duct is sometimes provided for it to be brought in from outside the room. Where discharge of a lot of wet air outside the room is difficult, an air-cooled condenser can be made part of the system. The water taken out of the air will then have to be emptied out at intervals or drained out.

Michalski has pointed out that the possible loss of lithium chloride, a corrosive salt, into a metal collection should predispose conservators to the silica gel type[107].

2 REFRIGERANT DEHUMIDIFIER

Because the *refrigerant* type of dehumidifier is nothing more than a refrigerator cooling unit rearranged, the ordinary refrigerator will first be described.

The device which keeps the domestic refrigerator cool consists of a long closed tube in which are a compressor and an expansion valve (*Figure 44*). Sealed inside this tube is the refrigerant, a gas so chosen as to be easily liquefiable by pressure alone without cooling. Both ammonia and sulphur dioxide have been used for this purpose in the past, but fortunately for museums (in case of leakage) these have been displaced by chemically inactive fluorinated hydrocarbons. When a gas is compressed into the liquid state it gives out a quota of its energy as heat (called the latent heat of vaporisation). When it evaporates the reverse happens and it has to draw in heat from its surroundings. All we have to do, then, is to cause the liquid refrigerant to evaporate inside that part of the tube which is in the refrigerator, thus drawing heat from the interior of the refrigerator. The refrigerant gas is then compressed back to a liquid in the part of the tube outside the refrigerator at the back, and this causes it to give up into the room the heat it took from the refrigerator.

Turning the refrigerant from liquid to gas is performed by the expansion valve, and turning it back to liquid is performed by the compressor. Thus the tube is divided into a low pressure section following the expansion valve, which contains liquid turning into gas and drawing heat from the inside of the refrigerator, and a high pressure section following the compressor, which contains gas turning back to liquid and giving up heat at the back of the refrigerator.

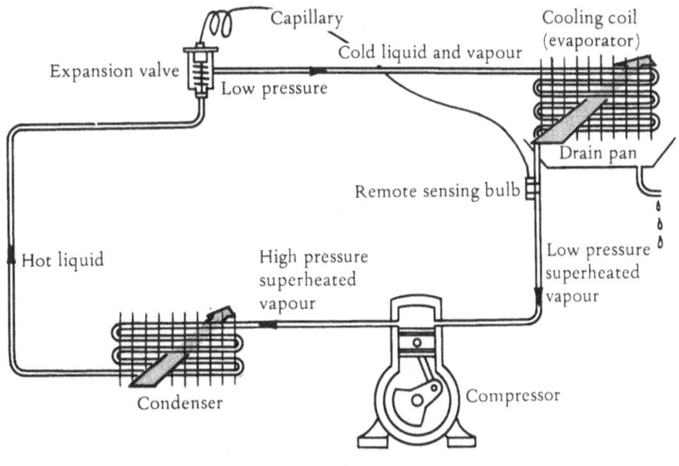

Figure 44
Vapour compression refrigeration system[108]

Thus, in the domestic refrigerator, as fast as heat leaks into it, this heat is transferred out again by coming into contact with the cooling coils inside. The device generates neither heat nor cold (except insofar as all machinery generates some heat), it merely transfers heat from one place to another: from inside to outside. This is why engineers call it a heat pump.

To turn this refrigerator system into a dehumidifier we only need to take it out of the refrigerator, put the high pressure coils directly in front of the low pressure coils and force the room air through them both with a fan (*Figure 45*). The following then happens. The room air is first cooled by the low pressure coils, and it must be cooled below its dew point so that water condenses from it and drips into a sump to be drained away. The air then passes directly to the high pressure coils which give back the heat which the cooling coils took from it. The air that leaves the apparatus is therefore drier and unaltered in temperature (actually slightly warmer), which is what we required.

CHOICE OF DEHUMIDIFIER

The desiccant type of dehumidifier is the choice for cold climates, and the refrigerant for warmer conditions. It is not hard to see why this is so. Desiccant materials hold less water when warmed and are therefore more effective in cool conditions. Conversely in cool conditions frost will form too readily on the cooling coils of the refrigerant dehumidifier thus blocking its action in removing water.

The break-even point, above which one would favour the refrigerant type and below which the desiccant, is probably about 10 °C.

Just as humidifiers cool, dehumidifiers have a heating effect. The figure is the same: 700 W for every litre evaporated or condensed per hour.

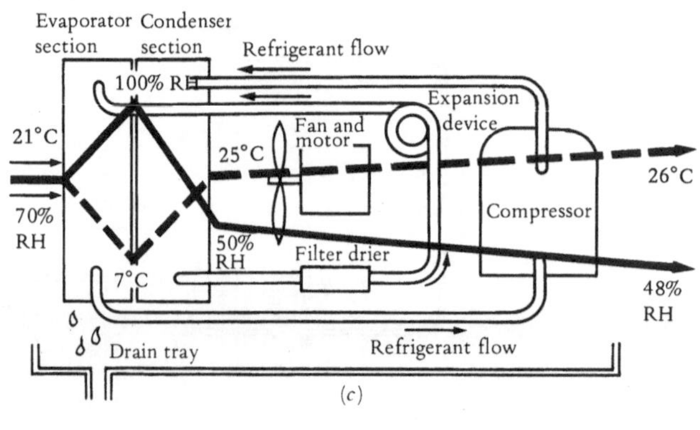

Figure 45
Westair refrigerant dehumidifier in diagrammatic form. Temperature and RH of air entering at 21 °C 70% RH and passing through the dehumidifier are shown in colour

ROOM RH CONTROL: MAINTENANCE AND AIR CIRCULATION

No machine will operate indefinitely without maintenance. The manufacturer's advice must be followed, and a regular procedure of maintenance set up in the museum.

It is amazing how reluctant air can be to mix. A great disadvantage of room humidity-controlling equipment is that we cannot ensure good mixing by just putting them in convenient places. As soon as the equipment is installed and operating, a survey must be made round the room to find out whether there are any areas or pockets which remain at the wrong RH through lack of air circulation or because of a draught of uncontrolled air[109]. Such problems can be solved either by resiting the equipment or by installing fans at suitable points.

Be sure that the air coming out of the equipment does not play directly on any exhibit before it has been diluted with room air.

Though perhaps unsightly in the eyes of the curator, the equipment will be the most interesting thing in the room to some visitors, especially to small boys. Control knobs and removable parts must be made secure from interference.

But the chief problem to the new conservator, or to the conservation adviser, remains: how do we eliminate acute danger and reduce deterioration without air conditioning?

At the risk of repeating something of what has already been said, it might be useful to run over the stages of appraisal in this kind of situation, perhaps in a country house or small museum.

STAGE 1. TYPE OF MATERIAL AND ITS NEEDS

We are here concerned with RH control, and therefore with wood, archival material, textiles, paintings and metals. Any mixed collection is certain to contain members of these groups, but the needs of metals are different from the needs of wood (p. 82), and climatic factors must also be borne in mind (p. 87).

STAGE 2. RECORDS

Can any RH and temperature records be produced? If these are of the eight-day record kind, look for evidence of calibration checks. If these are absent, the records should be politely ignored. Of far greater use are local meteorological tables[103]. These, together with observation and sling hygrometer readings on the spot, will show whether the conservator has a humidifying or a dehumidifying situation on his hands.

STAGE 3. APPARATUS REQUIRED

Perhaps the trouble resides in the misuse of existing equipment – most likely excessive dryness caused by heating so that no special apparatus need be acquired. Otherwise this stage of the advice consists in naming the equipment which should be used. What is available will vary from country to country. ICCROM (International Centre for the Study of the Preservation and the Restoration of Cultural Property, 13 Via di San Michele, Rome 00184) holds lists of equipment. The number of units and their cost will also need to be known. Here are two examples intended to help estimate the number of units required, the first for humidification and the second for dehumidification.

Humidification required

The usual situation is that a room is ventilated with air which has too little moisture in it. If we can estimate the rate of ventilation and measure the temperature and RH of this air, then the hygrometric chart will tell us the rate at which water should be added.

We have an exhibition or store room of volume 500 m^3, and the air in it and in neighbouring rooms is at 20 °C, 30% RH. We require 55% RH. We can assume a ventilation rate of one air change per hour, which is a normal sort of rate when doors and windows are kept closed, but without any refined draught exclusion. Reference to the Hygrometric Chart (*Figure 33*) shows that there is a difference of about 5 g/m^3 moisture between the 55% RH we wish to attain and the room air at 30% RH (both at 20 °C). Therefore every hour the room has to acquire 5 × 500 g of water (equivalent to 2½ litres).

But since we don't want our humidifiers to be working at full capacity for 24 h a day, we should double this requirement to at least 5 l/h.

At 50% RH, 20°C an evaporative humidifier will deliver up to about 12 l/h/kW of electrical input. So we need here 5/12 kW or about 400 W. Most room units are rated at less than this. The Defensor 4000-V, for example, has an output (with heater off) of 150 W. Our requirement would then be for three such units.

Dehumidification required

In this situation, we can picture an unheated basement storage room, again of 500 m^3 volume. It has very little ventilation and is generally damp (specific sources of damp must, of course, be identified and dealt with). Its conditions are 10 °C, 80% RH. Its ventilation rate is unlikely to be known, but we can work on one change every six hours. The Hygrometric Chart shows that, to bring 80% RH down to 55% without heating the room (10 °C), about 2½ g of moisture must be removed from every cubic metre of air every air change (six hours), or (500 × 2½ × 4) g (equivalent to 5 litres) per day. This is not too big a load for a fairly small desiccating dehumidifier.

If the high RH is due to ventilation by consistently damp air, as in the humid tropics, then the calculation of humidity requirements follows the same lines as for humidification.

At 50% RH, 20°C a dehumidifier, whether desiccant or refrigerant (refrigerant would be the choice in the tropics) can be expected to remove 10–15 l/day/kW.

But damp in basement storerooms is not easily reduced to figures. If a storeroom is poorly ventilated, quite a slow ingress of moisture through the structure will produce high humidities, and by the same token, a small dehumidifier will handle the task.

Most dehumidifiers are rated at more than 1kW, so one unit per room should suffice, at least as a first trial.

STAGE 4. OTHER MODIFICATIONS

Our investigation will inevitably widen to embrace the general conservation situation, including lighting (p. 22). Things interrelate, and we should check on excessive ventilation leading to extravagant energy demands, insulation, condensation on windows and skylights (double glazing?), keeping lower temperatures in stores (p. 45) etc.

STAGE 5. MAINTENANCE AND RECORDS

All apparatus requires maintenance, following the manufacturer's directions, and someone must be specifically delegated for this job. He should also set up proper climate records, e.g. an eight-day recording hygrometer calibrated every month, with recalibration particulars written on the record paper.

Summary

A humidistat must be used for any kind of room humidity control. This instrument may or may not be separate from the rest of the equipment. For humidification, an evaporative type of humidifier should be chosen. For dehumidification in warm climates use a refrigerative type of dehumidifier, and in cold climates a desiccant type. Proper air mixing and regular maintenance are very important in room humidity control. In advising on room humidification we must first survey what type of material is on display and then find out what we can of the local climate from records. Examples are given in this section on how to estimate the number of room humidity controllers required.

PACKAGED AIR-CONDITIONING UNITS

Until recently there was a big gap between room humidity controllers and central units supplying air to all rooms through ducting. Today this gap is being bridged by 'packaged' units, which may very likely be the most economical way of converting an existing building, a set of rooms at a time, to full air conditioning.

For the packaged unit the air must be delivered and returned through ducting, just as with central units. But the ducts lead to a unit close by, perhaps on the roof, and therefore do not involve extensive work on the building. A packaged unit can be made to do everything to the air that a central unit will do, removing dust and air pollution as well as maintaining RH and temperature.

A point to consider is that the cheapest form of heat is usually not electricity. Thus one can expect to have to pipe heated (and perhaps chilled) water from the central temperature control system, but this is not an operation on the same scale as ducting air. Unhappily this entails piping water above the collection, which is generally to be avoided for fear of leakage.

These packaged systems for full air conditioning are not to be confused with the coolers jutting out of every room in the richer parts of the tropical world. These are little more than refrigerators, and though in warm humid countries they can do much to reduce interior humidities to reasonable levels they are not capable of controlling RH to a set level since they operate on a thermostat not a humidistat.

DUCTED AIR CONDITIONING

The complete answer to climate control is a central unit distributing fully conditioned air through ducts to all parts of the building or at least to all exhibition and store rooms. Fully conditioned air means air whose RH and temperature are fixed and from which dust and polluting gases have been removed. New museums should give it very high priority.

Whereas the free-standing room units we first dealt with may be under the control of the conservation or curatorial staff, a central unit, and indeed a packaged unit, requires the attention of a qualified engineer. It would be inappropriate to enter this realm of specialisation in a book for museum staff, but an outline will be given for general guidance.

The objectives of the system must be clearly defined. For this purpose a Specification must be provided by the museum, such as the one suggested in the Appendix.

Figure 78 (p. 220) shows an air-conditioning system in diagrammatic form. It is perhaps surprising that moisture can be either added to or subtracted from air as it passes through a water spray according to the temperature of the spray system. This often consists of a set of nozzles which spray water onto pipes through which cold water may flow. For control reasons we require that the air leaves the spray in a near-saturated state (p. 214), but it may have been caused to release some of its moisture by the cooling, and when it is re-heated its RH may consequently be reduced.

Each duct leading to a set of rooms has its re-heater and humidifier for final temperature and RH adjustment. Air normally enters each room through spaced grilles at ceiling level and is extracted at several points near the floor. In this way dust raised in the room tends to be drawn out of the system.

Water for the spray cooling coils, must be provided by a refrigeration unit. This can be dispensed with in a cheaper system if we only want to control winter humidity. In winter we do not expect to need (a) a reduction of temperature* or (b) removal of moisture from the air. Conversely in the Tropics one would envisage an economical system without heating but with cooler and water spray.

Air-conditioning installations in museums could well have fail-safe switches, for example set at thresholds of 65% and 45% RH, so that if through failure of a component the RH goes outside these levels the machinery is automatically closed down.

Many conservation activities involve solvent vapours. The air from areas should not be recirculated.

*This may not be true if the heat from lights is itself enough to cause a rise in temperature.

CHEMICALS ADDED TO WATER USED IN AIR CONDITIONING

Chemicals may be added in low concentrations to the water in spray washers. These may be to prevent growth of fungi and bacteria or to reduce corrosion. They should not normally get into the air stream to affect exhibits or people because there is a large excess of water in the washers: the water cleans the air rather than vice versa. But an atomising humidifier disperses the total contents of the water into the air (see p. 95) and should on no account be used.

Another situation to avoid is the dispersion of volatiles by steam humidification. A case ocurred in a museum where *2-Diethylaminoethanol*, an anti-corrosion agent, was found to have appeared on surfaces in the museum. It has been suggested[110] that this deposit may have been an involatile product of reaction between the *2-Diethylaminoethanol* and acid pollutant gases. It appears to have caused skin and eye irritation, even though its concentration in the air (50 $\mu g/m^3$) was far below the Permissible Exposure Limit of 50 000 $\mu g/m^3$ [111].

To end this section let us emphasise the obvious. A museum is empty of people at night, so that in the old days it was the custom to shut down the heating. A modern museum is air-conditioned primarily for the safety of the exhibits, which live in it night and day. Therefore air conditioning must be continuous.

RH CONTROL IN A CLOSED CASE – BUFFERS

We have considered controlling the climate in a single room and in a whole building. There are also limited possibilities, not by any means fully explored, for the control of individual exhibition cases. The technology involved also applies to packing-cases, so that these also will be mentioned.

There are two distinct non-mechanical ways of reducing RH changes in a closed container. The first is by using a 'buffer' and the second is by using certain salts or salt solutions. We will first take a look at buffers.

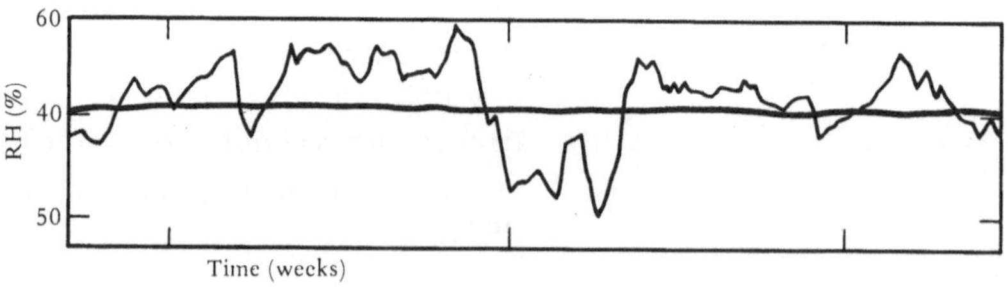

Figure 46
RH in room and exhibition case. The RH in the case is shown in colour. The case contained a wooden musical instrument and cloth-covered cork base, and so constitutes a fairly well buffered situation[112]

The RH fluctuations in an ordinary closed case, provided it is not heated by sun or strong spot-lights, are much smaller than those that occur in the room, and very much smaller than the changes in the weather outside the building (*Figure 46*). Daily RH fluctuations may be more or less eliminated by putting exhibits in a closed case, and weekly fluctuations much reduced. But this simple expedient will *not* protect against seasonal changes – dry European indoor winters or tropical monsoons.

Consider first a case which is sealed and empty. We must further specify that the case is made of non-water-absorbent material such as metal. No moisture can enter or escape. But the RH can still be changed – by a change in temperature. *Figure 36* shows what happens when the temperature rises: the RH goes down, and of course vice versa. If there is a non-moisture-absorbent material in the case, such as a metal bowl, cooling could result in condensation on it.

But museum cases are not usually hermetically sealed in this way. In fact we can expect the air in a normal case with a well-fitting lid and no warps or cracks to be replaced several times a day[97]. This implies that the RH of the air in an empty non-absorbent case would approximately follow the average daily RH in the room.

However *Figure 46* shows a much greater stability than this. The reason must lie in the presence of buffering material in the case. The term 'buffer' is used here to denote any material which resists or helps to buffer a change in the RH of the air surrounding it, whether this change is caused by a leakage of air at a different RH or by a change in temperature. If the RH of the air falls, then the buffer, in order to keep in equilibrium with the air, will give out some moisture. But this moisture will cause the RH to rise, thus counteracting the change, and vice versa.

The reader may by now have recognised that any of the moisture-absorbent materials which we have been discussing – wood, paper, cotton, etc. – do just this, so that all of these can themselves be used as buffers, and furthermore an exhibit made of such material will buffer the RH in the case – but not forever.

Imagine a metal museum case with some small wooden sculptures in it, in an uncontrolled gallery. Everything is about right, with the RH between 50 and 60%, and then comes the annual monsoon. Air of higher RH leaks into the case. The wooden objects, in order to keep in equilibrium with this higher RH, take in some moisture, and this lowers the RH of the air. So far so good: the wood has been a successful buffer. But in doing so it has taken in a little water, it has raised its own moisture content, so that it is now in equilibrium with a slightly higher RH. This continues throughout the monsoon, and very likely long before the end of it the wood will have become buffered to the monsoon humidity not to the dry season humidity.

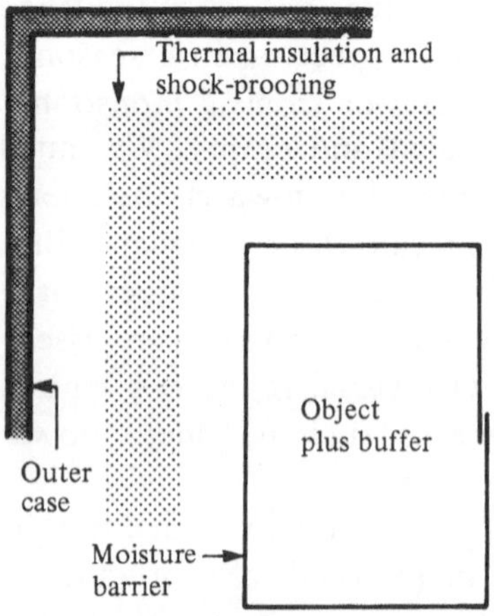

Figure 47(a)
Diagram of a packing case. The case must protect its contents from shock and from RH change. Preventing RH change involves preventing moisture entry or exit and preventing temperature change. For buffering see text p. 106. The moisture barrier could be polyvinylidene chloride, polyethylene, or a metallised plastic foil. For both thermal insulation and shock-proofing there is aerated packing material, e.g. foamed plastic. The outer case should in itself be moisture-resistant, and should on no account be made of butted planks. This diagram is for conveying principles. It is not a design drawing

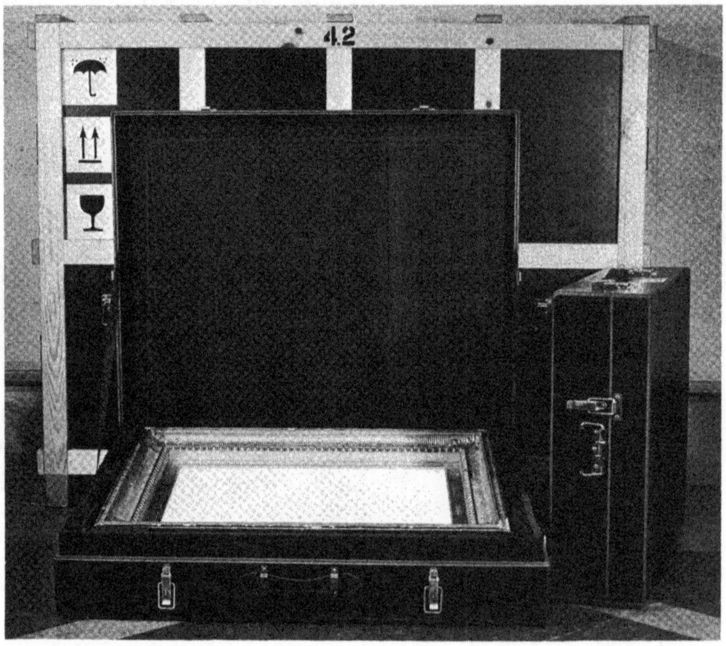

Figure 47(b)
Packing cases, National Gallery, London (EPS Dristyle). Constructed of plywood faced with phenolic resin and bonded together with an extruded polyurethane elastomer. The large case at the back is strengthened with wooden battens. Closures on sides and front can be padlocked. The cases are lined with a polyethelene foam, and pads of the same foam are cut to fit the painting to be packed. The painting is wrapped in polyethylene film. Glazing, if used, is acrylic sheet

Two paragraphs back, the statement was made that an exhibit can itself buffer the RH. We are in danger of losing sight of our objective: the moisture content of the exhibit is to remain unchanged. We can usually shift this goal to RH constancy, but we should be aware of what we are doing. The exhibit contains so much more water than the air surrounding it that any change in its moisture content will be revealed by a big RH change. So a constant RH should indicate that all is well.

Silica gel, first suggested by Toishi[113] and later advocated by Stolow, especially for the severe climate of Canada[114–117] is a good buffer. This fact has caused much confusion, for silica gel is widely used for its drying properties. The situation is that silica gel has no irreversible drying property like some chemical desiccants. Indeed it behaves very much like wood with regard to moisture. If it is in a very dry condition it will take moisture out of the atmosphere, and if it is very moist it will do the reverse. It is a better buffer than wood because it does all this faster, has a larger water reservoir and is chemically inert and non-flammable.

To use silica gel in the orthodox way as a drying agent it is heated to between 110 and 250 °C, whereupon an indicator incorporated in the granules changes from pink to blue. As its drying capacity becomes exhausted during subsequent use the colour changes back to pink.

Used as a buffer there is no need for the indicator colour. In fact the most convenient form for the silica gel is in small paper or cotton sachets. Either the granules themselves or the paper sachets are laid out in shallow layers at the required RH for as long as possible, at least a fortnight. During this period they will aquire a moisture content corresponding to the RH, and they can then be used as a buffer for this RH.

To test whether a bag of silica gel has reached the required RH, put it in a polyethylene bag together with an accurate dial hygrometer which can be read through the bag, and leave it for 30 min.

Another good buffer, better than silica gel according to the Japanese, is Kaken Gel, a clay impregnated with silica gel and sold as Nikka Pellets[288,291].

SILICA GEL IN PACKING CASES

Cases containing moisture-absorbent material which are not expected to undergo severe temperature changes do not require silica gel. At the other extreme, silica gel can protect metal from high humidities which might otherwise result in condensation.

Obviously the ability of a buffer to maintain RH within close limits depends on the relative proportions of buffer and air. There is very little air in a packing case compared to the usual exhibition case. Speed of response is also important. 1 kg of wood in the form of fine sticks in 1 m^3 of air will rapidly compensate – in fact slightly over-compensate – for any change in temperature[118]. But the same quantity of wood as a solid block would be much slower to change its moisture content. So, for packing cases travelling in arduous conditions it is better to leave RH control to a quick-acting buffer such as silica gel. For such a case the amount of buffer need only be 1 kg/m^3 of case volume.

Following the same principle, all the material used in the packing-case must also be allowed to come to equilibrium with the RH required in the case. Ideally all the wood used in the packing case, and of course the finished case itself, should be kept permanently at the correct RH. One cannot be sure that a pre-climatisation of even a month in the correct RH will be enough. Shorter periods may result in the objects packed being surrounded by a very different and possibly dangerous RH. This is a common mistake.

Rather than go into details of construction we need only note here that the general principles of packing for constant RH are dictated by two principles: (a) Moisture must not enter or leave the case. (b) Heat transfer must also be minimised because heating an enclosed volume of air lowers its RH and vice versa. The principles[121,122] are shown in diagram in *Figure 47(a)*.

The traditional concern of packing is, of course, still with us: protection against shock and vibration. If all the awful accidents to works of art in transit were published, if all the minor damages were photographed for all to see, a lot more would be spent by museums on packing. Advances have been made. Plastic foams, granules, etc. can perform three functions in one: (1) thermal insulation, (2) moisture barrier (together with the polyethylene sheet round the contents), and (3) protection against shock and vibration. For the latter purpose the foam must be adjusted to its loading, that is to say, it must neither be so flexible that it squashes flat under the load of the contents, nor so rigid that it transmits shock and vibration.

Some recent tests on cases using polyethylene foam are described in refs 119 and 120 (*Figure 47(b)*).

Of course we should not stop here. Conditions around the packing-case during transit should be corrected as far as possible in the direction of median RH values and room temperatures. For example until recently the holds of aircraft suffered violent extremes in both temperature and humidity.

It would be indiscreet of the author to reproduce photographic evidence of some disastrous damages which have occurred during transit to valuable works of art through lack of RH control. Packing cases can be successfully buffered against RH change. What about exhibition cases?

EXHIBITION CASES

The purpose of putting silica gel in a packing case is to protect the contents from an RH change caused by a *temperature* change. The case is assumed well enough sealed that there is very little air exchange between case and surroundings during a short journey. 1 kg of silica gel per cubic metre will suffice.

For the exhibition case we still need to protect against temperature-caused RH change. But we need to do much more. We need to keep RH within safe limits for a whole year. We are now working against leakage of air at a different RH into the case. For this we need much more silica gel: *20 kg silica gel per cubic metre of case volume* is suggested.

For a metal and glass case with corners sealed and a gasket round the door, this should be more than enough. For cases with cracks no more than 1/2 mm wide, twice this quantity would not suffice.

The normal exhibition case is certainly not a sealed case. Case air is freely exchanged with room air even through fine cracks. If the room air is much drier it will slowly but surely steal moisture from the contents of the case and hence slowly reduce its RH. Of course, we are now talking in terms, not of a short journey by air, but of months in a museum.

A buffer can never confer constant RH in a case if the air surrounding the case is on average of a different RH. But if the buffer is present in sufficient quantity in a well-made, leak-free case, it can reduce RH changes over a year to unimportance.

Theory and practice suggest that 20 kg of silica gel per cubic metre of case volume should be sufficient to carry its contents safely through a winter-heating period or a monsoon. (A temporary exhibition may not require 20 kg.) But the case should be so constructed as to allow circulation of air between the silica gel and the exhibition compartment, and should have a true-reading hygrometer in the case visible from outside it. If the average annual RH is very different from what is required in the case, there may have to be an annual replacement of freshly conditioned silica gel[123–125]. A buffer block of porous plastic with silica gel embedded is being developed[126]. Replacement of such a block would be quick and easy.

We need to examine whether there is any truth in the suggestion that a work of art, by some Pygmalian-like transsubstantiation, needs to breathe! Breathing is, of course, the means whereby living things burn their food internally to acquire energy. An oxidative process of this kind is the last thing we wish to happen to a work of art.

At the back of peoples' minds is probably the danger that stagnant damp air may be trapped so that moulds flourish. Indeed, if a case is permeable to moisture and if it is placed in contact with a source of moisture then ventilation will improve the situation. But allowing such a thing to happen in the first place is bad practice. Examples are a packing-case left standing on damp ground for customs clearance, or a painting hanging on a damp outside wall. Again, where humidity is generally very high, as in the humid tropics, and there is no mechanism for keeping it within safe limits, very much can be done in the museum building by ventilation and air movement.

As a third less usual example, an exhibition case containing wood, textiles, paper, etc., might be heated by direct sunlight or overpowerful spotlights to the extent that moisture is driven out of the wood and other materials into the air of the case. If the heating period is succeeded by strong cooling (lights switched off, fall in room air temperature,) there is the possibility of condensation behind the glass and even on objects close to the glass.

But these examples are all the result of bad housekeeping. In a healthy situation exhibition cases do not have to be ventilated, and certainly not properly constructed packing-cases. Needless to say, works of art do not breathe. Pollution in cases is reviewed on p. 154.

Summary

A buffer is here a moisture-containing solid which, when the RH rises, absorbs moisture, and gives it out when the RH falls. It thus helps to stabilise the air, but over a long period will drift towards the conditions surrounding it. The best buffer is silica gel, not in its dry form, but brought into equilibrium with air of the required RH.

THE BUFFERED CASE: TOWARDS A PRACTICAL SOLUTION

(1) For stability throughout the year 20 kg buffer per cubic metre of case volume is suggested for a case of one air change per day. (A temporary exhibition may not require 20 kg.)

(2) This involves a case not hermetically sealed but crack free. Metal and glass are the best construction materials, though acrylic or polycarbonate sheet could be substituted for glass. Wood should have a vapour seal such as aluminium foil. Corners should be sealed and openings gasketed.

(3) The buffer should be conditioned to the desired RH as tested in a polyethylene bag with a hygrometer. A fortnight at

the desired RH should be sufficient. Any wood in the case or in its construction will require much longer. Two months is a lower limit.

(4) A reliable and newly calibrated hygrometer should be placed in the case so as to be read from outside.

(5) However well made and buffered the case, its RH will eventually approach the annual average for the room. If this is very different from the desired RH the buffer will have to be changed periodically. This might be an annual job.

(6) There must be free air interchange between the buffer and exhibition compartments: fabric stretched over a perforated shelf would comply, or narrow slits front and back.

(7) The buffer should be laid out, either in bags or as free granules, as thinly as possible: not more than 2 or 3 cm thick.

(8) Trouble should be taken, especially with metal objects, to ensure that the case contains no polluting materials (p. 154).

(9) If diurnal RH fluctuations are eliminated but a monthly drift persists, then the case is probably insufficiently leak-proofed.

RH CONTROL IN A CLOSED CASE – USE OF SALTS

In simple theory if we put a bowl of water into a closed case water will evaporate from the bowl until 100% RH is reached in the case. In practice this figure may only be reached fairly close to the water if the water surface is small.

Any salt dissolved in the water will restrict evaporation, but the capacity of salts to do this varies. It also depends on the strength of the solution, and so we can simplify matters by always using saturated solutions. A saturated solution of a salt is one that holds as much as it can in solution. The usual way to make a saturated solution is to warm water and salt while stirring until no more will dissolve. On cooling some of the salt will crystallise out, and this is allowed to remain as visible evidence that the solution is saturated.

Saturated solutions of magnesium nitrate[127] and of sodium bromide[112] have been used to maintain RH stability inside showcases. At 20 °C magnesium nitrate gives, in practice, about 50% RH and sodium bromide about 58%. Although the humidity maintained varies somewhat with temperature the great advantage of this method lies in the large water reservoir, simplifying maintenance. But one thing prevents it from being more widely used. The salt crystals, going into and out of solution, tend to creep over the edge of the container and into the case, though it has been claimed that magnesium nitrate in stainless steel containers does not do this. Further, the presence of salt crystals in a case is not desirable for metal objects which could be electrolytically corroded.

The earliest use of salts for RH stabilisation in a museum case was a little different. Not a saturated solution but a dry salt containing water of crystallisation, zinc sulphate, was used. Like many other salts zinc sulphate takes into its crystal structure a certain number of molecules of water. These become part of the crystal structure and cease entirely to be liquid water. Some of them are loosely bound and will leave the crystal in conditions of low RH forming water vapour, to condense again into the crystal structure if the RH rises above a certain level. Here we have conditions for stabilisation.

An important painting by Hugo van der Goes, known as the *Trinity College Altarpiece*, was rehoused in 1936 in a closed case containing dry zinc sulphate crystals in the Scottish National Gallery[128]. The temperature of the Gallery in winter was then about 15 °C, and at this temperature the change-over point between zinc sulphate containing six molecules of water per molecule of zinc sulphate and seven molecules of water occurs at 55% RH, the required relative humidity. If the RH rises above 55% all those molecules of zinc sulphate containing six molecules of water will take up one molecule of water, and the reverse will occur if the RH falls below 55%.

This system, like all the others, has a limited reservoir of water (*see* p. 212), or capacity to take up water. If, for instance, in conditions of high RH the salt takes up all its complement of water so as to be converted 100% to zinc sulphate with seven molecules of water it can do nothing further to reduce the RH. The capacity of the system is much less than that of a saturated salt solution, and yet, though to a lesser extent, the danger of corrosion remains. There is no danger of corrosion with silica gel, nor of any other chemical reaction.

MECHANICAL RH STABILISATION IN CASES

Installation of a mechanical system for each case or group of cases, taking into account capital cost and maintenance, is not likely to be the cheapest option. But where special conditions are required for only a few cases in a general exhibition, containing perhaps unstable metal or glass, this could be a good solution for imposing low humidity just in these cases or for dust control.

Michalski has devised an economical and safe mechanical conditioner for a group of cases. This solution seems a truly viable alternative to non-mechanical buffering[129,130]

FUTURE DEVELOPMENT OF EXHIBITION CASE STABILISATION

Without ducted air conditioning, the curator now faces a situation of conflicting alternatives for case stabilisation: (a) should he construct low-leakage cases for use with silica gel and take care to avoid pollution from case materials? (b) should he install a miniature air-conditioning system serving one or a group of cases? or (c) if case RH control is secondary, should he ventilate the case by passing air through a filter against particulate and gaseous pollution, with or without a fan?

If the primary need is for reduction of, say, sulphur dioxide, as with precious archival material, and the room has some humidity control, then a filter and pump on the case could be the best answer. But even here a room filter (p. 254) might be a better solution.

There is much in favour of a system with no mechanical moving parts, because this is usually where failure occurs. Regular maintenance is required, possibly skilled. Silica gel systems also require maintenance, but at less frequent intervals and of a kind that can be performed by any conservator.

This might be the place to emphasise the need for development of a system to maintain stable RH over a season, whether a dry season (European winter heating season) or a wet season (tropical monsoon). The less mechanical the system the cheaper it is liable to be both in cost and maintenance. This would bring immense benefit to the less well-endowed museums of the world. The difficulties are practical rather than theoretical (*see* Part II, leakage rate, p. 232), but the work will inevitably be slow because, in the absence of large climate-controlled test rooms, testing must proceed through several seasons, meaning several years.

Summary

It is possible to stabilise the air in a container within limits by the use of silica gel or other buffer, salts, or a miniature mechanical system. But all require some knowledge and care in maintenance. We can expect development towards simple maintenance-free methods.

RH IS OFTEN A MATTER OF COMPROMISE

It can be seen from *Table 8* (p. 88) that there is a pretty strong tendency towards middle-of-the-range values for RH. We might move away from the centre for two reasons: (a) because the material exhibited requires a special RH, or (b) because the RH in an air-conditioned museum should reflect average indoor RH at the locality. The old objects of local origin will have been made in this climate and will have adapted to it, and also keeping near the local RH results in economy of energy use.

However most museums these days have temporary exhibits from time to time and so on occasion will borrow and lend. For the larger museums this implies exchange between countries, possibly of very different climate.

The lending museum may very properly demand conditions reasonably close to its own. This will strengthen a trend towards median RH values (50 or 55%).

Another similar tendency is as follows. Supposing we find from local records that 45% RH represents average indoor conditions, and specify 45% RH for the new air conditioning. In practice this means 45% ± 5%, in other words the range 40 to 50%. But 40% is bordering on the unwise for certain objects (parchment, leather, imported wooden objects), so to avoid this we go a little above average conditions and opt for 50% ± 5% RH. We cannot go very far from median RH conditions. But in temperate winter climates there are two situations which might make 55% too high an RH to handle. Firstly, a single layer of glazing in roof or windows will cool in winter to the extent of causing condensation. In England, not the coldest country, an attempt to maintain 55% at 20 °C is likely to lead to condensation-water dripping from skylights unless these are double glazed. In such cases it would be wise not to exceed 50%. There are troublesome examples of hot moist air rising and condensing under cold poorly insulated roofs. The result: water dripping down walls or on to exhibits plus corrosion and rotting of roof materials[131].

The difficulties are vastly increased for museums in the harsh winter climates of Canada, the northern parts of the U.S.A. and north-eastern Europe (*Figure 40*). To maintain 50% RH in a Canadian winter without condensation, triple glazing becomes necessary[132]. Furthermore the danger of frost formation within the walls of the building is great. There are both a temperature gradient and a moisture gradient in the walls of a building, the inner and outer surfaces being in two different climates. Thus there will always be some moisture movement in or out to reduce the difference. During a winter with high RH maintained indoors together with a high temperature the indoor moisture content will be much higher whatever the outside RH, and so the movement will be outwards. With the outside temperature well below freezing this moisture will freeze before it reaches the surface, i.e. inside the wall. Expansion caused by freezing creates space by pushing masonry apart, and the process is cumulative because after every thaw there is more space for moisture to penetrate and extend the crack.

It is possible to construct a new building for 50% winter RH, but special insulation is needed in the walls and condensation trouble in and around windows forces the design towards a window-less building. Rogers[132] quotes an extra 14% on the basic structural cost solely for humidification to 50%.

Another solution has been proposed by Amdur[133]. It is that the museum should be a building within a building. Only the inner building, containing moisture-sensitive exhibits and also storage space, would be maintained at 50% RH, while the space between inner and outer walls would be limited to the display of dry-loving material such as metal, glass, and salt-free stone and ceramics, and thus could be at a lower RH.

A third solution is now being studied by the Canadian Conservation Institute[104] for general application. It involves a controlled submission to low winter RH. For example, the midsummer RH is set to 50 ± 3% RH (47–53%), but the set point is lowered at a steady 2% per month from 50% to 38% in midwinter. The next six months to midsummer reverse the process, so that the RH again climbs to 50%. Thus the ideal of constant RH is jettisoned in favour of a change which is made gently and steadily. We know so little about the relation between actual physical damage and RH that it is difficult to pre-judge this proposal. Most exhibits would probably settle down to the cycle, but there would certainly be wooden objects which could be critically strained and might therefore require special conditions. The tests necessary to establish the safety of this routine should teach us much about the actual RH tolerance required in a museum.

It can now be seen that RH control involves consideration of the building as much as of the exhibits it contains. The architectural side of the problem is of great importance and merits the separate study given to it by Feilden[134].

HOW MUCH RH VARIATION CAN BE TOLERATED?

There could not be a better question than this to expose the inadequacy of the quantitative data so far collected on the effects of climatic factors on deterioration.

It is easy enough to test an experimental exhibit to destruction or to illustrate the response of a wooden panel to changes in RH. The panel warps in the predicted manner[84,135]; the response of a thin panel to a single breath can be measured[136]. But warping is not the same as damage, which is difficult to measure under real conditions.

There is impressive general evidence, for example in the records of the National Gallery, London, that transferring paintings to an air-conditioned environment very greatly reduced the need for treatment of detached paint. The adverse response of wooden objects to low RH is now universally recognised, especially of musical instruments where warping and shrinking affect sound as well as appearance.

But the question of how constant RH needs to be to ensure that no physical deterioration will occur at present remains unanswered. The standard specification of ± 4 or 5% in RH control is based more on what we can reasonably expect the equipment to do than on any deep knowledge of the effect of small variations on the exhibit.

Consider what is involved in discovering this information. In the first place the sledge-hammer technique of testing under exaggerated conditions is useless: we know what happens under exaggerated conditions. But in this case there is probably a threshold of humidity variation for each category of exhibit below which damage will not occur.

To find this threshold we might subject a group of exhibits to the most minute scrutiny while changing the RH round them in a controlled manner. When we find no damage whatever what can we conclude? We can conclude nothing. On the contrary we realise that, even though our particular samples suffered no damage, we cannot be sure of other samples. We would have to test a very large sample indeed. The whole thing escalates beyond practical experiment.

As a matter of interest this same impasse arises in other situations: attempts to assess the effects of vibration, the sonic boom or various anti-theft devices, for example.

A statistical approach is clearly essential. Hopefully there may be a day when many galleries keep careful records of damage occurring, treatment given, and environmental conditions. These records should be as quantitative as possible and exchangeable between museums. It seems to me that this approach is as much a basic duty of museums as a specialised scientific investigation.

Summary

Choice of RH level depends on several factors but cannot go too far from 50 or 55% RH. An exception may be found in the very low winter temperatures of Canada and north-eastern Europe where attempts at humidification to this level may endanger the building. The tolerance usually quoted of ± 4 or 5% RH is based more on what can be expected of an air-conditioning plant than on what exhibits can actually stand without deterioration, which is not known in any detail.

HISTORIC BUILDINGS CLOSED IN WINTER AND CHURCHES

It will not be possible to humidify certain buildings up to first-class museum standards.

Energy is costly and cannot be wasted. Were it not for the danger of condensation, sound buildings could safely – indeed advantageously – be left to cool to just above freezing point. Condensation occurs whenever air is cooled below its dew point, usually because the air meets a colder surface.

The main danger from condensation is in the spring, when the building is open to ventilation by warm moist air and interior surfaces are still cold from winter. Mould growth, dormant during winter, re-awakens. There is nothing against using windows for airing provided this fact is borne in mind: the air should not be damp.

Internal sources of moisture in historic buildings, such as cooking and washing activities, may cause trouble at any time of the year, especially if the damp air travels some distance and condenses in a cold, stagnant hidden corner.

But it is not only the cost of energy that limits winter heating. The reader must by now be tired of the reminder that major damage is caused by heating cold air without humidification. This has been well documented by Beck and Koller for Austrian churches. They recommend that winter temperatures should on no account be caused to rise above 12 °C by heating[137]. Inland Europe is much colder in winter than its coastal areas, yet Briggs suggests the same sort of figure for historic houses in England: 10 °C[138]. Thus there is some agreement that artificial heating should not cause winter temperatures to rise much above 10 °C.

Churches face another danger. If they are heated only for weekly services, the congregation bring in, even on a dry day, quite a load of moisture (something like 50 ml/h, *see* p. 230), which is readily absorbed by the warm air. The congregation leaves, the church is closed and its air temperature falls. Water condenses on any surface below the dew point of the air, which itself may be near saturation.

These surfaces may be decorated with murals. While the church is heated, water from the wall will evaporate. The water inevitably carries salts, which then crystallise as efflorescence[137–147].

Efflorescence may be on or beneath the paint surface. The latter is the more dangerous situation and will prevail as the wall dries, carrying the evaporation zone inwards. Condensation can dissolve air pollutants and redissolve efflorescence, carrying both through and below the surface. Each cycle causes further damage.

Thus cyclic situations of heating and cooling are to be avoided. Wall paintings are safest when dry, or at least when not transpiring moisture either in or out.

The best approach is to heat only moderately and to heat only those parts where heat is needed for human comfort, for instance with radiant heaters. Bordass gives good advice[148].

IMPROVISATION AND RH CONTROL

Supposing that there is no specialised apparatus for humidity control available or that in an emergency time is too short to await their arrival. What can be done to prevent humidity falling or rising to dangerous levels? Since improvisation depends on what is to hand one cannot cover the subject in a textbook, though a few guide-lines might be useful.

The first thing to note is that we may indeed be able to sense whether things are getting too damp or too dry, but that it is dangerous to rely only on our senses. The means of measuring temperature and RH *must* be at hand. A good electronic hygrometer (*Figure 30*) will do both jobs, or any form of wet-and-dry-bulb instrument.

Of course the conservator newly arrived in his post may discover a situation of emergency where none was suspected, or his post might have been created to solve a problem of deterioration. In such cases, though not true emergencies, his first actions will be improvisations, to be followed by permanent improvements.

In an emergency, such as flooding, involving objects susceptible to water damage, the flooded building may be rendered uninhabitable for the exhibits. These would have to be moved to drier surroundings, ideally where heat and humidity could be controlled[149]. The conservator would then have to tackle three problems simultaneously: drying, fungicidal treatment and restoration first-aid. Detailed restoration would obviously come last, but treatment such as applying facing tissues to paintings may be necessary to prevent things falling apart. Even if the weather is so cold as to discourage moulds (though after the Florence flood frosty weather did not discourage moulds), drying by heat will alter the situation, making it necessary to precede drying by some fungicidal treatment.

Just because of this mould problem, and also because water-soluble constituents may migrate, the rate of drying from the wet state will often have to be as fast as conditions will allow. But fast drying, which may be safe enough for individual sheets of paper and textiles, is dangerous for wood, bone, ivory, leather and composite materials.

There are two reasons for ensuring that an object is not dried too fast (or moistened, but this is less critical). Firstly, drying sets up internal forces aimed at contracting the object. A block of wood dries first on the outside, causing the outer layers to attempt contraction. But the bulk of the wood has not dried, and so resists this contraction. The outcome may well be a general stretching of the surface layers and the formation of surface cracks. Drying then penetrates to the core, which in its turn tries to contract. But the surface layers, having been stretched, are now too big for the core. Internal voids may be set up to resolve the situation. All this can be prevented if the rate of drying is slow enough to prevent a big difference in moisture content between surface and core.

The second reason for drying slowly is that a very close watch must be maintained on objects whose moisture content is undergoing a radical change. Provided the drying is not too fast it can be stopped if any trouble is detected and before any major damage occurs.

Maximum rates of safe drying cannot be tabulated since they vary too much according to type of object and strength. They can vary from an hour or two for textiles and prints of good strength to a few months for bulky wooden objects, and even longer for buildings where the plaster must be preserved from surface damage by efflorescence.

To dry the wall of a building is to draw water through it to the surface, where it evaporates. All salts dissolved in the water are deposited on or just beneath the surface. Formation of salts in this way can exert considerable pressure, and may destroy a fresco. In such a situation every effort must be made to evaporate from other surfaces than the painted ones.

Incidentally the effective RH at a painting hanging on a damp wall, or even a wall slightly less than room temperature, can be very much higher than the RH in the room. In such circumstances mould is often found on the back of the canvas, and this is the one situation where covering the back of the frame with a protective board can do more harm than good, since this reduces the chance of the damp air being ventilated away. The answer to cool and/or damp walls is to build a false inner wall of light construction with an air space between it and the true wall, and vents near floor and ceiling for air circulation. Pictures can safely be hung on this false wall. In a milder situation cork spacers between picture and wall would suffice.

Floods are rare, but damp in buildings is the first result of neglect. The source of the trouble must be clearly identified, and though repair is a builder's job, the conservator who is thoroughly familiar with the structure, may be as competent as the builder in finding this source.

Damp in buildings can originate from roof, foundation, exposed wall or plumbing failure. Less obviously it can arise from a temperature fluctuation or from visitors. In an unheated building the seasonal change from winter to spring finds a building cold from the winter filled with warm damp spring air. Condensation and mould growth may then occur on the coldest walls. Condensation may also occur on a daily rather than a seasonal cycle, though the example that follows is from a domestic rather than a museological situation. In a private home where both husband and wife go out to work they will be likely to reduce the heating during the unoccupied part of the day. But each morning there may be sufficient production of water vapour in kitchen and bathroom for condensation to occur later in the day and to accumulate every day. Museum equivalents might be found.

As for condensation caused by visitors, it has already been mentioned (p. 120) that the stones of a building can take up considerable quantities of human moisture. The situation is common in churches which are heated for services, but not long enough for the fabric of the building to warm[140].

The weather of the moment can help or hinder an emergency. After a hurricane there may be plenty of sun for drying things in the open and for drying a building outwards rather than inwards[150], though warm weather encourages moulds. In the humid tropics and at night in other climates natural ventilation is not going to get us very far, since outside air may very well be at 80% RH[151]. But even here, continued movement of air by fans is a great prophylactic against moulds. Whatever the situation it will obviously pay the conservator to use the weather in any way he can. It requires no specialist knowledge to bring in the laundry when the sun gets low in the sky, but in less definite situations work with wet and dry bulb will serve to determine whether the available air will be effective for drying or moistening. We can certainly dry if the outside air is drier but cooler, for the slight heating it will sustain if brought into a warmer shelter will make it drier still. If the outside air is dry but warmer than the air in the building, the cooling which it gets on entry will drive up the RH. In this case the only thing to do is to follow progress carefully with a sling psychrometer or equivalent.

Generally speaking, we open windows for our comfort, usually because we are too hot. If the interior of a building is hotter than the outside it is likely also to be drier, perhaps too dry. Opening windows also ventilates, and provided that the change of conditions brought about by the fresh air is not too harsh, ventilation is good. But, of course, windows must not be left open for rain or for cold night air. So, for rooms without any kind of humidity control, some good and little harm can come from judiciously opening windows onto good weather.

Apart from an intelligent use of climate by opening and closing windows supplemented by using fans, artificial heat of some kind for a damp situation is usually available. Electrical heating is straightforward. Oil (paraffin, kerosene, etc.) and bottled-gas burners add moisture to the room if they are of the type where the exhaust gases mix with the room air. They would never normally be used in a museum.

Up to this point we have dealt mainly with drying. In the contrary situation, the over-dry situation where moisture must be added to the air, a better method than spraying water on the floor and hanging damp textiles round the place – especially if the weather is cold is the use of electric immersion heaters to boil off water vapour which is then dispersed by fans. In all cases fans are important to keep the air mixing. Textile hangings can certainly help to form a reservoir of moisture. The conservator must be on the lookout for cold surfaces where water may condense. Pictures hung on such surfaces could suffer.

All in all the improvisation situation involves much running around, watchfulness and ingenuity. As soon as possible the specialised apparatus for humidification or dehumidification should be brought in and operated by humidistat. Watchfulness must still be maintained but can now be more relaxed.

Summary

We cannot reliably sense the RH and therefore cannot safely deal with an emergency humidity situation without the means of measuring RH, the best being a good electronic hygrometer. In flooding and other high humidity situations three problems must be tackled concurrently: drying, fungicidal treatment and restoration first aid. Drying must not be too fast, especially for bulky wooded objects. The outdoor climate can be used by ventilating through open windows when it will help the situation. Heating is a simple way of lowering RH. Dry conditions can be counteracted by the use of electric immersion heaters and fans.

HUMIDITY CONTROL IN ARCHAEOLOGY

Delicate organic material such as textiles and flesh can be amazingly preserved in three contrasting situations:

1 Extreme dryness as in desert sand.

2 Water logged anaerobic conditions as in acid bogs.

3 Temperatures which never rise above freezing point as in the permafrost regions of northern Russia.

In all these cases the reason for preservation is absence of biodeterioration, and the reason for absence of biodeterioration is that neither rodents, insects, moulds or bacteria can flourish in dry sand or in water which contains insufficient oxygen, or in constant freezing conditions.

In parenthesis it is interesting that in tombs or tumuli closed for many centuries yet not devoid of oxygen, microbial growth seems often to have come to an end. Arai has made the interesting suggestion that anti-microbial material has been generated which might consist of alkylamines formed during the putrefaction of proteinaceous matter such as flesh[152].

However as soon as the man with the spade comes along he destroys a precarious equilibrium.

Journalists are fond of the story of the archaeologist who bursts into a tomb only to find the most precious treasures turning to dust before his eyes. Today we must assume that the falling to dust does not happen in his hands or in the hands of his workmen, at one time a very likely event. However where great changes do take place quickly on excavation these are likely to be caused not by oxygen in the air nor by light but by moisture change. A common situation is a tomb near the water table, which has been at a constant 100% RH for a very long time, but on opening is exposed to the lower humidity of the outside air. This can cause rapid drying of material which has, in the course of its long entombment, become extremely delicate. Although we speak of marvellous preservation, large changes will have occurred and things that were once strong may now be very weak. There is no wind in a tomb, and temperature and RH are very constant. But biodeterioration may have occurred in the early life of the tomb and water and chemical change may have been at work. For example the pigment layer on a tomb painting is likely to have lost any organic adhesive and may curl up in flakes as soon as its front surface is subject to drying. Other objects may only have sufficient strength to retain their integrity without movement of any kind[153].

The first thing is to establish an airlock *before* opening. Nothing elaborate is required in the way of pressure doors or hermetic sealing: merely an effective means of reducing interchange between inside and outside air. Thus the airlock might consist of nothing more than a double door constructed of wood, plywood and polyethylene. Where there is a considerable difference between ground and air temperature, thermal insulation ought to be added. The space between the two doors should be just large enough to ensure that one door can be closed before the other is opened. Humidifying equipment may be required.

The maintenance of high humidity plus light plus the presence of people within an airlock brings the danger of mould and algal growth. Because of toxicity risks it is probably best not to take preventive action but to keep a careful watch for first signs, and only then to disinfest.

Inflatable air-domes are coming to be used at excavations[154] for protection against weather. However the constant ventilation which is a feature of these domes is likely to increase the rate of drying on the site of excavation. This can only be prevented by humidification.

For major cave or tomb discoveries with wall decoration there are dangers associated with opening the chamber to visitors. These can be summarised under three headings:

(1) Movement of water through the paintings caused by evaporation/condensation.
(2) Microbial activity.
(3) Change in carbon dioxide level.

(1) Records of RH before interference should have been made. If, as commonly, it was near saturation, it should be made to remain at this level. Otherwise evaporation and efflorescense will occur. At the other extreme it can be generalised that a dry wall painting is a safe one. If at a later stage subterranean water can be successfully diverted and the walls allowed to dry inwards, perhaps by surrounding the chamber with an air gap or moisture barrier or drainage channels, then the RH can be allowed to fall. Thereafter dewpoint conditions must be avoided by ensuring that the walls are never much cooler than the air in the chamber.

(2) High RH, as above, unfortunately favours microbial activity: algal growth in the more brightly lit areas for example. In Lascaux *la maladie blanche* (efflorescence) was succeeded by *la maladie verte* (algal growth). Visitors furnish the small amounts of nutrient needed. Disinfection may be required[155,156].

(3) Dissolved carbon dioxide gives rainwater a pH of 5.6. When rainwater falls on soil it picks up further carbon dioxide from plant root activity:

$$\begin{array}{llll} \text{water} + & \text{carbon dioxide} & & \text{carbonic acid} \\ H_2O + & CO_2 & \rightleftharpoons & H^+ + HCO_3^- \end{array}$$

This equation is reversible: if the rainwater, collected into a stream, returns to the open air, some of its carbon dioxide will be given up and the equation will move to the left.

However, if the CO_2-rich rainwater seeps down into chalk ($CaCO_3$) or $CaCO_3$-containing sediments, the carbonic acid it contains is capable of dissolving some of the $CaCO_3$, to form calcium bicarbonate:

$$CaCO_3 + H^+ + HCO_3^- \rightleftharpoons Ca^{++} + 2HCO_3^-$$

These two equations can be summarised:

$$H_2O + CO_2 + CaCO_3 \underset{\text{deposited}}{\overset{CaCO_3 \text{ dissolved}}{\rightleftharpoons}} Ca^{++} + 2HCO_3^-$$

In a limestone cave there will always be an excess of $CaCO_3$ present, but the level of carbon dioxide can vary. By a well known chemical law, the law of mass action, already assumed above, the equation will go to the right if the concentration of material on the left increases, or if material on the right decreases, and vice versa. Thus, to dissolve $CaCO_3$, raise the CO_2 concentration. To deposit $CaCO_3$ as efflorescence, lower the CO_2 concentration.

In a closed cave the CO_2 will long ago have risen to the level where $CaCO_3$ is neither deposited nor dissolved. Things will be in equilibrium. But if we open the cave and ventilate it with fresh air, the CO_2 level will fall, causing the deposition of $CaCO_3$ on paint and other surfaces.

If, on the other hand, no ventilation is allowed, but people enter the cave, the CO_2 level will rise to the point where the water is capable of dissolving $CaCO_3$ from the cave surfaces. The water will have become mildly corrosive. Incidentally, the people will also suffer from the rise in CO_2. To compound the danger, even with moderate CO_2 levels, condensation water can be corrosive because it is absolutely free of Ca^{++} ions, thus driving the equation to the right. Condensation water trickling down the surface may wash off the paint, as has been observed in Lascaux[157].

Whereas control of (1) RH and of (2) microbial activity is simple in principle, control of carbon dioxide is not. The only answer may be to close the cave except to a very limited number of people (5 per day at Lascaux) or to have it viewed through a window, as with the Japanese Takamatsuzuka tumulus[158].

Air Pollution
Part One

Air Pollution

THE PROBLEM

'In considering the position of the National Gallery, our attention was drawn to the vicinity of several large chimneys, particularly that of the Baths and Wash-houses, and that connected with the steam engine by which the fountains in Trafalgar Square are worked, from which great volumes of smoke are emitted. In the neighbourhood also the numerous chimneys of the various club-houses are constantly throwing out a greater body of smoke than those of ordinary private residences.'
Michael Faraday and Charles Eastlake[159].

The problem of solid dirt in the air of cities is not new, and complaints date back much further than Faraday and Eastlake's investigation of air pollution in the National Gallery, London. The acid vapours which belched out of furnaces with the smoke were also known to Faraday. However the automobile has produced a distinctly modern form of pollution known as the oxidant type (p. 149).

Air pollution is associated with towns and industry and is almost entirely caused by the burning of fuels.

To deal with air pollution we must be able to measure it, and it is simplest to adopt a universal system applicable to both solids and gases, for gases can be weighed just like solids. We therefore measure the concentration of a pollutant by finding the weight of it in a given volume of air. The unit commonly used is *micrograms* of pollutant *per cubic metre* of air, abbreviated $\mu g/m^3$ or $\mu g\ m^{-3}$. A microgram is a millionth of a gram. If we write the words 'Air Pollution' firmly with a soft pencil on a piece of paper we might add about 100 μg to its weight.

Concentrations of gases are often quoted in parts per million (ppm) or parts per billion (ppb). Note one billion equals one thousand million. One part per million of sulphur dioxide in air simply means that if we could separate the sulphur dioxide from the air its volume would be one millionth of the volume of the air it came from.

To convert from $\mu g/m^3$ to ppm or ppb the molecular weight of the pollutant gas must be known:

TABLE 10. Conversion from ppb by vol. to $\mu g/m^3$

1 ppb (0.001 ppm) of sulphur dioxide, SO_2	= 2.6 $\mu g/m^3$
1 ppb (0.001 ppm) of nitrogen dioxide, NO_2	= 1.9 $\mu g/m^3$
1 ppb (0.001 ppm) of ozone, O_3	= 2.0 $\mu g/m^3$
1 $\mu g/m^3$ SO_2	= 0.38 ppb
1 $\mu g/m^3$ NO_2	= 0.53 ppb
1 $\mu g/m^3$ O_3	= 0.51 ppb

General formula: 1 ppb = $0.0409M$ $\mu g/m^3$
where M is the molecular weight of the pollutant gas

In town museums the suspended dirt in the air gives rise to an obvious problem. Some of these particulates are heavy enough to settle in still air, but our major concern is with those that are too small ever to settle under their own weight, and therefore enter into the furthest corners of buildings. Dirt on museum objects, especially textiles, not only becomes unsightly as it accumulates, but sooner or later necessitates the risky operation of cleaning.

Within the last half-century various gaseous pollutants have become a world-wide hazard to the health of both people and antiquities.

PARTICULATES

Two terms are currently in common use for the solid particles suspended in the air: particulates and aerosol. The former will be used here. Other terms, particularly 'smoke' which is sometimes limited to the products of incomplete combustion, are too imprecise. Particles that settle in still air can be referred to as dust, or sometimes grit, whichever seems most appropriate.

The sizes of particulates are conveniently scaled by quoting diameters in microns (1 micron (abbreviated as μm) = 1 thousandth of a millimetre). In what follows a size of one micron means a diameter (not radius) of one micron.

The particulates, though they might all seem too small to measure, vary hugely in size. Particles larger than 15 or 20 microns settle near their source of origin or at worst on window sills. Particles smaller than this remain suspended until trapped on some surface. The bottom end of the scale can be taken as

0.01 micron, a hundredth of a micron (*see* Part II p. 247), and a good deal smaller than the wavelength of visible light (½ micron). This gives us a range of three orders of magnitude in diameter, or 1000 to 1. Size distribution is important when trying to find the right air filter for the job.

Since the particulates produced outside the building arise largely from the burning of fuels in power stations, and from vehicles and heaters there is a lot of sooty and tarry material in them. They are also usually acid from adsorbed sulphur dioxide and often contain traces of metals such as iron which can catalyse deterioration.

Inside the museum the furnishings and the human occupants provide a quota of textile fibres and fragments of skin. These may be chemically harmless but are biologically attractive as food sources. Smoking provokes a real problem in those museums which open their exhibition rooms for receptions, and should not be allowed where important works are on display. Experienced picture restorers are familiar with a tarry brown stain soluble in water characteristic of paintings hanging in club rooms[94].

PARTICULATE CONCENTRATIONS TODAY

In the U.K. and probably in most of Western Europe the situation was – literally – at its blackest long before the passing of the Clean Air Act in 1956. Smoke levels at Kew, a suburb of London, were falling when records were started in 1922[160]. Average winter levels were near 1000 $\mu g/m^3$ in London in the early 1950s[161], whereas the average winter level near the National Gallery (one of the smokiest areas of London) had fallen to 160 $\mu g/m^3$ by 1967[162]. The yearly average level for London is now only 40 $\mu g/m^3$. This can be taken as an average sort of figure for towns in Western Europe.

There are not now any pockets of pure country air left in Europe and North America, industrial pollution having spread over the whole area to give background levels of around 20 $\mu g/m^3$ [163].

Apart from urban pollution, natural particulates produced by forest and plant cover, including volatile material given off by them and polymerised by sunlight, may contribute up to 15 $\mu g/m^3$ [164], but usually much less.

NEW CONCRETE BUILDINGS

Toishi[165,166] has focussed attention on the possible danger of putting works of art into new concrete buildings. New concrete and cement gives off a dust of ultrafine particulates at the low end of the range (0.01 microns), which will readily pass the usual air filters. These particles, according to Toishi, are alkaline enough to cause damage to hardened oil paint, silk and certain dyes and pigments. Their presence can be detected by bubbling the air through slightly acid water[167].

Alkalinity of this kind can be dealt with by facing the interior concrete surfaces or by sealing with a suitable paint or varnish. If this is not done, up to two years may pass before the alkalinity drops to a safe level.

It should be emphasised that the above recommendations depend entirely on the work of Toishi and his colleagues, but it seems that they should be taken seriously.

REMOVAL OF PARTICULATES

While we should not neglect the simple ways of keeping things dust-free, such as well-made cases in exhibition rooms and storage in polythene bags, we cannot positively clean the air in this way. Nor can particulates be removed effectively by standing a fan and filter in the corner of a room so that the air that passes through is cleaned. Removal of particulates involves full ducted air conditioning so that the whole of the room air passes through the filters. The only alternatives are to deal with individual showcases (p. 114) or to install room air cleaners (p. 254). The same applies to the removal of gaseous pollution.

The main question at issue then becomes: what should be the efficiency of the air cleaning?

The more efficient an air filter is in removing dirt the more pressure is needed to force the air through it and the more maintenance it requires.

The most efficient filters of all, often called Absolute Filters, find uses in operating theatres and radiochemical laboratories. They are used where specific toxic particles must be removed with great efficiency. Their efficiency would be wasted if their installation were not coupled with other measures to prevent entry of those particles, e.g. through doors and on people. Clearly the standard to be aimed at in museums is something less than this.

It is of very little use to quote the efficiency of a filter as the proportion by *number* of particles removed since a particle of ten microns will weigh a billion (10^9) times as much as a particle of 0.01 microns diameter. It is very little better to say that the filters are '99% efficient down to two microns' when we know that more than 20% of the mass is likely to be below this size (p. 252). What we need in general is the proportion by *weight* of the total particulates in the air which are prevented from entering the exhibition rooms. The degree of soiling is more or less related to the quantity of dirt, assuming that colour is independent of size. Paint manufacturers know that to grind down a pigment to finer particle size is to increase its covering power. We could extend this principle to air pollution and suppose that a given weight of small particles makes things blacker than the same weight of larger particles. But this relation breaks down near the wavelength of light (½ micron), and we have to deal with particles both smaller and larger than this. So we can ignore the effect and simply deal with the total weight of particulates.

Two filter tests are now standard: Eurovent 4/5 in the EEC area, and ASHRAE 52/76 in the USA. They use similar test dusts and are approximately interchangeable.

Eurovent 4/5 comprises two tests. The first, termed 'Arrestance', will not be used here. It comprises a synthetic test dust made up of grit, carbon black and cotton linters. We shall use the second test, the Eurovent 4/5 Efficiency. This test measures the amount of an atmospheric dust penetrating the filter under test. The measurement is repeated several times, meanwhile loading the filter with the Arrestance dust so as to simulate a filter under actual use partly filled with dust. An average is then taken.

It is suggested that the main filters in a museum air-conditioning system should have a Eurovent 4/5 Efficiency of 85%. Other filters used in the system do not need to reach this level (*see Table 11*).

Figure 48
A bank of eight bag-type particulate filters: Vokes Univee Grade II. The air flow is from the right and the V-section of each bag ensures a large surface of filter in contact with the air. The bags are of composite layers comprising both synthetic fibres and glass wool, and have an efficiency of 80–85%

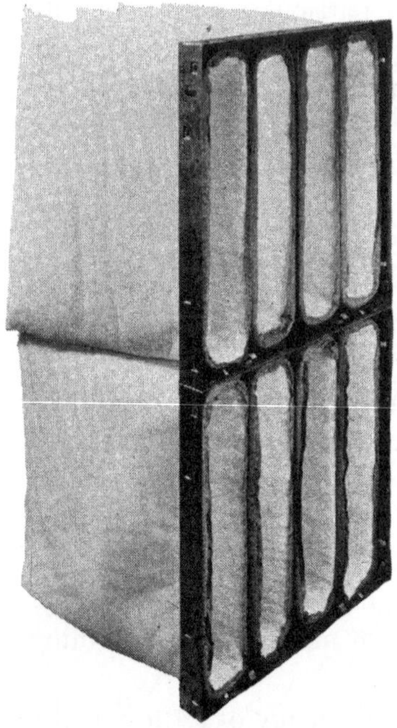

TABLE 11. Recommended Eurovent 4/5 Efficiencies[168]

Main filter	85%
After-filter (after activated carbon filter)	45%
Pre-filter or rough filter	25%

Because most of the air passes through the filters at least twice by recirculation, the effect of installing such filters is to reduce the proportion of dirt reaching the exhibits to well below 5% by weight of the outside levels (*Figure 48. See also* p. 252).

The maintenance of a ducted air-conditioning system is not a job for the curator or conservator: specialised staff must be employed. However the curator is interested in its efficiency. While he is never likely to have to change a filter himself the following information is relevant.

A pressure gauge is mounted across the bank of filters so as to measure the pressure drop caused in the air stream by its passage through the filters. The more dirt that is picked up by the filters the harder it becomes to push air through them, and so the pressure rises with use. Thus the reading on the pressure gauge gives an indication of when the filters need to be changed. If this change is not carried out with care some of the dirt adhering to the filters will fall into the duct, to be swept into the galleries as soon as the fans are restarted. Also filters which are forced incorrectly into their mountings will create gaps for the free passage of unfiltered air.

Today there are many kinds of particulate-measuring apparatus and most countries have air pollution laboratories which can undertake surveys. Such a survey carried out in the exhibition rooms from time to time is a valuable check on the functioning of an expensive system.

ELECTROSTATIC PRECIPITATORS (ELECTRO–FILTERS)

An efficient way of removing particulates from air is to pass the air stream through an array of wires charged to a high positive voltage. The particles are thereby given a positive charge. Collector plates charged negatively then attract and hold the particles further downstream. This is the principle of the electrostatic precipitator. Unfortunately the charging voltage is high enough (around ten kilovolts) to ionise and thus to initiate chemical reactions on a small proportion of the air molecules, producing in particular ozone[169]. Since ozone is a very undesirable contaminant (p. 149) electrostatic precipitators *should not be used in museums*. For good measure they also aid the conversion of sulphur dioxide to sulphuric acid[170].

GASEOUS POLLUTION

Like the particulates, gaseous pollution is caused overwhelmingly by the burning of fuels in power-stations, factories, domestic buildings and automobiles. Before the 'age of the car' it appeared there were two main types of gaseous urban air pollution: acidic and oxidant. Acidic pollution was associated with coal burning, oxidant with cars. With cities all over the world full of cars the two types have merged. Now the three main pollutant gases everywhere in the industrialised world are sulphur dioxide, nitrogen dioxide and ozone.

SULPHUR DIOXIDE (SO_2)

The gas sulphur dioxide appears in the air when we burn any of the fossil fuels (coal, coal gas, petroleum, oil and natural gas). All of these contain some sulphur (*see Table 12*).

TABLE 12. Weight Percent of Sulphur in Some Fuels Used in the U.K.

Fuel oil	2.2–2.6
Coal	1.3–1.5
Smokeless fuels	1.0
Diesel fuel	0.35
Motor-vehicle petroleum	0.1
Natural gas	not more than 0.002

When the fuel is burnt the sulphur combines with oxygen in the air to form sulphur dioxide. Sulphur dioxide itself is only a mild acid. However it quite readily combines with further oxygen to form sulphur trioxide, SO_3. As soon as this is formed it immediately combines with the ever-present water molecules to form sulphuric acid (*see* below). Sulphuric acid is a very strong and corrosive chemical. Furthermore once it appears on a material it cannot be removed by cleansing the air because of its involatility.

$$S + O_2 \longrightarrow SO_2$$
sulphur (in all fuels) + oxygen → sulphur dioxide gas

$$2SO_2 + O_2 \longrightarrow 2SO_3$$
sulphur dioxide + more oxygen → sulphur trioxide gas

$$SO_3 + H_2O \longrightarrow H_2SO_4$$
sulphur trioxide + water → sulphuric acid

The sulphur dioxide in the air is far from all man-made, indeed it may come as a surprise to learn that roughly half of the sulphur dioxide in the air is from natural biological sources[171, 172]. But this is only sufficient to give a background concentration all over the globe of less than 1 $\mu g/m^3$ [173]. The trouble with man-made sulphur dioxide is that it is produced in high concentration in the industrial parts of the world, just where most of the

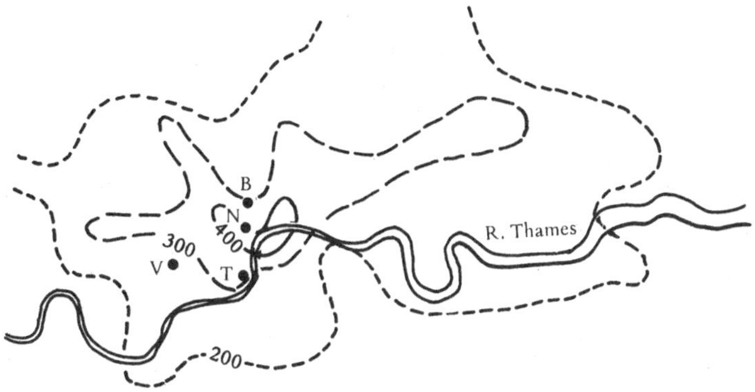

Figure 49
Contours of sulphur dioxide pollution in London for winter 1966–67 in $\mu g/m^3$. The major museums and galleries are all sited near the area of maximum pollution. B, British Museum; N, National Gallery; T, Tate Gallery; V, Victoria and Albert Museum. (Adapted from Nat. Survey[162] p. 173.) In 1980 concentrations had dropped to just under half these values[175].

museums and monuments of antiquity are also concentrated (*Figure 49*). Over 90% of the sulphur dioxide within Europe is from man-made sources[174]. In north-west Europe, for example, concentrations are in the 50–100 $\mu g/m^3$ range for industrial areas (*Figure 50*).

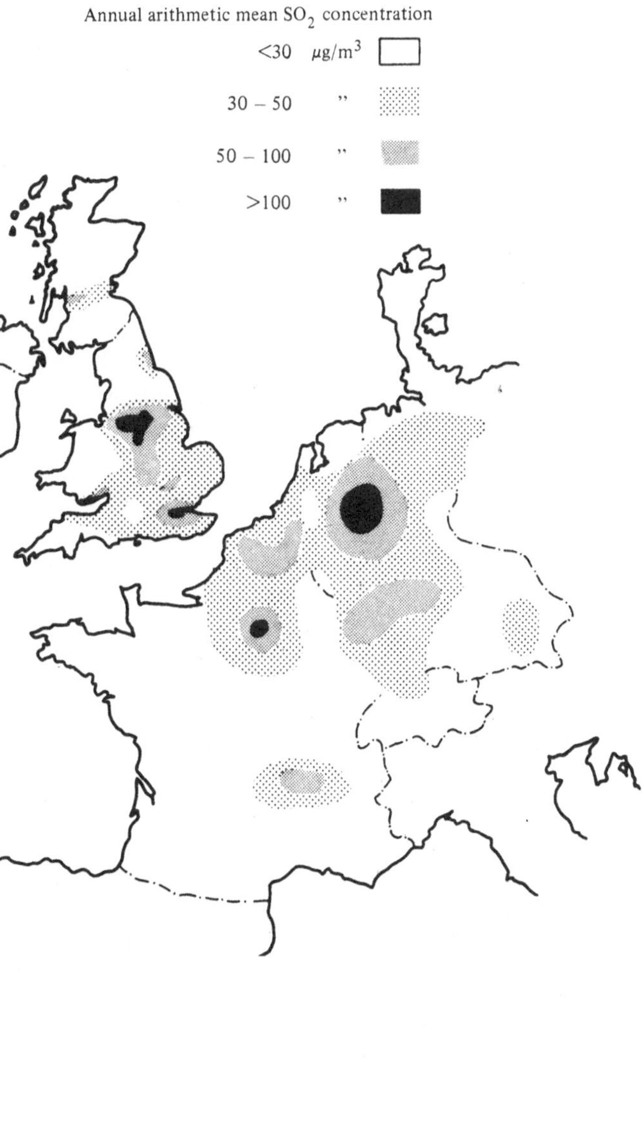

Figure 50
Distribution of sulphur dioxide over Western Europe, 1980[176]

Even in the European countryside levels will not fall much below 20 or 30 $\mu g/m^3$. Because sulphur dioxide is quite quickly adsorbed on many surfaces, particularly fresh plaster, its concentration indoors is always lower than it is outside. In a review of sulphur dioxide in the Tate Gallery, London, Hackney found indoor/outdoor ratios mostly within the range 15–70% in rooms without carbon filters. The ratios were consistent within each room and probably depended on 'the state of the roof, the airflows within the building and the proximity to filtered spaces'[177]. External sulphur dioxide levels can change quite rapidly, e.g. falling during rain. The response of an exposed exhibition room can be quite rapid on a windy day: a matter of minutes rather than hours. As a rough rule, for older buildings, with windows closed, we can assume that the indoor concentration is about half that outside[112,178].

50 μg of sulphur dioxide will give 0.04 microlitres of sulphuric acid. This may not seem a lot, but if we take an exhibition room of 100 m^3 and give it the average ventilation rate of one air change per hour, 3 ml of concentrated sulphuric acid will have passed through it in a month, and some of it will stick.

Understandably the public cares a great deal more about its health than about its works of art. It is therefore fortunate from our particular point of view that sulphur dioxide is a health hazard. This means that there is effective pressure to reduce its presence in the atmosphere. This can be done at source by washing the exhaust gases or passing them through an alkaline absorbent. However the commoner approach has been to build chimney stacks higher so that the concentration near the source is reduced. The output is unchanged of course. One result is that, in the prevailing wind system, some of the sulphur dioxide found in Scandinavia originated in western Europe and the U.K.

DAMAGE CAUSED BY SULPHUR DIOXIDE

CALCIUM CARBONATE MATERIALS: CHALK, LIMESTONE, MARBLE, FRESCOES AND ALKALINE SANDSTONES

Chalk, limestone and marble are all geological variations of calcium carbonate, $CaCO_3$. Sandstone may also contain calcium carbonate.

Buon fresco, or true fresco, is a technique of wall-painting wherein the pigment mixed with water is laid on wet lime plaster. Both lime water and plaster contain calcium hydroxide, $Ca(OH)_2$. The carbon dioxide in the air slowly converts this to

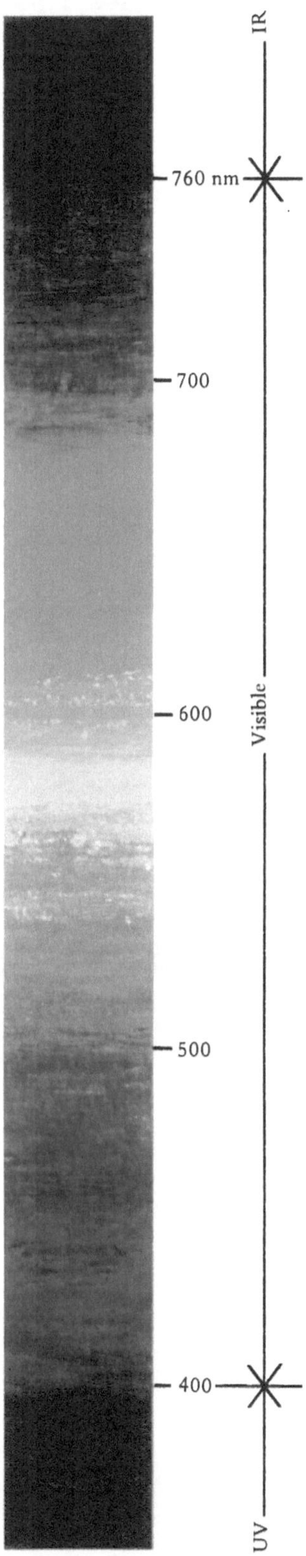

Plate 1
The wavelengths of ultraviolet (UV), visible and infrared (IR) radiation. (The spectrum is hand-coloured to minimise colour-shift in reproduction.) The wavelength scale is in nanometres (nm). 1 nm = 10^{-9} metres, or a thousand millionth of a metre. A normal varnish film on a painting might be 10 000 nm thick. 1 nm = 10 Ångstroms. The UV is actually considered to extend to 1 nm wavelength (below which the waves are called x-rays), but radiation of wavelength shorter than 300 nm is not present in daylight and in any case cannot penetrate glass. The IR range extends by convention to 10^6 nm (1 million nanometres or 1 mm), beyond which the waves are called Hertzian or radio. The sun's IR emission extends to about 22 000 nm

Plate 2
Green and purple water-colours mixed to produce a warm grey. In the original the purple is alizarin plus Prussian blue, and the green is viridian (hydrated chromic oxide)

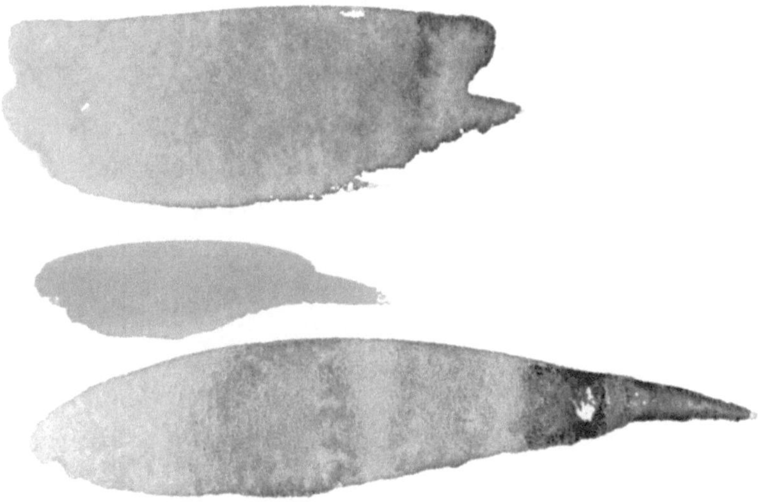

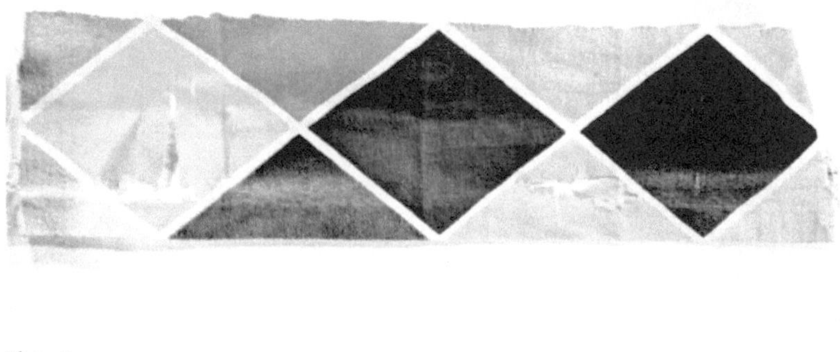

Plate 3
The two effects of UV and visible radiation: colour change and weakening. The changes have occurred predominantly on the edge of this cotton curtain where it was turned to the light. Whereas the black dye has been bleached though its textile support is still strong, the yellow dye is unchanged but has passed on the absorbed radiant energy to the textile, causing its destruction. (This destruction was not apparent until after cleaning.)

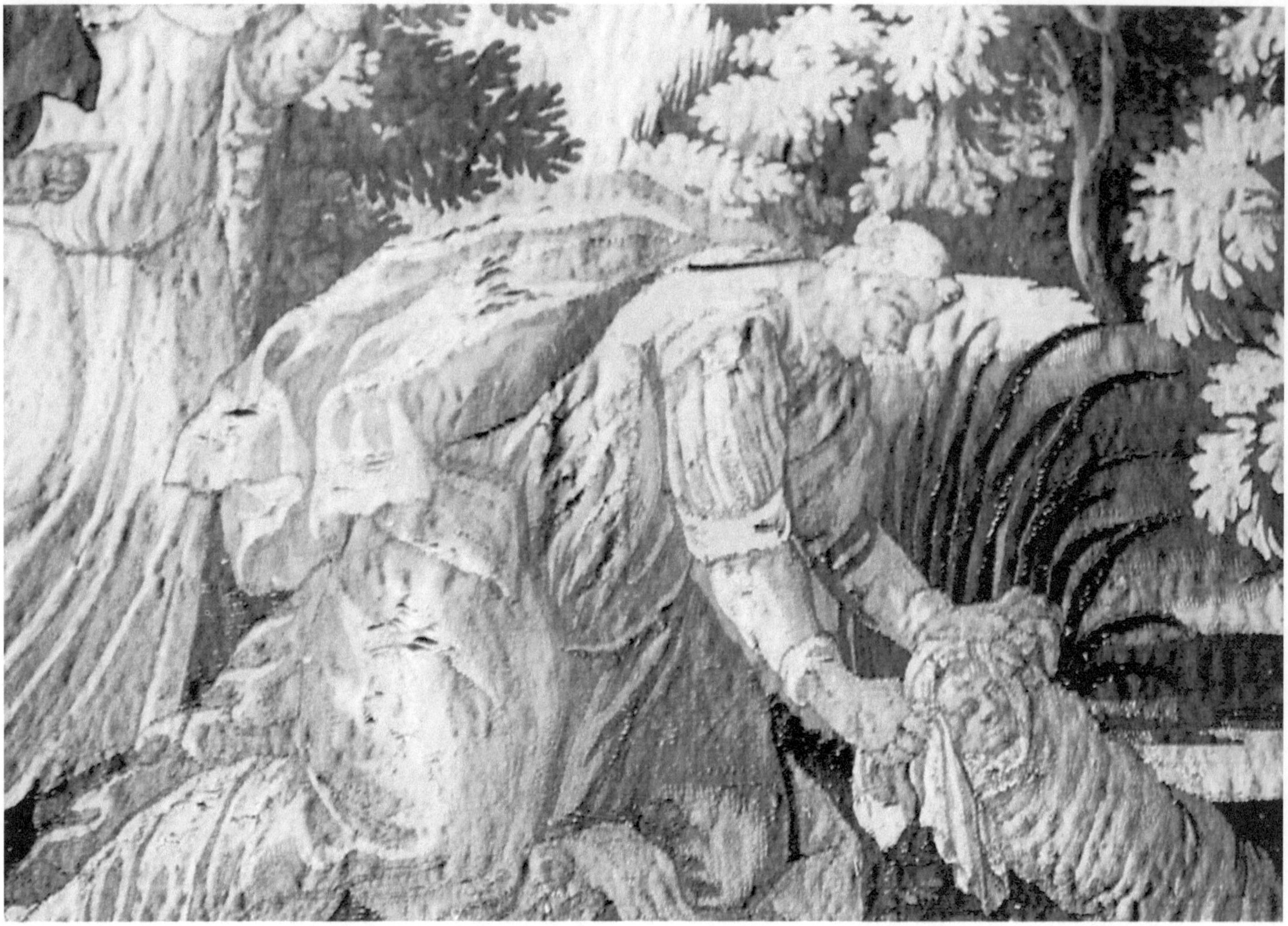

Plate 4
The effect of light on a tapestry. All colours except blue have changed to various shades of brown. Light has also weakened the weft threads to the extent that holes are appearing

Plate 5
Details from a Turner water-colour which has been protected at the edge by the frame from damage by light. The blue in particular has suffered. (By courtesy of the Trustees of the British Museum)

Plate 6
Sunset Landscape by Giorgione.
To give an illusion of depth Giorgione graded his tones from warm, in the foreground, to blue in the far distance. The blue has remained unchanged, but copper-resinate green has become brown, destroying this aerial perspective. (N.G.6307. By courtesy of the Trustees of the National Gallery, London)

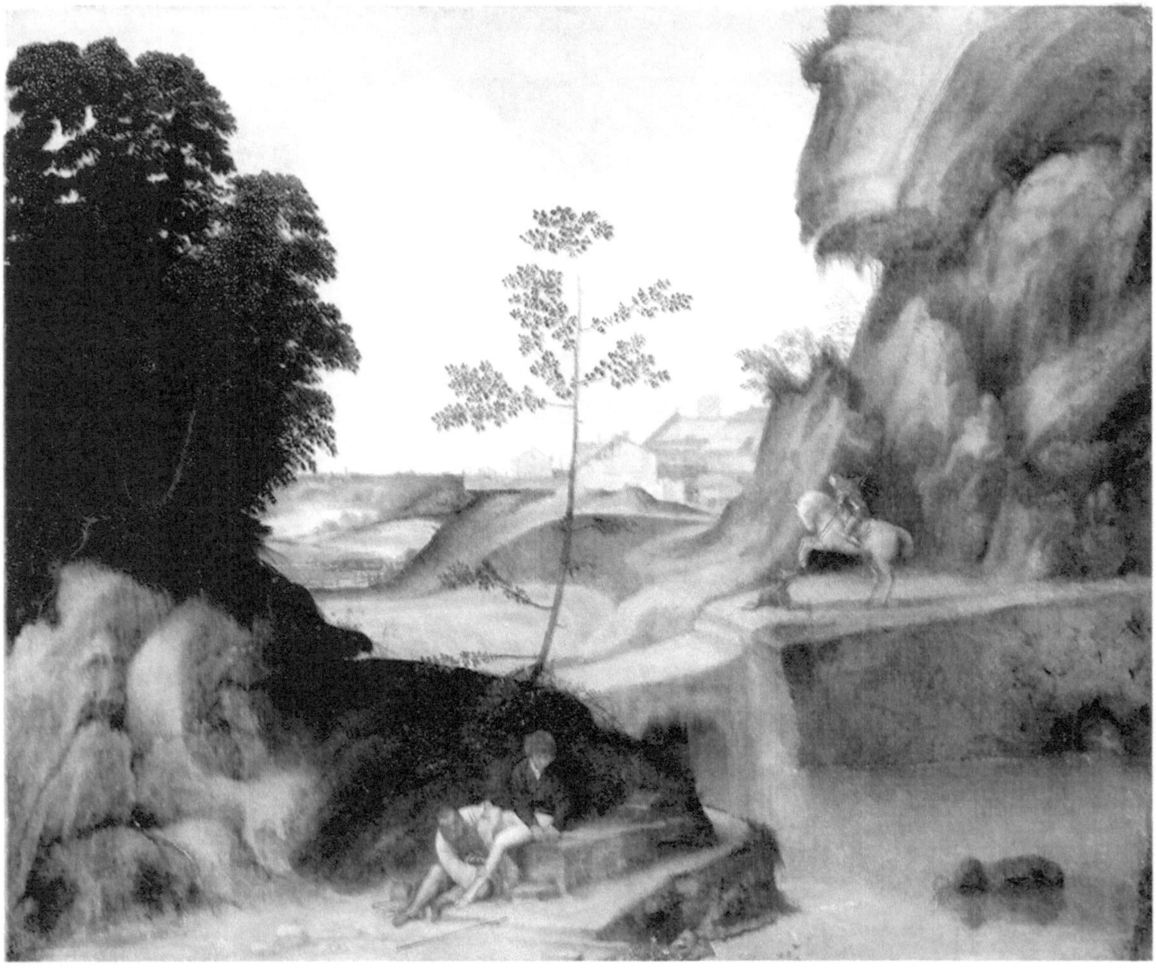

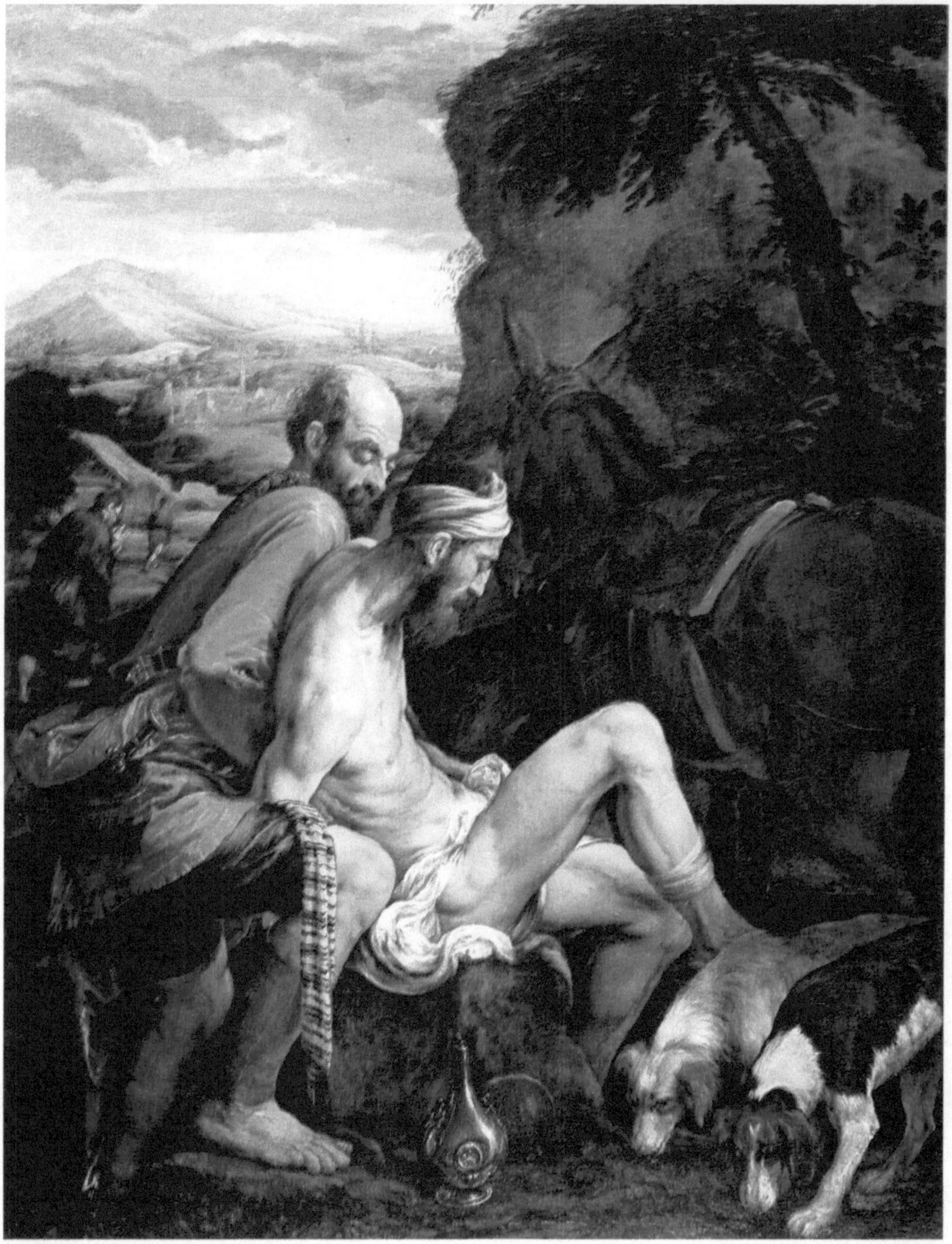

Plate 7
The Good Samaritan by Jacopo Bassano (active c. 1535) (N.G. No. 277. By courtesy of the Trustees of the National Gallery, London.)

Plate 8
Cross-section of a tiny fragment of paint and ground from one of the apparently brown stripes on the scarf lying across the Samaritan's knee. The upper layer of pigment has turned brown while much of the lower remains green. Pigment: copper resinate. Magnification X 330

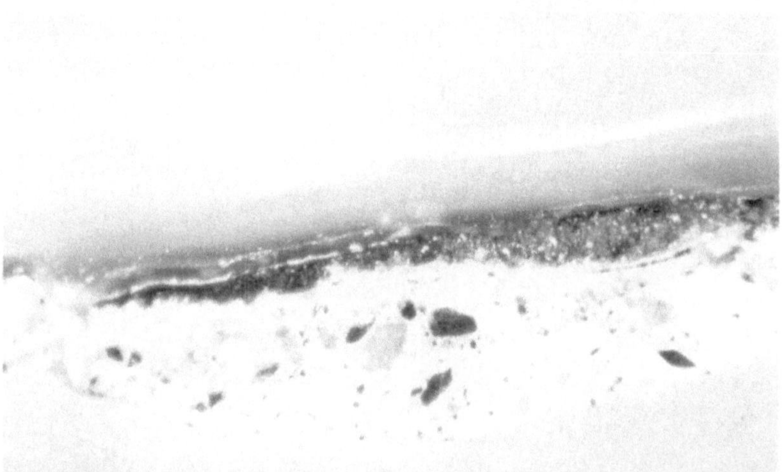

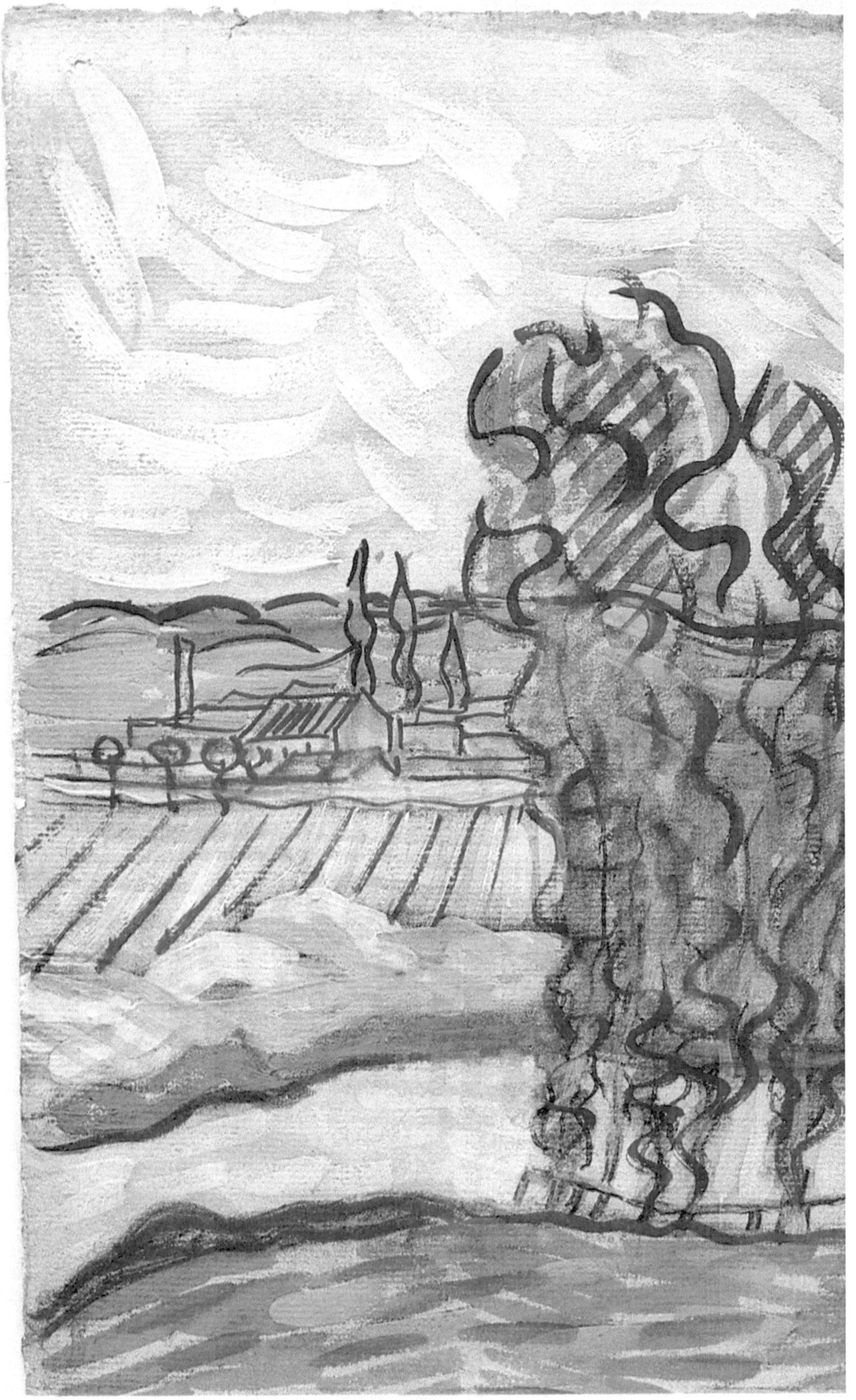

Plate 9
The Oise at Auvers by Van Gogh. The frame has protected a red lake in the sky from fading. (T.4714. By courtesy of the Trustees of the Tate Gallery, London)

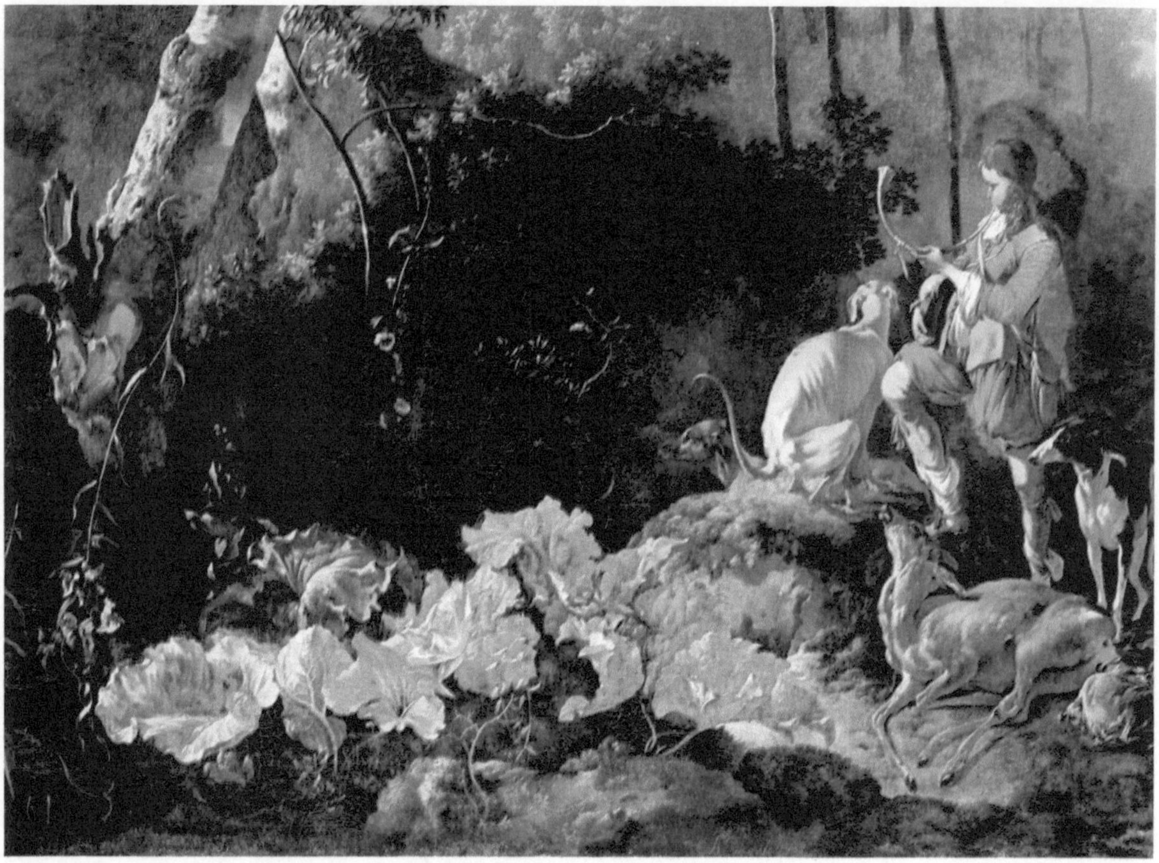

Plate 10
Landscape with Sportsmen and Game by Adam Pynacker (1621–73), detail. The large leaves, now blue, in the foreground, were painted with blue underpaint glazed or mixed with yellow, giving green. The yellow glaze has disappeared, leaving the blue. (By permission of the Governors of Dulwich College Picture Gallery.)

Plates 11 and 12
Lighting as in *Figure 14, Plate 11* direct and *Plate 12* diffuse. The darker samples in this set of glossy colours appear deeper in direct ligh because a general veiling light has been superimposed on them in the diffuse situation by reflections from all the light surfaces round the room. Thus the diffuse component of the lighting should not be too pronounced for glossy surfaces such as varnished paintings and metalwork, nor for showcases

Plate 11

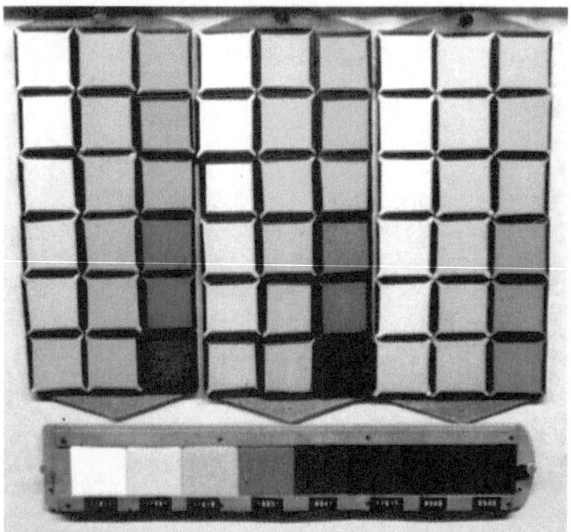

Plate 12

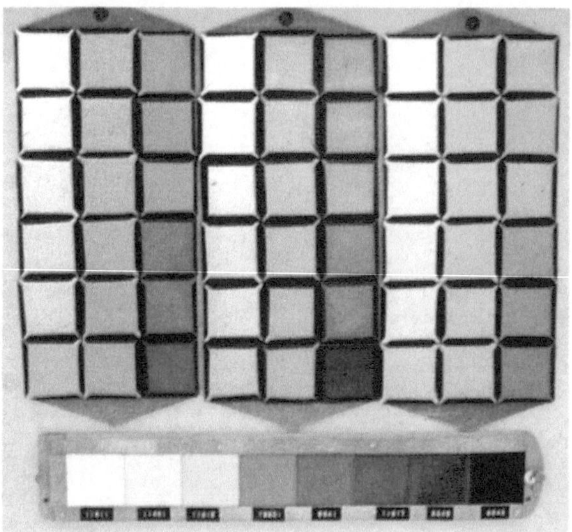

Plates 13, 14 and 15
The camera and colour rendering. A group of objects in the snow including paper flowers, ceramics and an encrusted glass bottle from the sea-bed.

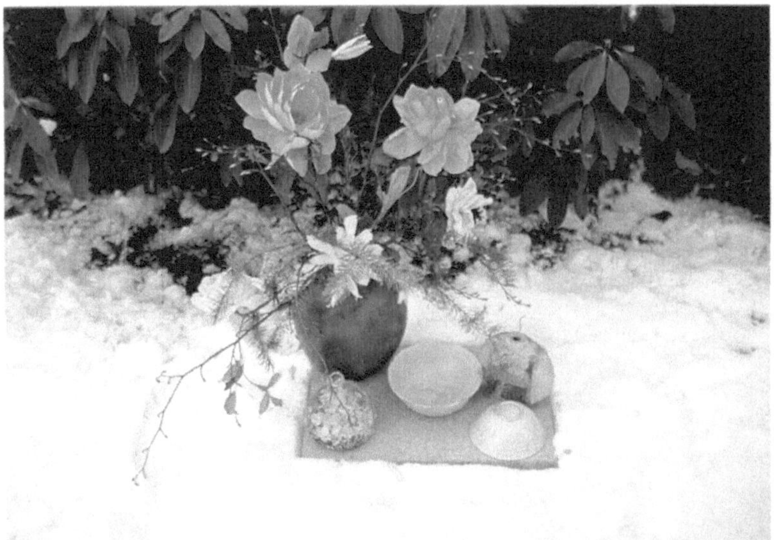

Plate 13
Artificial-light film. Since the film is adjusted to the much redder light of tungsten, everything has a strong blue cast, to the extent that the rhododendron foliage in the background looks blue rather than green.

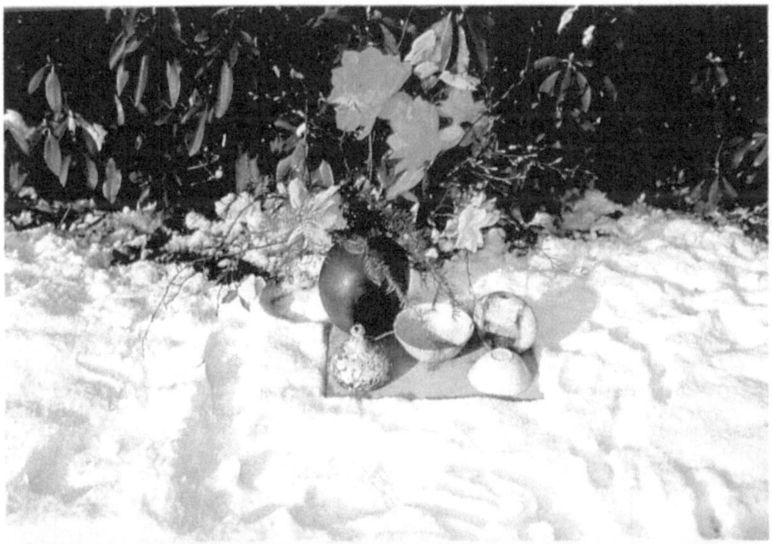

Plate 14
Daylight film. Watery sun. The film is correctly chosen for the colour temperature of the light and so the snow looks white. Even the shadows are neutral since the sky is not too blue. Consequently colours are correctly rendered.

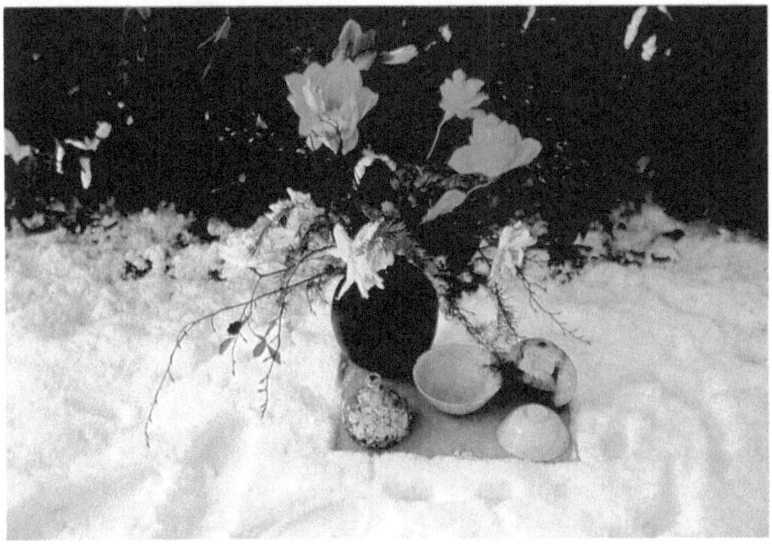

Plate 15
Daylight film again. The sun has now moved round so that the objects are in shadow. The sky is blue enough for the colour temperature of the light to be well above that of the film, and so once again a blue cast appears, but not by any means as strong under these conditions as in *Plate 13*. (Films: Kodachrome II reversal.)

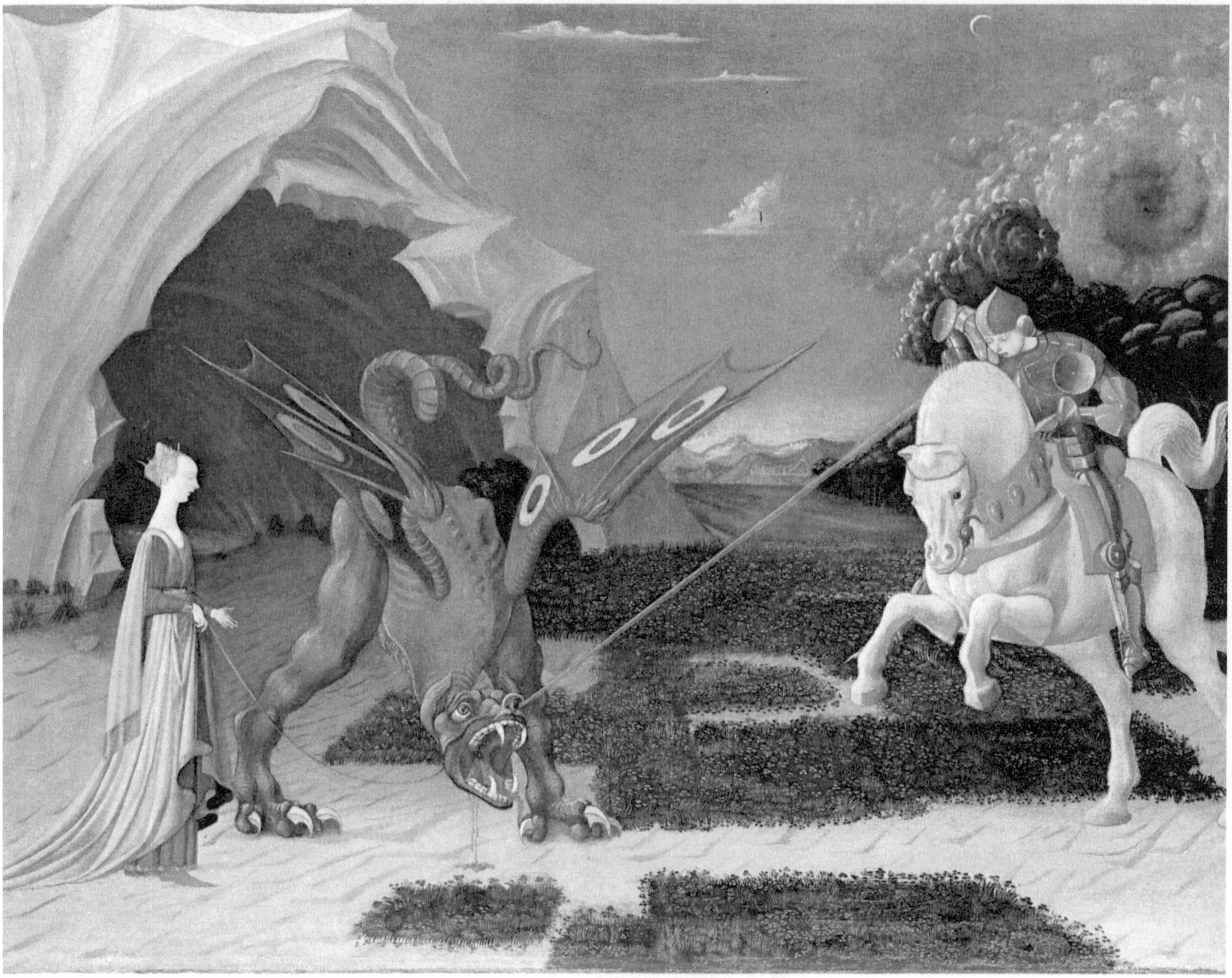

Plate 16
Uccello's *S. George and the Dragon* (N.G. No. 6294). The reflectance curve of *Figure 67* was taken at the spot marked X. (By courtesy of the Trustees of the National Gallery, London.)

calcium carbonate, which thus forms a solid paint medium holding the pigment particles securely in place. More commonly frescoes have been made on lime plaster by painting in colours mixed with some other medium such as glue. But the support is still of calcium carbonate.

If sulphuric acid is dropped onto calcium carbonate in any form there is a violent fizzing as carbon dioxide gas is evolved, and the calcium carbonate is converted into a powder of calcium sulphate, gypsum. We have seen that the sulphur dioxide in the atmosphere is converted into sulphuric acid, and so it is not surprising that sulphur dioxide pollution is a major enemy of calcium carbonate in all its forms.

We need to realise that there is much more carbon dioxide gas in the world's atmosphere than sulphur dioxide in even the worst smog (average carbon dioxide concentration 600 000 $\mu g/m^3$ or 0.03% by volume). This carbon dioxide is also capable of dissolving calcium carbonate by converting it into the water-soluble calcium bicarbonate, $Ca(HCO_3)_2$. In this way rain-water containing dissolved carbon dioxide has permeated limestone rocks all over the world, forming characteristic tunnels and caverns, but also reforming calcium carbonate as stalagmites and stalactites in places where the carbon dioxide can evaporate from the water.

However there are two differences between the subterranean solution of limestone by dissolved carbon dioxide and the solution of calcium carbonate in buildings by sulphur acids. Firstly the sulphur acids make rain in industrial areas more strongly acid than natural rain (*Figure 51*). Dissolved carbon dioxide gives rain a pH of 5.6. Town rain today may be as acid as pH 4 and is often pH 4.6[172,173]. The acid strength (hydrogen ion concentration) at pH 4 is 25 times that at pH 5.6. This is

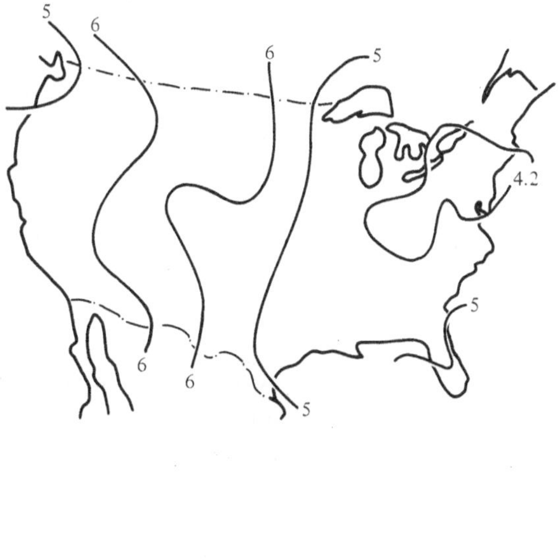

Figure 51
pH of rain falling in the U.S.A. during 1976–9 (ref. 173, p. 41). Neutrality is pH7. pH below 7 is acid. One digit down on this scale represents a 10-fold increase in acidity. Thus pH4 is 10 times as acid as pH5. In both the eastern United States and in Europe, acid in rain consists of about 65% sulphuric acid and 30% nitric acid[174]

because sulphuric acid, though in much lower concentration, is a very much more powerful acid than dissolved carbon dioxide[179].

Secondly, rain passing through the soil picks up more carbon dioxide from plant roots and decaying vegetation, to the extent that it may gain five times as much dissolving power by the time it reaches buried limestone rocks. In any case this geological tunnelling takes millions of years.

The Palace of Fatehpur Sikri was built by the Moghul ruler Akhbar in 1569. *Figure 52(a)* shows a photograph of a tomb in the palace grounds taken in 1872. The tomb is made of intricately carved marble. With this photograph in my hand I examined the tomb in 1972 (*Figure 52(b)*) and could not see that any wear had occurred during the intervening century. Indeed the structure looked to be in pristine condition, though it was said by conservators present that some of the marble of the cupola, perhaps where rain had collected, was becoming brittle and crumbly. Fatehpur Sikri is situated in a very unpolluted part of the world. In the absence of sulphur dioxide the deterioration of sound marble properly erected will not be disturbing even after several centuries have passed.

How different from Fatehpur Sikri is the situation in the towns of Europe, North America and Japan (*Figure 53*). The rainwashed parts of alkaline stone are the worst affected, since the sulphate formed is continually washed away to expose a fresh surface for attack (*see* p. 257). Sheltered parts are protected to a much greater extent, but blackened with a coating of soot and sulphate. Rate of attack is widely variable, but if we work back from measurements of the rate of deposition of sulphur acids on marble, a rate of about two or three mm per century may fairly represent an average[180]. For half a century of erosion examine any limestone or marble memorial monument to the First World War, for example that in the Parc du Cinquantenaire in Brussels.

Limestone buildings in polluted atmospheres soon acquire visible signs of deterioration. The stone surfaces can be divided quite sharply into black, white and grey areas[182].

The grey areas are quite distinct from the other two. They are sheltered from rain and unattacked. They are grey because of an adherent dust coating.

Those areas directly exposed to rain are white. This does not mean they are unattacked, as close examination will show. They have been attacked by acid rain and the products of the attack, sulphates, have been washed away. There is likely to be a

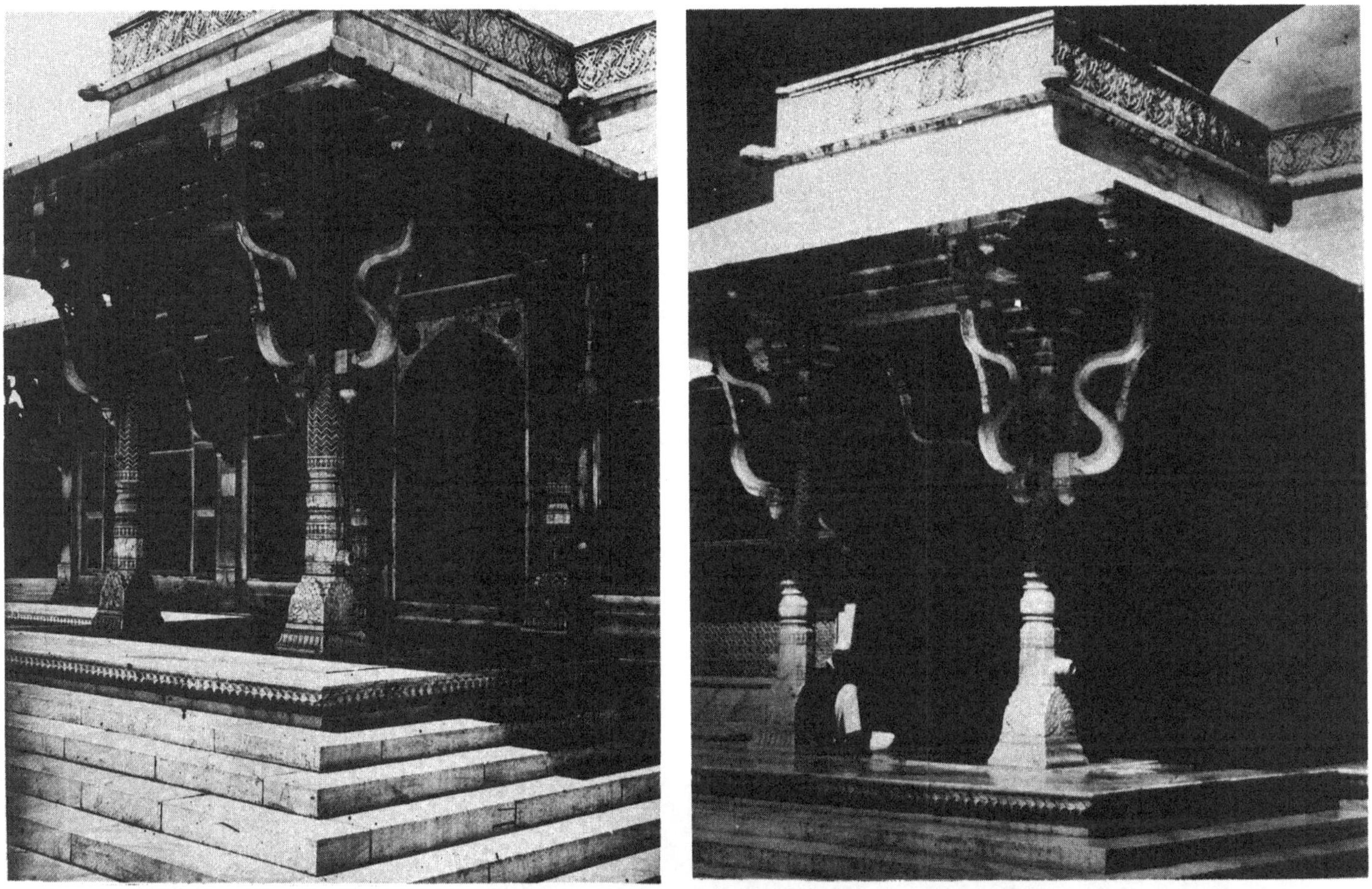

(a) *(b)*

Figure 52
(a) Marble tomb in the Palace of Fatehpur Sikri, near Agra, India. Photograph taken in 1872. (By permission of the Director of the India Office Library and Records). *(b)* The same in 1972. (Author's photograph)

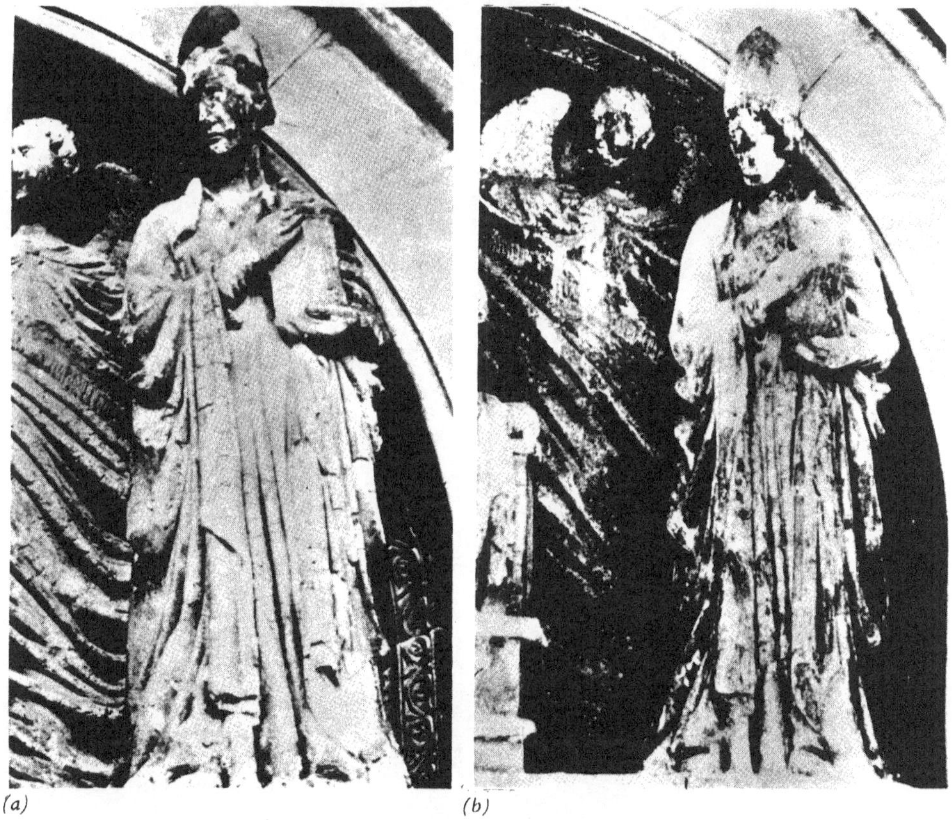

(a) *(b)*

Figure 53
Sulphur dioxide attack. Lunette on the cathedral at Arrezzo, Italy: a recent photograph (*b*) compared with one taken 30 years earlier (*a*). (By courtesy of Prof. U. Baldini)

surface layer of reprecipitated carbonate. These are the white areas.

The black areas are more sheltered but still wet by rain, through percolation, not washout. The black surface consists of a mixture of black carbonaceous particles (products of fuel burning) embedded in gypsum (calcium sulphate) and recrystallised carbonate. The black particles are those above 2 microns diameter, the fraction most effectively washed out by rain (*see* p. 249). There is a strong suggestion that these particles catalyse the conversion of carbonate to sulphate, and also that soot from petroleum is more effective than that from coal[183]. Bacteria may also be involved. Gauri supposes that it is sulphur dioxide, not sulphuric acid, which is responsible for most of the attack. Sulphur dioxide converts damp limestone or marble to calcium sulphite, which then becomes oxidised to sulphate[184].

Figure 54
Sulphur dioxide attack. An area of the Andrea del Sarto frescoes in the Chiostro dello Scalzo, Florence, where deterioration through conversion of calcium carbonate to gypsum has occurred[181]

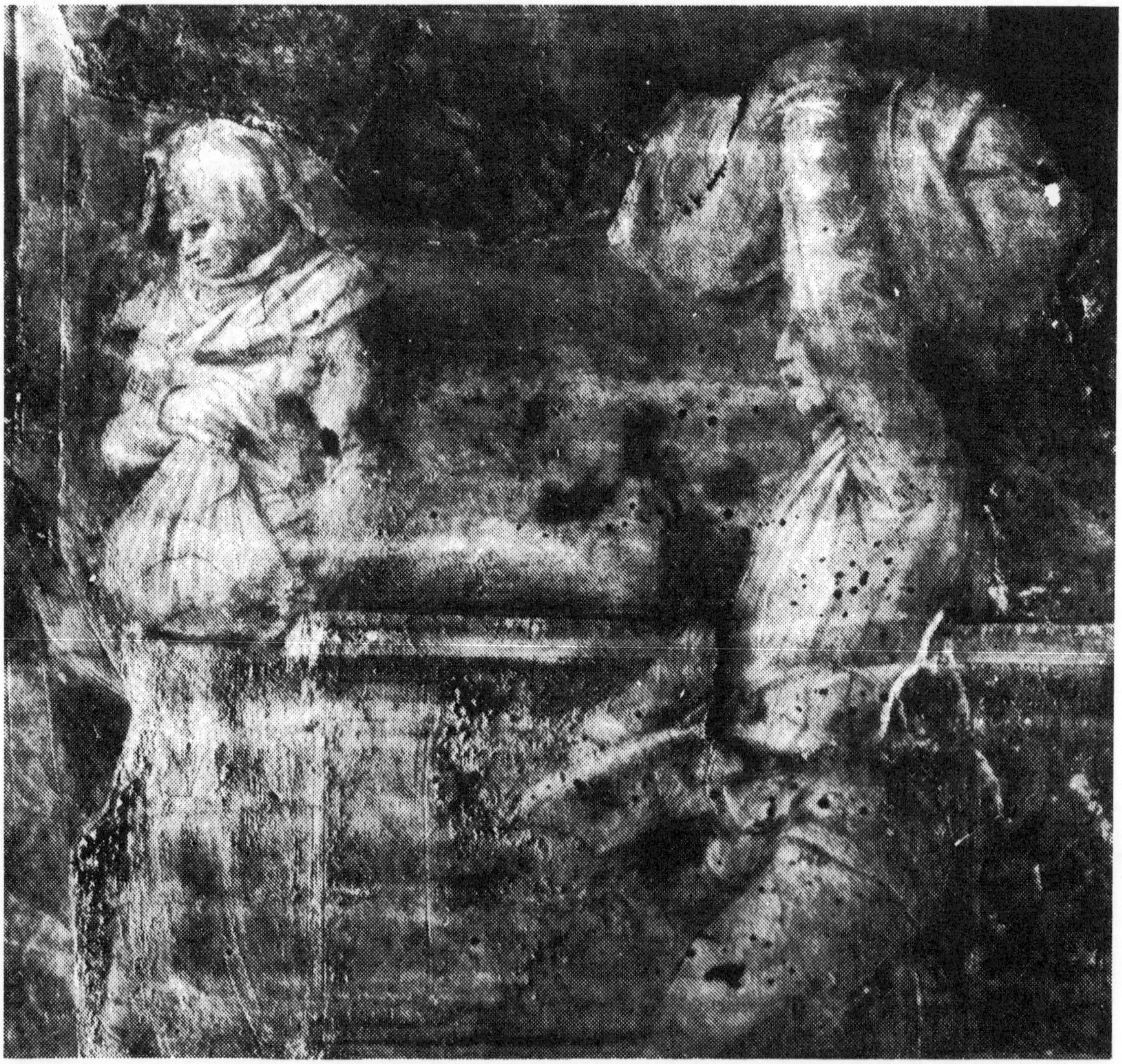

Many frescoes in Europe are beginning to show the tragic depredations of sulphur dioxide. The Giotto frescoes in Padua and some others (*Figure 54*) have been carefully examined. The pustules which had started to form were shown by analysis to be composed of calcium sulphate. Attack was particularly advanced where moisture had been allowed to seep down the fresco surface.

CELLULOSE MATERIALS: PAPER, COTTON, LINEN

The structural material of all plants is made up of cellulose, a polymeric substance whose sub-units are sugars. Wood, though cellulosic, is not included in the heading because in its bulk it remains unaffected by sulphur dioxide, but a thin veneer might prove an exception. Paper has been made from all kinds of plant fibres, though the bulk of newsprint and other cheap papers is from conifers. Many of the synthetic textile fibres are immune from sulphur dioxide attack, but rayon, being reconstituted cotton, is as susceptible as any other form of cellulose.

The oxidation of sulphur dioxide to sulphuric acid occurs both in the air and on surfaces (p. 256). Most of the sulphuric acid now to be found on exhibits arrived as sulphur dioxide and was converted on the spot, in many cases aided by traces of iron and other catalysts already present. Though all cellulose materials are attacked, the rate of attack varies widely, and this depends on other factors as well as the presence or absence of catalysts.

RH certainly influences the rate of attack, a very high RH speeding the reaction. But in the middle of the range, from 30–70% RH, we cannot do very much to influence things.

Textiles which hang in the light are, of course, affected by it whether or not sulphuric acid is present (p. 10), and light is the main factor in their decay[4]. But a combination of light and sulphur dioxide does more damage than either separately[9,185,186].

This applies also to exposed paper. However most archival paper is not exposed to light. Here the presence of certain impurities acts powerfully. It is well-known that centuries-old rag paper (paper made from selected pieces of waste cotton textiles) remains in good condition, even in urban environments. Yet a cheap book shows signs of browning and brittleness round the borders of the pages, where sulphur dioxide has diffused in, within less than ten years (*Figure 55*)[187]. Readers had better examine the pages of this book after it has lain on their shelves for some years.

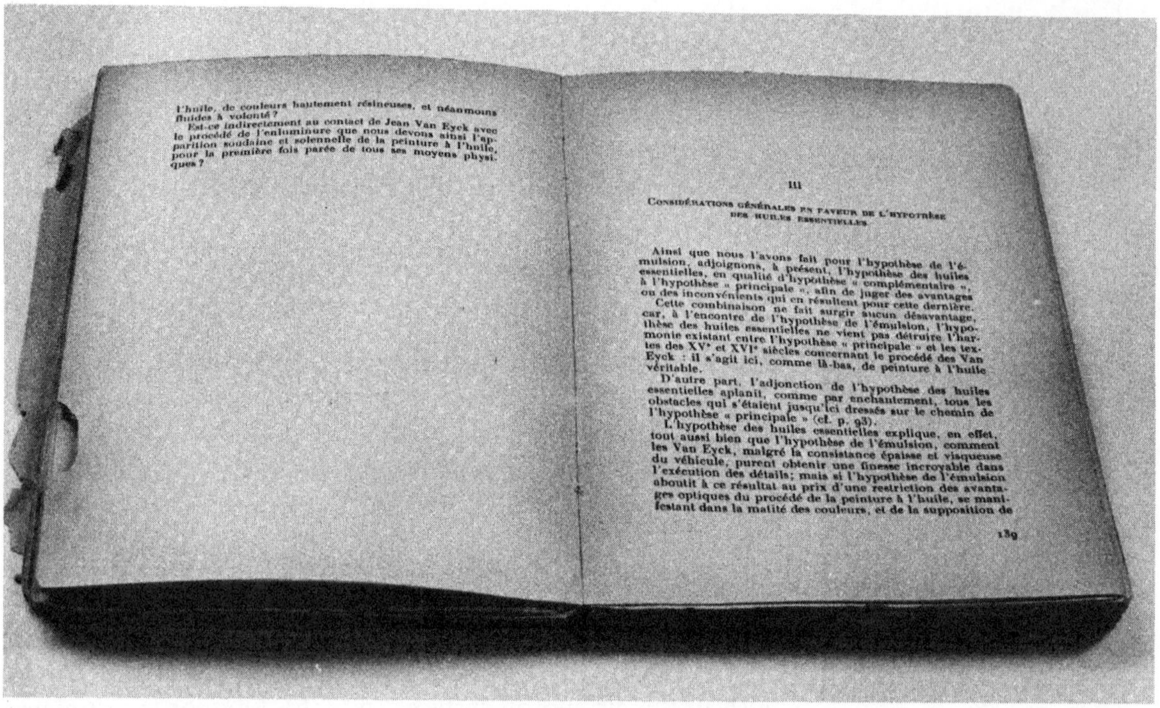

Figure 55
Sulphur dioxide attack. Browning and embrittlement of book paper, worst at the edges where sulphur dioxide has diffused into the closed book

This difference in rate of attack may be ascribed to impurities acting in two distinct ways. Firstly, several of the materials added during manufacture are acid, notably alum which is used in the size[188,189]. For this reason such paper becomes weaker all over the page. But this will not explain the extra damage and acidity round the edges of each page, caused by sulphur dioxide. But the damage is greater on poor quality papers.

Such paper contains a relatively high proportion of lignin. As has been said above, all plant tissues have cellulose as their main structural component. Lignin is the material that gives plant tissues the added character of woody tissues. The analogy has been given of a wet fabric frozen stiff, where the fabric represents the cellulose and the all-pervading ice the lignin[190]. Lignin is an aromatic polymer which breaks down to compounds with a strong affinity for sulphur dioxide[191]. Hence more sulphuric acid is to be found on high-lignin papers than on rag papers at the same location.

Other factors such as length of cellulose fibre and degree of crystallinity within the fibre also play their part.

Summary

All cellulose, whether paper, cotton or linen, is attacked by sulphuric acid from sulphur dioxide. Light, and even more, UV radiation, increases the damage. Poor quality paper deteriorates more quickly than rag paper, firstly because it contains more acid materials introduced during manufacture, and secondly because it absorbs acid gaseous pollution more strongly because of the lignin also present.

PROTEIN MATERIALS: SILK, WOOL, LEATHER, PARCHMENT

Animal fibres, as opposed to plant fibres, are proteins, also polymeric but made up in this case of amino acids. Animals which spin produce protein fibres, which we call silk. Natural textile silk is from the cocoon of the silk moth, *Bombyx*. The cross-hairs on an old scientific instrument have been made from spiders' web silk. All animal skins are proteinaceous.

Faraday discovered the connection between the rotting of the leather armchairs in the London Athenaeum Club in 1843 and sulphur dioxide from gas lamps[192]. Badly affected leather can be reduced to a powder by gentle rubbing. Vegetable-tanned leather (skins processed with tannin from bark, wood, etc.) is particularly susceptible to 'red rot', a form of deterioration caused by the sulphuric acid from sulphur dioxide[193]. Parchment is affected like leather, but is not usually so exposed.

The effect of light on the breaking strength of silk has been shown to depend on its acidity[15]. Wool is also affected, but, even when the fineness of most silk yarns is taken into account, silk is a great deal more susceptible to both light and sulphur dioxide than wool[4].

DYES, PIGMENTS, PAINTS AND SYNTHETIC TEXTILES[194]

A few modern dyes and pigments have been found to be specifically sensitive to sulphur dioxide. Effects on natural dyes have not been fully investigated, but they are probably minor.

Nylon stockings were found to 'run' much more quickly in the London smogs of the 1950s[192], and sulphur dioxide does indeed attack nylon[186]. Recently nitrogen dioxide has become another likely culprit[195, 196].

Sulphur dioxide eventually attacks most other synthetics to some extent, but with a few exceptions the effect will be small compared to the normal deterioration which occurs in pure air and light, so that it can be ignored. Notable exceptions are certain elastomers (synthetic rubbers) whose flexibility can be destroyed. Paints are affected[197].

METALS

Iron is the chief metal to suffer from the presence of sulphur dioxide. Iron corrodes to rust electrolytically. This means that both moisture and an electrolyte must be present on the iron surface. All water-soluble salts, acids and alkalis form electrolytes. Those electrolytes which attract moisture, form soluble corrosion products and are involatile, are the most corrosive. Sulphuric acid and the ammonium sulphate to which it is often partly converted fulfill all these conditions admirably. As a result rusting will commence in urban atmospheres at about 60% RH[198]. Thereafter it is no good merely transfering the iron to a clean atmosphere: it must also be washed free of electrolytes[81].

Ancient patinated bronzes typically have stratified corrosion layers, the innermost of which is red cuprous oxide (cuprite, Cu_2O) and the outermost either a basic copper carbonate (green malachite or, less commonly, blue azurite) or a basic copper chloride (green $CuCl_2$, atacamite or paratacamite). If chloride is present, there may be a waxy layer of pale grey cuprous chloride next to the unchanged metal. This is unstable when exposed to moist air, since the green cupric chloride then forms by re-attacking the metal, and the bronze is said to be suffering from 'bronze disease'[81,199].

Indoor bronzes with stable patinas need not cause the curator any anxiety about air pollution, unless the museum is close enough to the sea (within a mile or so) to receive wind-blown salt crystals (p. 153).

Outdoor copper or bronze statuary and roofing material quite often has a basic cupric sulphate patina composed of the green mineral brochantite ($CuSO_4.3Cu(OH)_2$)[200]. This is normally stable. However certain bronze alloys may become deeply pitted. This is especially true of bronze with a high lead content, for the lead is removed by rain as sulphate[201,202].

Even bare bronzes will not be attacked by sulphur dioxide in the shelter of a museum unless the RH rises above 60 or 70%.

Silver is specifically attacked by hydrogen sulphide (H_2S). The concentration of hydrogen sulphide in the air is variable, but frequently in the range 5–30 $\mu g/m^3$ (Junge[203], p. 3). It is just as likely to be present as a natural product of organic decay in the country as in towns. The natural level is sufficient to tarnish silver slowly[204]. Rapid tarnishing of silver in museums is much more likely to be due to some local production, perhaps by the materials in a showcase (p. 154). RH above 70% greatly accelerates tarnishing[92]. Silver is not attacked by sulphur dioxide in the concentrations likely to be present.

The chief danger to lead is not sulphur dioxide, though sulphate may form, but organic acids, particularly acetic acid. Again local production should be suspected (p. 154).

Pure gold is immune to sulphur dioxide and to all other forms of air pollution.

GLASS AND SULPHUR DIOXIDE

It has more than once been suggested that sulphate layers identified on old stained glass point to attack by sulphur dioxide. Newton, in an important review[205], inclines to the view that the first cause of glass corrosion is the breakdown of the silica network by alkali leached out of the glass itself in damp conditions, for example when water is drawn between panes of glass stacked together (there were examples of this in the wartime storage of stained glass). The water leaches out alkaline hydroxides, which will at some stage combine with the carbon dioxide in the air to form carbonates. These would in turn be converted to sulphates by any sulphur acids in the air. But the sulphur acids would not be contributing to decay.

The chief cause, then, of glass decay is condensed moisture. A small amount of water is more damaging than a large amount because only within a small amount will the concentration of alkali leached out of the glass be high enough to attack its own silica network. Those working in the Brussels Institut Royale du Patrimoine Artistique will be very familiar with the results of this reaction since the inside surfaces of the double glazing on their building were attacked in this way, giving the glass a milky appearance. Free-flowing rain does little damage since any alkali dissolved by it is not concentrated on the glass. But modern glass will deteriorate rapidly if kept moist and warm for only a few days[206].

Summary of sulphur dioxide attack

The chief materials to suffer from sulphur dioxide pollution are calcium carbonate (marble, limestone, frescoes), cellulose (paper, cotton, linen), silk, iron and steel. Leather, parchment and wool are also attacked. Many other materials, including certain bronze alloys, synthetic rubbers, dyes and textiles may be affected.

EFFECTS OF SULPHUR DIOXIDE ON LICHENS AND MOSSES

Air pollution affects plants as well as people and antiquities. Particularly sensitive to sulphur dioxide are lichens and mosses, to the extent that the centres of large towns in the industrial world have become deserts to these plants.

The attack on plants by air pollution, including ozone, is not the concern of antiquities conservators, but the misfortunes of lichens and mosses can be made use of as sulphur dioxide pollution indicators. In Europe and North America it is possible, by noticing the presence or absence of certain species, actually to estimate the average yearly sulphur dioxide level in the air[207].

If we examine the trunks of deciduous trees we will find two extremes. Where pollution is greater than about 170 $\mu g/m^3$ there will be no lichens at all, though a green powder of *Pleurococcus viridis* algae may be found.

At the other extreme, in pure country air only, long tufts of *Usnea* and other hanging (fruticose) types will be growing freely (*Figure 56*). Nowadays these are limited to the West coast of England, but I have seen long tufts of lichen on trees 200 km inland in the south of France.

In zones of intermediate pollution, the concentration of sulphur dioxide can be followed by the succession from crustaceous lichens (crusts close to the surface of, e.g., *Lecanora conizaeoides*) at about 125 $\mu g/m^3$, through the foliose types (leafy lichens such as *Cladonia saxitilis*) at 40–60 $\mu g/m^3$, to the

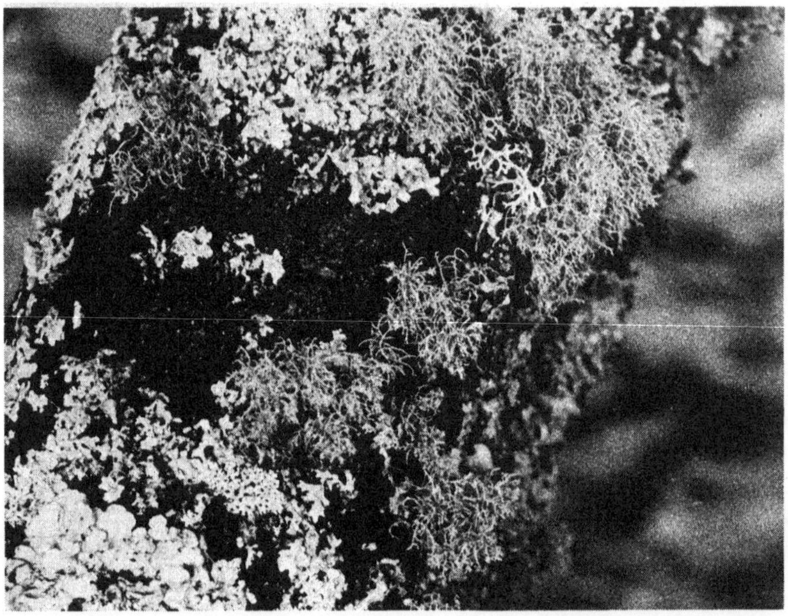

Figure 56
Lichen as an indicator of air purity. An assemblage of foliose (leafy) and fruticose (hanging in tufts) lichens characteristic of unpolluted air. Photographed near the coast of North Wales

fruticose types in pure air (*Usnea, Evernia* and *Ramalina* species).

Alkaline surfaces such as asbestos roofs and calcareous stonework will neutralise the acid in the air enough to support close-growing types. Characteristically the orange lichen *Xanthoria parietina* is found on such surfaces at and below about 60 $\mu g/m^3$, and round clumps of the moss *Grimmia pulvinata* at and below about 45 $\mu g/m^3$.

OZONE

The idea lingers that ozone is healthy and that you find it at the seaside. Ozone is a poison found at highest concentration in polluted towns.

Ozone can arrive in the museum from three sources:

1 From natural production in the upper atmosphere.

2 From the effects of sunlight on car exhaust gases (photochemical smog).

3 From certain kinds of lamp and electrical equipment which might be used indoors.

1 THE NATURAL BACKGROUND OF OZONE

Ozone is produced in the stratosphere at a height of 20–30 km by the action of UV radiation on oxygen. The UV concerned here is of much shorter wavelength than ever penetrates to the surface of the earth (*see* Part II p. 167). In fact the reason why there is no UV of wavelength shorter than 300 nm at ground level is that the ozone in the stratosphere effectively absorbs it. The destruction of this ozone shield would have serious consequences for life on earth, and there is current argument about whether, on the one hand, supersonic aircraft, and on the other, fluorocarbon aerosol propellants, may be capable of doing just this.

Mixing between the upper and lower layers of the atmosphere results in a ground level natural background of about 20–60 $\mu g/m^3$ of ozone[208].

2 PHOTOCHEMICAL OR OXIDANT SMOG

The London 'pea-souper' fogs, now fortunately a thing of the past, were the first form of air pollution to be profitably studied. As we have seen, they were characterised by acid sulphur gases. Wherever fuel is burnt in quantity this kind of pollution continues to be a problem.

But the next city in air pollution history was Los Angeles with a different kind of smog. Although it also contains sulphur dioxide it is chiefly notable for abnormal concentrations of ozone, a powerful oxidising agent. Therefore this type of pollution has come to be known as oxidant as opposed to acidic. Because sunlight has to be plentiful, it is also known as photochemical smog. Besides sunlight the other necessary ingredient is a high concentration of motor vehicles.

The new metropolis, wherever it arises, is jammed with cars, and ironically efforts to get rid of particulate pollution have increased the strength of sunlight in towns. Therefore oxidant or photochemical pollution is now a world-wide problem.

The production of ozone in this way is chemically complicated (*see* p. 258), and we need here merely to note that it is accompanied by the production of nitrogen dioxide (NO_2).

Ozone levels in a bad photochemical smog may exceed 1000 $\mu g/m^3$.

3 ELECTRICAL EQUIPMENT

Electrical equipment can generate ozone and oxides of nitrogen in two ways, firstly through the formation of strong electric fields, and secondly from light sources which emit UV radiation of wavelength shorter than 300 nm.

An example of the first is the electrostatic precipitator, which has received adverse comment on p. 135. It should not be used in museums.

Mercury vapour lamps with quartz envelopes are an example of the second. Currently these are used in certain types of photocopier. Therefore these should be kept well away from any exhibit.

Both ozone and nitrogen dioxide have characteristic smells. If they can be smelt they are in high concentration. Lower, but still very undesirable, concentrations cannot be smelt.

EFFECTS OF OZONE

Ozone has a specific and complete action on unsaturated organic compounds, that is to say it will break every double bond on a carbon chain with which it comes into contact. This destroys the material. In this way transverse cracks appear on rubber bands which then snap when stretched.

However attack by ozone does not end with rubber bands. It

is a powerful oxidant, that is to say destroyer, of almost all organic material[197, 209, 210]. The effect of ozone on certain materials such as cellulose may be due to its partial conversion to hydrogen peroxide by reaction with water.

The reader will hardly need to be reminded that paintings, textiles, archival materials, furniture, biological specimens, leather, fur, feathers, etc., are all made wholly or predominantly of organic material, and that therefore ozone is extremely dangerous in the museum. Ozone also increases the rate of oxidation of silver and iron and of sulphidation of silver and copper[211]. Hopefully one may suppose that the visitors may in this case help conservation, since every inhalation is likely to destroy all the ozone in the breath.

NITROGEN OXIDES

Chemistry textbooks give a long list of nitrogen oxides: N_2O, NO, N_2O_3, NO_2, N_2O_4, N_2O_5, NO_3 and N_2O_6. Fortunately the situation in polluted air is much simpler than this.

N_2O_3, N_2O_4, N_2O_5, NO_3 and N_2O_6 dissociate almost completely in the atmosphere into simpler oxides, and N_2O is too stable to be a worry for antiquities (Junge[203] p. 81–3, Leighton[212] p. 189). Neither is NO reactive enough to constitute a threat, but nevertheless it must be considered for the important part it plays in the formation of ozone (p. 241).

With N_2O and NO present but harmless, of the long list of nitrogen oxides we are left with just one: *nitrogen dioxide (nitrogen peroxide),* NO_2. But nitrogen dioxide is a worry. Like sulphur dioxide it is soluble in water to give, eventually, a strong acid, in this case nitric acid:

$$2\,NO_2 + H_2O \longrightarrow \underset{\text{nitrous + nitric acids}}{HNO_2 + HNO_3} \xrightarrow{\text{air}} \text{nitric acid}$$

The nitrous acid formed along with the nitric acid is oxidised by the air to more nitric acid.

For the formation of nitrogen dioxide in polluted air *see* Part II p. 259.

EFFECTS OF NITROGEN DIOXIDE

Since NO_2 dissolves in water eventually to form nitric acid, an acid as strong as sulphuric, and on top of that an oxidising agent, we can expect all the troubles that sulphur dioxide gives us plus a few more: corrosion of metals, hydrolysis of cellulose and attack on calcareous stones and murals.

However nitric acid is volatile and so, at the concentrations found in polluted air, the free acid formed from NO_2 cannot be entrained on dry surfaces, but must react on contact. Therefore it is probably the lesser menace of the two.

Certainly reports so far written have shown that sulphate rather than nitrate accumulates on paper, alkaline stone and frescoes in polluted atmospheres. But nitrogen dioxide concentrations are still rising with the increased use of motor cars, and so nitrogen dioxide is in future likely to be responsible for a proportion of the damage. It has been reported that both cotton and wool are susceptible to nitrogen dioxide[4].

Nitrogen dioxide attacks dyestuffs containing amine groups, especially on cellulosics and polyesters[213]. Indigo is such a dye. PAN (peroxy acyl nitrates: organic chemicals produced from motor car exhaust gases in oxidant pollution) may also be active, but as an oxidant and not as an acid. No investigations known to me have been made on its effects in the museum.

LEVELS OF OZONE AND NITROGEN DIOXIDE LIKELY TO BE ENCOUNTERED

It has been mentioned that the natural background of ozone which diffuses down to sea-level from its site of production in the upper atmosphere is usually 20–60 $\mu g/m^3$ [214], but may be higher[208]. On top of this there is now sufficient man-made ozone in summer in the industrial areas of Europe, America and Japan to push even rural levels up higher than this.

In London summers, peaks of 300 $\mu g/m^3$ are now quite often recorded[208] but this is still a good deal lower than the really efficient areas of production such as Los Angeles, which has occasional peaks of 1500 $\mu g/m^3$ [215], and which exeed 200 $\mu g/m^3$ by mid-afternoon on most days of the year.

The natural background of nitrogen dioxide is very low indeed. Junge quotes 2–3 $\mu g/m^3$ [203]. In towns (*Table 13*), however, it is produced along with ozone and reaches similar levels, e.g. recent London values in the range 80–200 $\mu g/m^3$ [175,216].

TABLE 13. Ozone and Nitrogen Dioxide Concentrations ($\mu g/m^3$)

		O_3	NO_2
Natural background		20–60	near zero
London: sunny summer afternoons			100–300
	peaks		over 300
Los Angeles: peaks			over 1500

Concerning indoor concentrations, no measurements are yet available for nitrogen dioxide, though we can expect it to be present at a somewhat reduced concentration, like sulphur dioxide.

Ozone is interesting. Though of course it can gain acess through open windows, in non-ventilated rooms it has a very short indoor life, being rapidly destroyed by organic materials[217], including presumably by human beings and exhibits! Its half-life in an unoccupied bedroom has been estimated at six minutes[218] (that is to say its concentration is halved in six minutes). Some measurements made in 1973 in the National Gallery, London, with a level in Trafalgar Square outside of about 100 $\mu g/m^3$, showed its presence to be barely detectable both in conditioned and in non-air-conditioned exhibition rooms (0.2 to 0.4 $\mu g/m^3$ throughout)[219]. But in other situations the indoor concentration has been found to be as much as half the outdoor (*see* p. 261).

CHLORIDES

Chlorides are a most dangerous contaminant for metals, and anyone who has lived in a house on the coast knows how rusty iron and steel becomes, even when painted, under such conditions.

A wave breaks, droplets are thrown into the air, and these may evaporate to sodium chloride crystals which are carried by the wind inland.

In fact only coastal museums will be affected. Published figures indicate that a level of up to 5 $\mu g/m^3$ of chloride may be found anywhere, even far inland[220]. Some of this may have come from the sea, the greater part of it more probably from industry, but the level is generally lower than that of sulphate unless some specific industry leaks it into the air. At coastal sites which are far away from any industry the concentration of chloride does not exceed this sort of level even 1 km inland, and one has to go to within 200 m or so of the shore to meet the sharp rise to 20–30 $\mu g/m^3$ characteristic of the coast, though strong onshore winds may temporarily drive this a little further inland[221].

POLLUTION THROUGH STORAGE CONDITIONS

Having taken a lot of trouble and expense to eliminate all undesirable gases from town air, it is alarming to realise that others may be introduced into the very air surrounding the exhibits by incorrect storage. Hydrogen sulphide (sulphuretted hydrogen), and formic, acetic and hydrochloric acids have been implicated in this way.

Organic acid vapours may be given off by a variety of timbers, particularly oak. Mr Paul Levi brought the attention of the National Gallery Laboratory to an interesting example. A pastel portrait on paper had been framed with a wood panel behind it in contact. On disassembly a reverse image of the portrait could be seen on the back of the paper bearing the pastel. Some material migrating from the wood panel, presumably an organic acid, had turned all the paper brown except where calcium carbonate pigment was present in the portrait. This had neutralised the acid and thus prevented it from staining the paper. Consequently the portrait on the reverse was an image of all the white areas on the portrait (*Figure 57*).

In this case not too much damage had been done, but lead objects can be rapidly destroyed by being stored in cases made of unsuitable wood, particularly oak and Douglas fir. Certain polyvinyl acetate emulsion adhesives may have the same effect, but not polyvinyl acetate solutions. Lead is converted by organic-acid vapours into lead carbonate – this is indeed how it was once made. Zinc and bronze have also been known to be affected, but not so strongly as lead.

Unusual tarnishing of silver (formation of silver sulphide) should also lead one to suspect the materials of the case. Here volatile sulphides are the enemy. These may be given off by rubbers, composite boards, and even by textiles (*Table 14*) if these have been dressed with sulphur-containing compounds. A great variety of synthetic materials may be implicated, including the adhesives used in plywood and veneers.

Silver is the image-forming material in the majority of black and white photographs, and therefore the above applies equally to them. Collins and Young describe a simple tarnishing test for materials used in the conservation of photographs[191]. If it is suspected that a silver collection is being tarnished with hydrogen sulphide from outside the showcase, it should be made more or less leakproof except for one hole into which is inserted either silver wool or crumpled lead acetate indicator paper (Plenderleith and Werner[81], p. 241). Cellulose nitrate, the first commercially successful plastic, fortunately finds little use

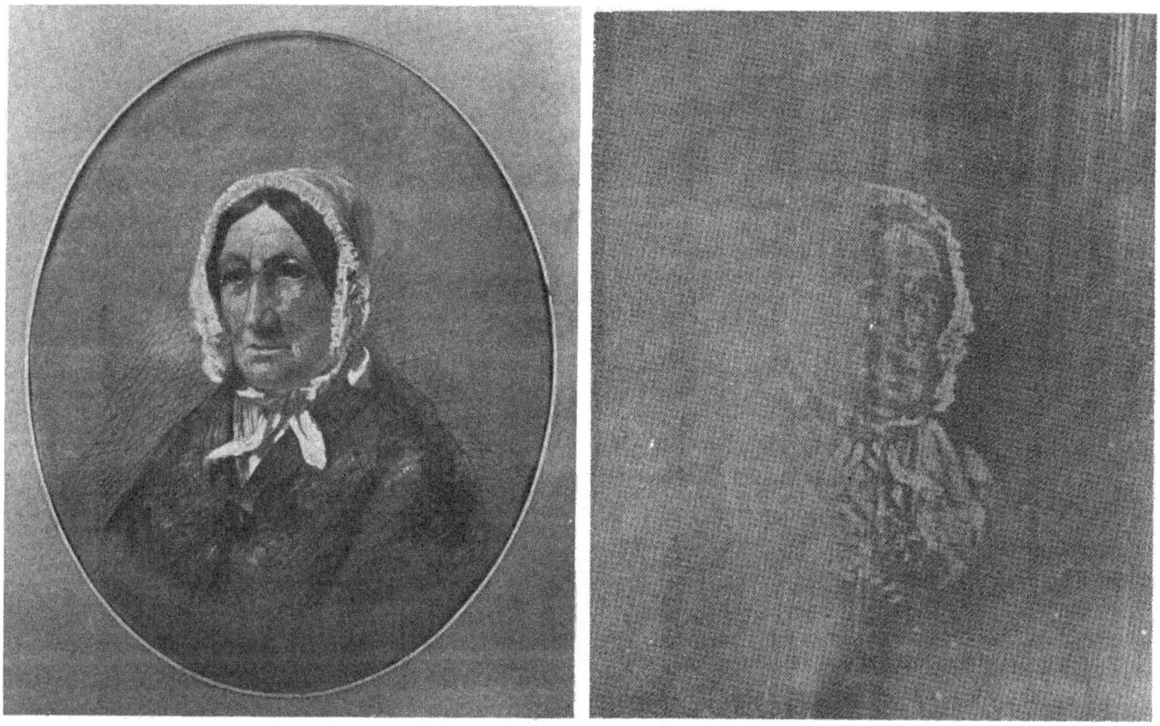

(a) *(b)*

Figure 57
(a) A pastel painting on paper, as framed. *(b)* The reverse side of the paper, which has received an image of the wood backing board except where calcium carbonate pigment is present in the pastel on the other side. The image of the wood has probably been formed by an organic acid which has migrated to the paper from the wood and turned it brown. Example provided by Mr Paul Levi

today for it is very unstable and inflammable, giving off nitrogen oxide vapours. Old cine film and early artificial leather (pyroxylin) were of cellulose nitrate and therefore pose serious conservation problems.

The degradation of polyvinyl chloride (PVC) by either heat or light is accompanied by the evolution of hydrogen chloride (hydrochloric acid). Though stabilisers have made PVC sheet fairly permanent it should not be used for storing museum material. Polymethyl methacrylate and most acrylics, polystyrene (protected from strong light), polyethylene, and polyvinyl acetate from organic solution should all be trouble free.

Oddy[222] descibes a simple general test which can be carried out without laboratory facilities where trouble is suspected. Activated carbon, perhaps in the form of a coating on paper, could be useful for mopping up pollutants in cases, though it might not be able to compete with silver in its avidity for hydrogen sulphide.

TABLE 14. Some materials known to release harmful vapours

Wood (particularly oak, birch, beech)	– Organic acids
Protein-based glues, wool	– Volatile sulphides
Vulcanised rubber	– Volatile sulphides
(Some dyes may release sulphur compounds)	
Cellulose nitrate	– Oxides of nitrogen
Polyvinyl chloride	– Hydrochloric acid
Polyurethanes are suspect	
Materials that appear to be safe	
Metals	
Glass	
Ceramics	
Inorganic pigments	
Polyethylene and polypropylene	
Acrylic, polycarbonate and polystyrene sheet	
Acrylic solutions (not necessarily emulsions)	
Polyester fibres	
Cotton and linen	

Based on ref. 223.
(Bradley gives a list of about 100 fabrics found safe for metals[224].)

REMOVAL OF GASEOUS POLLUTANTS

We require our filter system to remove primarily sulphur dioxide (SO_2), nitrogen dioxide (NO_2) and ozone (O_3). Other gases present in urban pollution may also be removed but are normally of minor importance.

There are two established methods for removing gaseous pollution:
1 Water spray
2 Activated carbon filters.

WATER SPRAY

A widely used method of air-conditioning involves passing the air through a water spray which is running the whole time with the temperature of the incoming air and of the water variable. Plain water, provided that it is never allowed to become too acid by absorption of acid gases, can be very effective against all such gases, including SO_2, and NO_2 which is even more soluble. In the early 1960s in the National Gallery the water sprays operating then reduced SO_2 to much less than one tenth of its outside value.

It has been suggested that the water would be even more effective if it was made alkaline[225], but problems of corrosion and clogging of nozzles may arise. Of course no volatile alkali should be allowed ingress.

Water spray is not effective against ozone.

ACTIVATED CARBON FILTERS

These are in packs, which can be mounted in the air stream like particulate filters, containing specially prepared charcoal (*Figure 58*). This presents a very large area for adsorption to gases flowing through it. Those gases most easily condensed are

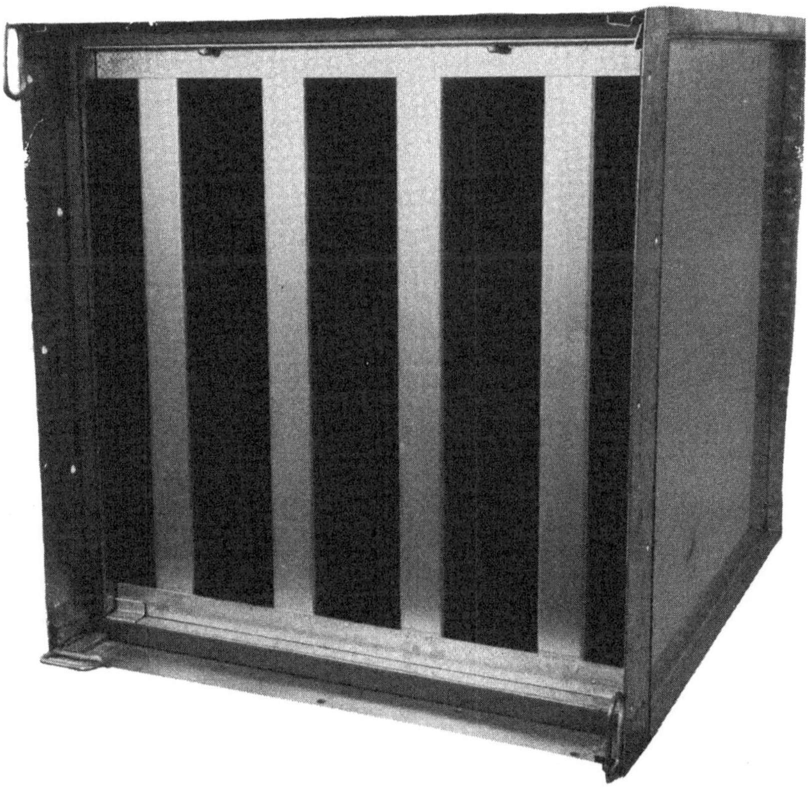

Figure 58
Activated carbon filter pack (Vokes). Air flow is into the front, and the carbon granules are contained in panels placed in V formation at an acute angle to the air flow, so that it impinges on them in the same way as in the particulate filter of *Figure 48*

adsorbed preferentially. SO_2 condenses at −10 °C and NO_2 at +21 °C, so that both should be quite well adsorbed.. However, tests show that, whereas activated carbon is satisfactory for SO_2, it will not always remove NO_2 efficiently. With most of the air recirculating more than once through the filter system, efficiency against SO_2 with a well maintained system is perfectly adequate (*see* Part II p. 252). A test of activated carbon filters in actual use in the northern extension to the National Gallery gave an SO_2 average over several months of just under 5 $\mu g/m^3$.

Ozone is removed with high efficiency by activated carbon, not through adsorption but by destruction, perhaps on the carbon surface itself, perhaps on traces of organic matter trapped there.

The activated carbon filters must be replaced at intervals, and there is no visible indicator to show when this stage has been reached. Usually the firm which installs the filters will take tests during the first period of use to determine a routine for their treatment at regular intervals.

A copper-impregnated activated carbon, trade name Sonoxcarb, is particularly effective and has been installed at the National and Tate Galleries in London[226]. Tests show that these filters should last for more than 5 years before renewal.

Activated-carbon filters are usually designed for mounting in ducts, but they could equally well be used in room units (*see* p. 254) or in show-case air purifiers. A Sonoxcarb unit providing a small flow of air into display cases in the Victoria and Albert Museum reduced sulphur dioxide to less than 3 $\mu g/m^3$.

TABLE 15. Effectiveness against Gaseous Pollutants

	Water spray	*Activated carbon*
Sulphur dioxide	yes	yes
Nitrogen dioxide	yes	variable
Ozone	no	yes

Note
Among other pollutants mentioned, water spray would be completely effective against hydrochloric acid and chlorides, and activated carbon against PAN. Neither could be relied on to take out nitric oxide, NO, but this is unlikely to be important.

TABLE 16. Suggested Top-grade Specification

Sulphur dioxide	not more than 10 $\mu g/m^3$
Nitrogen dioxide	not more than 10 $\mu g/m^3$
Ozone	reduce to trace levels (0–2 $\mu g/m^3$)

Note
These levels are readily attainable in the museum situation with present technology in air conditioning with recirculation. They will also ensure that other acidic gases and powerful oxidants will not gain significant access.

It is worth noting that the U.S. Department of Commerce recommends for archival storage not more than 1 $\mu g/m^3$ of sulphur dioxide and not more than 5 of nitrogen dioxide[227].

FIRE EXTINGUISHERS

Here the emphasis is on fire extinguishers as possible sources of pollution. As always, prevention and detection are more important than cure.

There are three types of substance which are likely to be used in fire-fighting equipment: water, carbon dioxide and halogenated methanes. Each is available in various types of equipment, from hand held extinguishers to automatic flooding equipment.

Halogenated methanes ('Halons'), such as ICI's B.C.F. (chemically CF_2ClBr) and Du Pont's B.T.M. (CF_3Br) are more effective than carbon dioxide, volume for volume.* Whereas carbon dioxide extinguishes fires by displacing the oxygen, the halogenated methanes operate by a positive physicochemical inhibition which is not completely understood. The result is that smaller amounts of the latter are required, which both makes them cheaper and reduces the chances of asphyxiation.

However though the halogenated methane as it comes from the extinguisher is relatively non-toxic to humans, small amounts of toxic gas may be formed on hot surfaces.

We must now consider the effects on antiquities of the three types of extinguisher. Unless a building has been completely gutted there will be objects which have survived the heat and flame. Will they also have survived the extinguishers?

High-pressure water by the very power of its delivery can destroy delicate objects, and steam is a potent hydrolyser of paint. Carbon dioxide is chemically harmless but, if released in error, can cause a savage drop in temperature.

The halogenated methanes, however, generate small quantities of the strong halogen acids, hydrochloric, hydrobromic and hydrofluoric acids at 500 °C. These could be damaging to a wide range of materials, especially metals.

We therefore have to balance up a number of factors in our choice of equipment.

For automatic equipment, e.g. in book and document stores not open to the public, a carbon dioxide system is probably best. In areas of the museum at higher than average risk, such as conservation laboratories, hand held halogenated methane extinguishers, extinguishing blankets and a fire-hose capable of reaching all areas should all be available.

Automatic sprinklers should not be installed in any areas containing exhibits. Water hoses cannot be dispensed with, but

*These are both formulations for minimum corrosiveness. B.C.F. is coded Halon 1211, where the numbers represent successively 1 carbon, 2 fluorine, 1 chlorine, 1 bromine

very close to each other.

The U.K. Chartered Institute of Building Services recommends a maximum of NR 35 for a museum exhibition room. Compare:

Broadcasting studio	NR 15
Suburban bedroom	25
Urban living room	35
Restaurant	45
Workshop	65

Light
Part Two

Light

SPECTRAL CURVES

The four kinds of spectral curve which the scientific conservator may need to use are the transmission curve (of a filter for example), the reflectance curve (of a surface), the sensitivity curve (of a light receiving mechanism such as the eye or a photocell), and the power distribution curve (of a light source).

The first two are straightforward. At each wavelength the y-axis presents the proportion of energy transmitted or reflected (*Figures 7 and 67*).

Spectral sensitivity is usually represented in relative units. For example the sensitivity of the daylight-adapted eye is at its maximum at 555 nm. If this maximum sensitivity is given the value 1 the sensitivity at any other wavelength falls in the range 0 to 1. We could, however, put a figure on this maximum sensitivity (*Figure 63 see* p. 177). The curve would then give, not relative sensitivity at other wavelengths, but an absolute value, as defined.

The scale of relative units is an easy way to avoid calculation, and so very often the fourth kind of spectral curve, the power distribution curve of a light source, is also represented in this way: the y-axis is labelled 'relative power' or 'energy'. Looking at the curve for a tungsten lamp, (*Figure 59(a)*), we can ignore the y-axis scale and, merely by measuring the height of the curve with a ruler at 400 nm and at 700 nm, we can say that this lamp emits about 13 times as much red light energy at 700 nm as violet light at 400 nm. But for certain purposes we have to go further. A scale of absolute units is arrived at as follows.

Light is energy. Therefore when light falls on a surface, energy is continuously being received by that surface. The amount of energy received at a moment in time is vanishingly small, so that we must measure the energy received in a standard time interval, conveniently one second. Energy is measured

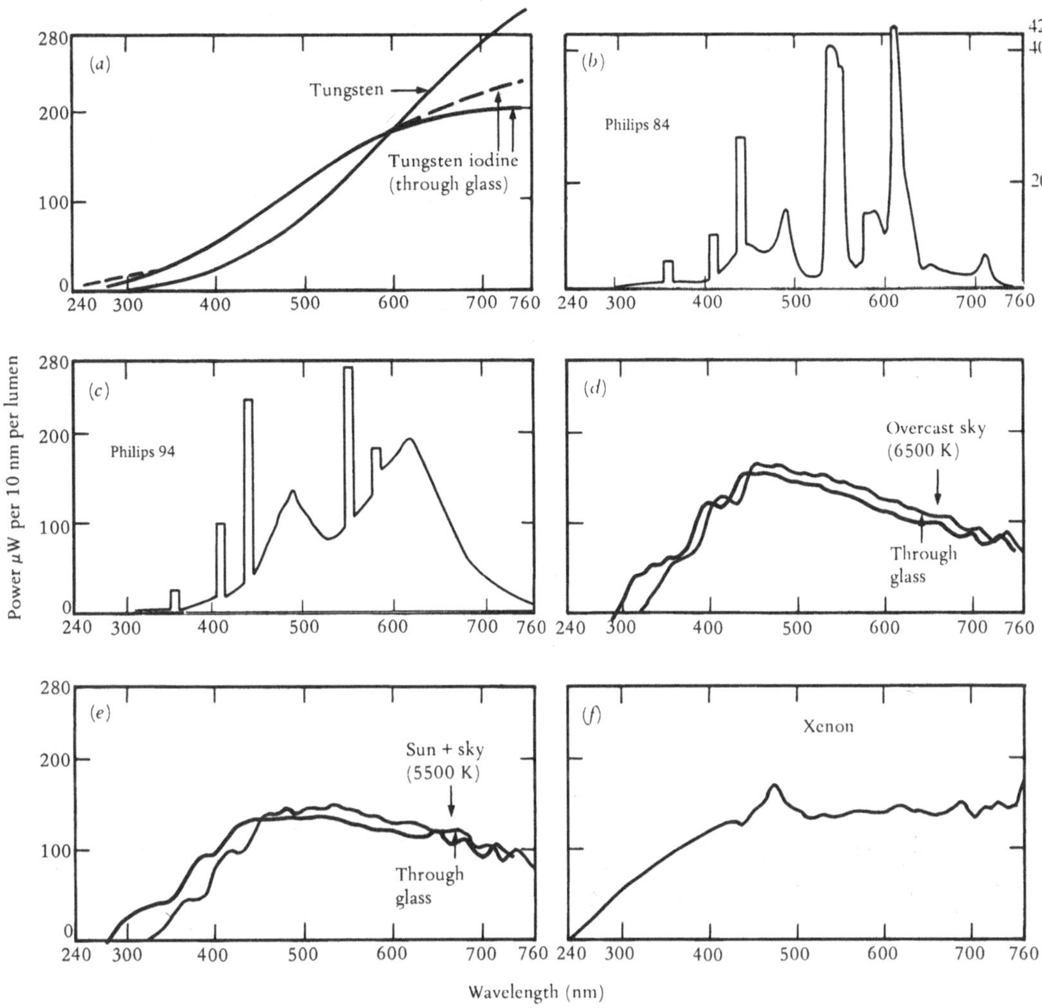

Figure 59
Spectral power distributions of some light sources in microwatts per 10 nm per lumen. *(a)* Black curve: Tungsten incandescent lamp (CIE Source A, 2854 K). Broken colour curve: Tungsten iodine lamp at 3360 K. Full colour curve: The same tungsten iodine lamp with glass shield. *(b)* Philips 84 'triphosphor' fluorescent lamp (note the different vertical scale). *(c)* Philips 94 fluorescent lamp. For the two fluorescent lamps the curve represents the radiation from the phosphors. On top of this curve are bars representing mercury-line radiation. For example, radiation in the band at 550 nm for the Philips 94 is composed of 94 μW from the phosphors plus 170 from the mercury line. *(d)* Colour curve: Radiation from an overcast sky at 6500 K. Black curve: The same through window glass. *(e)* Colour curve: Light from the sun and sky at 5500 K. Black curve: the same through window glass. *(f)* A typical xenon lamp. Note the considerable output of UV radiation. This necessitates the use of a UV-absorbing filter where museum objects are regularly photographed by (xenon) electronic flash. The lamp is too powerful for use inside the museum as a continuous source. If it were to be used, one would advise both glass and a plastic UV-absorbing filter

Note. Numerical data and further details for all these curves are to be found in *Table 17* . This table also includes a metal halide lamp of good colour-rendering quality, the Osram HQI/L 250W. Data for this lamp were not available below 360 nm, but it can be seen that a UV-absorbing filter (though not glass) would be very necessary. Metal halide lamps are in process of development, and present products vary widely in performance in joules, and joules per second are called watts, so, following the Système Internationale (SI), our measurement of the rate at which the surface receives energy, the power, will be in watts. But this energy does not fall on a single point. It is spread over an area, so that we must next specify the area over which our measurement is made. Let us use a square metre. Our unit is now watts per square metre, or W/m^2. This would do to measure the total power of the radiation: UV, visible and IR. But we are primarily interested in how this power is spread through the spectrum, how it varies with wavelength. Just as the energy received at a moment in time and at a single point is vanishingly small, so we cannot expect to be able to measure the energy at a single wavelength. We must use instead a band of wavelengths narrow enough to define the spectrum sharply, commonly 10 nm. We now have all we need to specify quantitatively the power distribution of the radiation. The *y*-axis could now be labelled $W\ m^{-2}\ (10\ nm)^{-1}$.

The only remaining trouble is that we do not yet know the illuminance of the light we are measuring. Increase the illuminance and all the *y*-values go up. We need to compare curves at a standard illuminance, say one lux. The *y*-axis now becomes 'watts per square metre per 10 nm per lux'.

Since lux are lumens per square metre those familiar with the manipulation of units will see that some cancelling out can be done, since 'per square metre per lux' is 'per square metre per (lumens per square metre)', which reduces to 'per lumen'. In standard notation '$W\ m^{-2}\ (10\ nm)^{-1}\ lm^{-2}\ m^{2}$' becomes '$W\ 10\ nm^{-1}\ lm^{-1}$'. In practice this yields very small figures, so we more often use microwatts (10^{-6} W, μW). But this is a matter of convenience on which no two graph-makers agree. So long as the *y*-axis unit contains a term for energy rate or power (e.g. microwatts), a term for bandwidth (10 nm) and a term for luminous flux (lumens) we have the information, which can be converted if necessary to other units.

Thus our final *y*-axis is in '$\mu W (10\ nm)^{-1}\ lm^{-1}$' or microwatts per 10 nm per lumen. See, for example, *Figure 59*.

There is a small complication with fluorescent lamps. The radiation from these lamps has two sources: firstly the continuous spectrum from the mixture of phosphors coated on the inside of the tube, and secondly the sharp lines which penetrate the phosphor layer from the mercury vapour. Thus the spectral power distributions of all fluorescent lamps consist of smooth hills with a few tall towers (*Figure 59(b) and (c)*). The smooth hills of the continuous spectrum are read as just explained. To add the mercury vapour energy we could just draw vertical lines at the wavelengths of emission and label them with the values of energy emitted, and this is sometimes done. But the general spectral power distribution curve has the very useful property that the area under any waveband interval of the curve gives the power emitted within that waveband. There is no area under a vertical line, and so we cannot follow this convention if we merely draw vertical lines for the mercury spectrum. Also vertical lines give no picture of this energy distribution. All we need to do to resolve the difficulty is to widen the lines into towers centred at each point of mercury emission, making the width of the towers the same as our chosen waveband (10 nm), and their heights equal to the power emitted. Usually the tower is shown on top of the hill, as here. The height from zero to the top of the tower represents the power from the phosphor plus the power from the mercury line. *See Figure 59(b) and (c).*

SUN AND SKY

The temperature of the sun's interior is very high, probably around 14 million degrees centigrade[230], but the surface visible to us is at no more than about 6000 °C. The sun radiating in black space comes near to being a perfect black body radiator (p. 49) though an outer atmosphere of gaseous elements selectively absorb some of the radiation to produce the missing lines of Fraunhofer. Little further modification of the sunlight occurs during its passage through space to earth. On arrival at our outer atmosphere its energy flux density is about 1350 W/m^2 [231] corresponding to a blinding 131 000 lux[232].

But between us and this incoming energy is the earth's atmosphere where absorption and scattering produce important changes in its composition and strength.

When all the vagaries of climate are combined the sun's radiation has no net heating effect on the surface of the earth. All the energy that arrives is re-radiated into space, but only part of it directly. Firstly, 40% of the incoming radiation is reflected from atmosphere, clouds, earth and sea without absorption. This is called the *albedo*. 15% is then absorbed by the atmosphere and the remaining 45% at sea level. All the absorbed radiation, including that used in life processes, is ultimately re-radiated as heat in the IR. These figures may be approximate but the balance is almost exact. It has been estimated that if the total energy radiated away from the earth were only 0.1% less than that which arrives the temperature of the earth's atmo-

TABLE 17. Spectral Power Distributions for Figure 59

	Power per 10 nm per lumen ($\mu W\ 10\ nm^{-1}\ lm^{-1}$)										
Wave-length (nm)	*(1) Tungsten*	*(2) Tungsten iodine*	*(3) (2) thru glass*	*(4) Philips 84*	*(5) Philips 94*	*(6) Overcast sky*	*(7) (6) thru glass*	*(8) Sun + sky*	*(9) (8) thru glass*	*(10) Xenon*	*(11) A metal halide*
250		1								7	.
60		1								13	.
70		2								23	.
80		3								31	.
90		4								43	.
300		6		0		0		0		51	.
10	0	7		2		5	0	3	0	58	.
20	3	8	0	3	0	28	1	16	1	66	.
30	4	10	2	4	1	52	11	29	6	73	.
40	5	13	7	7	1	56	27	34	16	78	.
350	6	16	12	8	0	63	49	39	30	85	?
60	8	19	17	42	22	65	59	43	40	90	84
70	10	22	22	7	1	73	69	48	46	95	125
80	13	26	24	7	5	70	65	46	43	101	57
90	16	31	30	5	9	76	74	53	52	107	98
400	20	36	36	8	13	115	114	85	85	114	115
10	24	41	41	85	95	128	127	96	95	119	162
20	28	47	46	27	18	130	128	100	98	122	186
30	33	53	52	47	22	121	118	95	93	125	121
40	39	59	59	244	229	146	145	120	119	126	104
450	46	65	64	67	44	163	162	137	137	139	112
60	52	71	71	62	67	164	163	140	140	145	126
70	59	78	78	52	92	160	160	140	140	170	121
80	66	84	85	68	114	161	162	144	145	148	114
90	74	91	92	125	130	151	153	137	138	142	114
500	82	98	99	51	116	152	154	141	143	139	127
10	90	105	107	18	98	150	152	141	143	132	121
20	99	112	114	12	85	146	149	140	143	131	113
30	107	119	122	20	78	150	153	146	150	135	222
40	117	125	127	388	83	145	147	143	145	136	174
550	127	132	134	354	265	145	146	144	145	136	118
60	137	138	140	42	105	139	140	140	141	139	126
70	146	145	146	32	120	134	135	136	137	139	140
80	156	151	150	119	175	133	133	137	136	139	118
90	166	157	156	122	158	123	122	128	127	136	140
600	176	163	161	85	166	125	123	132	130	136	144
10	185	169	165	422	178	125	121	133	130	142	139
20	195	175	170	182	184	122	117	132	128	145	124
30	205	181	175	115	170	116	111	126	122	139	106
40	217	187	178	26	147	117	110	129	123	136	109
650	225	192	182	34	123	111	105	124	118	133	97
60	234	198	185	26	100	112	104	126	118	131	115
70	244	202	189	18	79	114	106	132	122	136	96
80	252	206	190	16	61	109	100	126	116	142	113
90	261	210	192	19	47	97	88	112	101	150	82
700	270	214	194	17	35	97	89	116	104	131	70
10	278	217	194	51	27	103	92	119	105	142	48
20	286	221	194	8	21	86	75	98	86	142	48
30	295	224	196	4	16	97	85	111	96	148	38
40	302	228	196	3	12	105	90	119	102	142	43
750	309	231	196	3	10	89	75	101	86	142	63
60	316	234	197	2	6	65	55	74	62	170	66

sphere would rise by 6 °C in a year. This is why the rise in the carbon dioxide level is watched with such interest. This rise is due to the burning of fossil fuel and forests. The carbon dioxide in the atmosphere absorbs some of the infrared energy radiated from the earth so that a rise in carbon dioxide level will result in a rise in temperature.

Of vital importance to the preservation of both life itself and of the artifacts which man has produced is a layer of ozone 20 km and more above our heads. This uncompromisingly prevents all radiation of wavelength shorter than 300 nm from reaching us. Further modifications, mainly in the IR, result from absorption by water and carbon dioxide.

In overcast weather all our light arrives from the sky rather than the sun. But even a clear atmosphere scatters some of the light, predominantly blue, and the sunlight is consequently a little yellowed.

Lastly, between the exhibits and the daylight falling on the museum there is fortunately a layer of glass which removes a further slice of the UV in the region 300–325 nm (*Figure 1*) as well as much of the IR. If, instead of glass, some substitute totally transparent to UV had been used through the ages, we would today have many fewer antiquities.

Figures 59(d) and (e) gives the spectral power distribution of the light from the sun and sky which results from all these modifications.

The great difficulty in planning for daylight illumination of exhibition rooms is estimating in advance what the interior illumination is likely to be at various times of the day and year. This involves a combination of sun and sky calculations with model making. One cannot do it all from the figures, though talk of the *Daylight Factor* might lead the unwary to suppose that one can. In its place the Daylight Factor is a useful con-

---

Notes on the light sources

1 Tungsten: CIE Source A (2854 K).
2 Tungsten iodine 12V 100W (3360 K).
3 Tungsten iodine thru 6mm polished plate glass.
4 Philips 84 (3900 K).
5 Philips 94 (3800 K).
6 Daylight at 6500 K.
7 Daylight at 6500 K through 6mm polished plate glass.
8 Daylight at 5500 K.
9 Daylight at 5500 K through 6mm polished plate glass.
10 Compact source Xenon 250 or 500W (cct about 5500 K).
11 Osram HQI/L 250W metal halide (5200 K).

The number of photons per sec corresponding to any of these sources in a particular 10 nm band at wavelength can be obtained by multiplying the product of the power in the band and the wavelength by $10^{10}/1.99$. Thus at 1 lux the number of photons in overcast skylight in the band 545–555 nm

$$= 145 \times 550 \times 10^{10}/1.99 = 4 \times 10^{14} \text{ photons per m}^2 \text{ per sec.}$$

Mercury wavelengths are: 313, 365, 405, 408, 436, 546, 578 nm. Neon 334 nm.

vention. It is a measure at any point within a building of the proportion of light *from the sky* which falls on a horizontal surface at that point. It is thus a ratio of internal to external illuminance, but with the sun shaded off from the meter.

The reader may exclaim that this will give a reading which is badly wrong whenever the sun is shining. But this is not the purpose of the Daylight Factor. Its purpose is to find if there will be enough light in the room, whether or not the sun is shining, thereby ensuring that the illuminance is never too low. But we as conservators are just as interested in ensuring that the illuminance is never too high, and for this purpose the Daylight Factor is not applicable.

All in all our best help is a set of figures for average sun plus sky conditions. This will vary from place to place and so it will have to be worked out from local meteorological data (*Figures 60 and 61*).

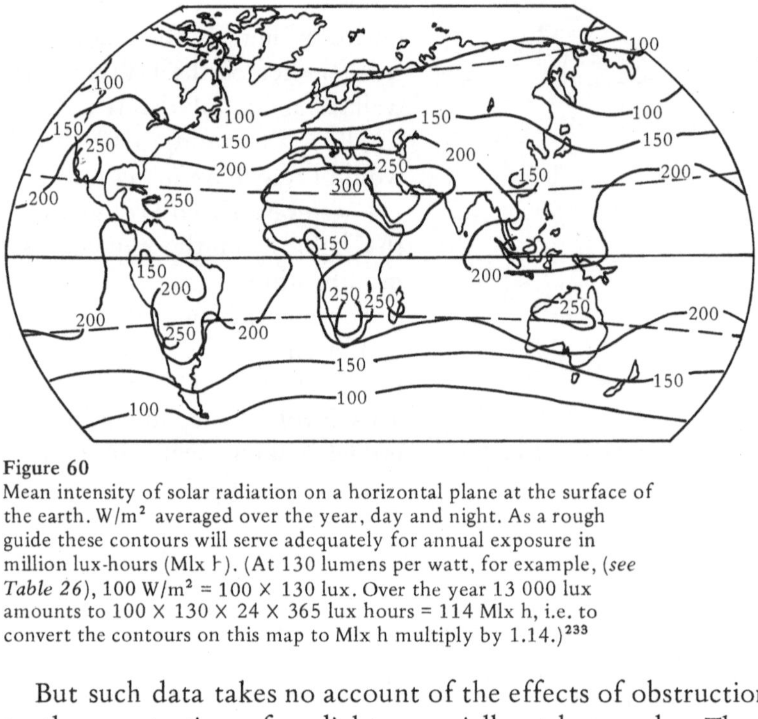

Figure 60
Mean intensity of solar radiation on a horizontal plane at the surface of the earth. W/m^2 averaged over the year, day and night. As a rough guide these contours will serve adequately for annual exposure in million lux-hours (Mlx ⊦). (At 130 lumens per watt, for example, (*see Table 26*), 100 W/m^2 = 100 × 130 lux. Over the year 13 000 lux amounts to 100 × 130 × 24 × 365 lux hours = 114 Mlx h, i.e. to convert the contours on this map to Mlx h multiply by 1.14.)[233]

But such data takes no account of the effects of obstructions to the penetration of sunlight, especially at low angles. Therefore a model should be built large enough for a photocell to be inserted and placed at various positions within the exhibition room. The model can then be taken out into the open, the higher the better to avoid obstructions from buildings and trees, and tested under different weather conditions. The photocell readings at the same moment outside and inside the model are recorded, the outside reading always being taken in the

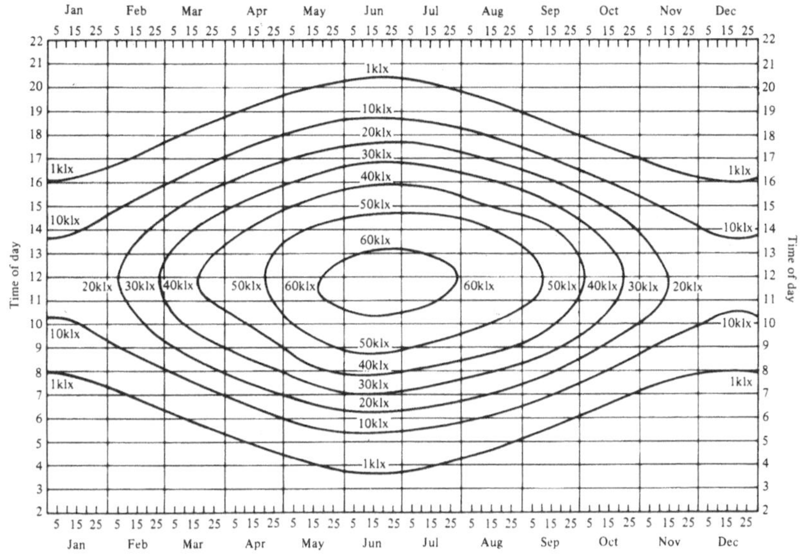

Figure 61
Illuminance throughout the year on a horizontal surface from both sun and sky under average weather conditions near London. The vertical scale reads Greenwich Mean Time. The contours are of illuminance in kilolux (1000 lux). Thus the illuminance at mid-day is expected to exceed 60 000 lux during most of June and July.[234]

same way with the photocell pointing upwards to the zenith and the observer ensuring that he himself does not shade off light from the photocell. Any time of the day and year can be simulated by tilting the model, and this becomes very simple if a suitable sundial is attached to its baseboard[235].

For the London area it will be found that, to maintain daylight at 200 lux on the exhibits, without supplementary artificial light, for 2/3 of the year, 2% of the daylight must reach the exhibits. This may not sound much, but it is quite difficult to achieve, especially in a top-lit picture gallery where the light has to be diffused evenly round the walls. Glazing, blinds and roof obstructions block out a great deal of light.

For other parts of the world, daylight availability should be plotted as in *Figure 61* and a correction made to my figures by comparing the two diagrams.

LAMPS AND CONTROL EQUIPMENT

The ordinary household electric light bulb and also most spotlights and miniature bulbs emit their light from a coil of tungsten wire heated by an electric current to about 2600 °C (*Figure 3*). Such lamps are called *tungsten incandescent lamps*. Their envelopes are made of glass, and they are filled with a mixture of inert gases which, for domestic lamps, is usually 90% argon and 10% nitrogen. Low-wattage lamps may contain a vacuum. Small low-voltage (low-wattage) lamp bulbs are often used for display. Since the source is compact the light beam can be more accurately controlled.

Tungsten incandescent lamps have a lifetime of about 1000 hours and are run directly from the mains electricity supply without control equipment, though low-voltage lamps naturally require a transformer to reduce mains voltage.

The luminous efficacy (lumens of luminous flux emitted per watt of electricity consumed) of a 100 W tungsten lamp is about 13 lm/W. Since most of the electrical energy consumed by such lamps is turned into heat and not into light (actually 94% for a 100 W lamp), considerable efforts have been directed towards improving the situation.

A chemical reaction has been ingeniously harnessed in the *tungsten–halogen* lamp to provide either an increase in efficacy or a longer lifetime, or some compromise between the two. These lamps are also very much more compact, which gives them advantages where the light beam has to be closely controlled, e.g. in film projectors. The higher the temperature at which a tungsten filament is heated the greater is its efficacy in light production. Also the light gets whiter, a natural consequence of the higher temperature. The temperature is limited in the standard tungsten lamp by the evaporation of the tungsten. To prevent loss of tungsten burning the lamp out in this way, a small amount of halogen vapour, usually iodine, is introduced; hence the alternative name of quartz–halogen, the envelope being made of quartz (fused silica) instead of glass. As tungsten atoms evaporate and travel away from the filament to a cooler part of the bulb they combine with the iodine vapour to form tungsten iodide, and remain in this form until by diffusion they find their way back to the filament. Here decomposition takes place, the tungsten is redeposited where it belongs and the iodine is returned to repeat its service.

There is a trade-off between luminous efficacy and length of life. If the bulb is run at 3250 °C the efficacy becomes 35 lm/W but the lifespan is not improved. At lower temperatures the lifespan can be increased sixfold.

The envelope of a tungsten–halogen lamp, being made of silica, is transparent to UV, even to a tiny amount of UV of wavelength shorter than any in sunlight (*Figure 59(a)*). This danger is completely removed by fitting a shield or housing of ordinary glass.

The great advantage of the ordinary (hot cathode*) type of fluorescent lamp is an efficacy (measured in lumens/watt) considerably higher than that of the tungsten incandescent lamp. The efficacy of a tungsten lamp is about 13 lm/W; that of a conventional (halophosphate) fluorescent lamp of good colour rendering is about 45 lm/W; whereas some of the newer triphosphate fluorescent lamps go up to 90. Thus even a low-efficacy fluorescent lamp uses less than a third of the electricity. Because fluorescent lamps emit much less heat, the reduced cooling load on an air-conditioning will also result in reduced bills in most countries.

The standard fluorescent lamp consists of a tube coated on the inside with a mixture of phosphors, and is available in a variety of sizes. At each end is a tungsten filament with two terminals. The tube contains a drop of mercury and a small amount of neon to aid conductance of electricity on starting. In operation electrons are given off by the filaments and these excite the mercury vapour atoms to emit radiation at certain fixed wavelengths, mostly in the UV. But most of this radiation, especially the UV, is absorbed by the phosphors which re-emit in the visible region.

Fluorescent lamps cannot be operated direct from mains electricity for two reasons. Firstly, during operation they are inherently unstable, with negative resistance characteristics such that a rise in voltage increases ionisation of the gas molecules, which lowers resistance and causes a surge of current. Secondly, they require special starting conditions to heat the filaments and initiate the discharge of electrons.

The control units come in several different patterns, but they must all include a choke ballast for stability, which is the largest and heaviest item. This is joined in series with the tube. The combination of tube and choke now has positive resistance characteristics so that a rise in voltage no longer increases the current.

Figure 62
A simple form of control equipment for fluorescent lamps

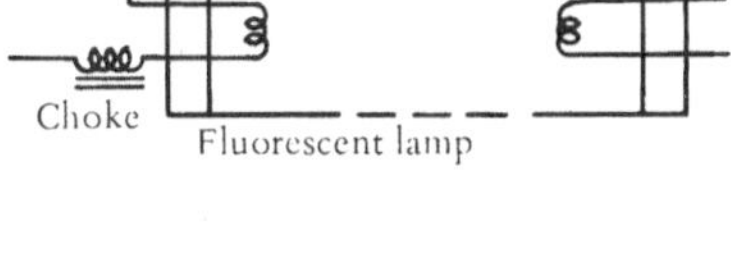

To start the lamp extra current must be provided to the filaments to heat them to the point when they provide sufficient electrons for the discharge to start. The simplest and commonest device is the glow-starter, a bimetallic temperature-sensitive switch enclosed in a bulb containing inert gas (*Figure 62*). When we switch the fluorescent lamp on, the discharge through the gas in the glow-starter heats the bimetallic switch which closes

*There is also a *cold*-cathode fluorescent lamp, but this is not normally to be found in interior lighting since it requires a starting voltage of around 12 000 volts so that special safety precautions are necessary. It has a very long life (about 25 000 hours), and is used for luminous signs.

it. In the closed position it immediately begins to cool and thus opens again in a few seconds. But while it is closed all the mains current flows through the filaments, bypassing the tube itself. The emission of electrons from the heated filaments thus begins and the tube lights up. If the first attempt is unsuccessful this simple device automatically initiates a second attempt. Repeated attempts can be recognised in the familiar flashing of defective lamps.

To prevent this flashing occurring lamp manufacturers point out that it is in any case more economical to change the whole batch of fluorescent lamps in one room at the same time, cleaning the fittings at the same time. The period between changes will be recommended by the manufacturers.

Of other lamps now being developed the metal halide lamps are the most promising. Some will undoubtedly emerge in the low-wattage range which will combine high efficacy with the good colour rendering required in museums. Metal halide lamps were developed to overcome the deficiencies in colour of the high pressure mercury lamp. It was found that the addition of certain metal halides, notably sodium, thallium and indium iodides, both filled in the gaps in the spectrum and about doubled the luminous efficacy. Other experimental lamps use the molecular radiation of aluminium halides. The lamp bulb is of silica, but there is likely to be an outer glass envelope, when the lamp is known as an MBI. If this glass is coated with fluorescent powders as additional colour modifiers we get an MBIF lamp.

TABLE 18. Some Lamp Designations

From the variety of codes designating lamp types and details the following are selected:

GLS	(General lighting service) The ordinary domestic tungsten incandescent lamp, of pear or mushroom shaped bulb
Crown-silvered	Pear shaped GLS with front half silvered. To be used with a reflector
ISL	(Internally silvered). A tungsten lamp shaped and internally silvered to form a diffuse-beamed spotlamp
PAR	(Parabolic). A sealed-beam tungsten spotlamp manufactured in two parts which are fused together: reflector and lens
MCF or TL	Tubular fluorescent lamp
MBI	Metal halide lamp with glass envelope
MBIF	The same with phosphors on the glass

Note
The GLS, ISL and PAR lamps are illustrated in *Figure 3*

MEASURING UV

On p. 166 we found that the y-axis on a spectral energy distribution curve could most conveniently be labelled 'μW 10 nm^{-1} lm^{-1}'. It is a small step from this to sum the whole of the UV energy in the waveband from 300 to 400 nm by adding together all the ten y-axis values at the centres of the 10 nm intervals (e.g. at 305, 315, 325 . . . 395 nm), in other words by measuring the area under the curve between 300 and 400 nm. This can be done on any of the curves in *Figure 59*, whereupon the unit of measurement becomes 'microwatts of UV radiation per lumen (of visible radiation)', abbreviated μW/lm.

In this way we are measuring a characteristic of the light source which is a constant for that particular light wherever it falls in the room. This is what we want. We want to know how high a proportion of UV is emitted by a particular fluorescent lamp, gets through a particular filter, is reflected from a white wall, etc. A UV monitor such as the Crawford Monitor (Littlemore Engineering, Littlemore, Oxford, England) is designed to answer such questions by direct measurement (*Figure 10*).

Of course one could alternatively measure the UV radiation falling on a particular surface, and there are incident UV meters available for this. Such a meter does the same for UV radiation as the light meter does for visible light. Its reading will depend on its distance from the light source. But usually this is not what the conservator needs. The conservator needs to make two checks on the lighting: *(a)* using the light meter, illuminance not to exceed, say 200 lux; and *(b)* using the UV monitor, proportion of UV in the light not to exceed that from a tungsten lamp (75 μW/lm, *see Table 19*).

To find the UV incident on a surface from the UV monitor reading, multiply it by the illuminance. For example a UV monitor reading of 300 μW/lm at 200 lux illuminance amounts to (300 × 200/1000) = 60 mW/m^2 of UV radiation incident on the exhibit.

TABLE 19. Microwatts of UV per lumen: Some Typical Values

Blue sky at 15 000 K	1600
Cloudy to overcast north sky	800
Direct sun	400
Fluorescent lamps	40–250
Tungsten–iodine through glass	up to 130
Normal tungsten	60–80

LUMINOUS EFFICIENCY AND THE LIGHT METER

The light meter measures illuminance in lux, the international unit. (1 lumen per square foot, lm/ft^2, also called 1 foot-candle = 10.76 lux). Illuminance is a measure, not of the energy flow which constitutes light, but of this energy flow as the eye sees it: not of energy flux but of *luminous* flux.

We can understand the difference by picturing the radiant energy flux as a set of lines coming from the light source. More lines striking a surface mean a bigger flow of energy (power) to that surface, a stronger radiation, a higher radiant flux density. But the eye does not respond to all these lines. It responds to some of them very well (yellow and green light), to some of them not very well (red and blue light), and to most of them not at all (UV and IR radiation). If in our imaginary diagram we rub out all the lines of energy flux invisible to the eye and reduce those only partly visible in proportion we will have turned our energy flux diagram into a luminous flux diagram. This is what the light meter measures.

The two kinds of measurement, radiation as energy and radiation as light which the eye sees, radiant flux and luminous flux, are related through an important curve known as the curve (or function) of Luminous Efficiency or Relative Spectral Sensitivity of the eye (*Figure 63*). Specialists refer to it as the V_λ curve (Vee lambda curve).

This curve shows the eye to be at its most efficient at 555 nm. The luminous efficiency of the eye at this point is 680 lm/W. This means that if light composed solely of this 555 nm yellow-green wavelength is made to fall on a light meter so that it registers 680 lux (680 lm/m^2) the energy flux or power will be 1 W/m^2. But at 510 or 610 nm the luminous efficiency has fallen to half, so that light at these wavelengths would have to register only 340 lux to give 1 W/m^2. Thus the luminous efficiency curve allows us to convert radiant flux into luminous flux and vice versa, provided that we have a picture of the spectral power distribution of the radiant energy like those shown in *Figure 59* (*see also Table 28*, p. 196).

For our purposes a good light meter should have the following characteristics:

1 Its luminous efficiency curve should match that of the human eye as closely as possible (*Figure 63*).

2 It should receive light from all directions as a bare surface does. This may seem simple, but in fact the protective glass over a bad meter prevents light at low angles from reaching it due to reflection. The meter must be, as it is termed, cosine-corrected (a reference to Lambert's Cosine Law which relates the illuminance of a surface to the angle which the rays make with the perpendicular to the surface). In practice there are two kinds of cosine correction. In the first the protective glass is replaced by

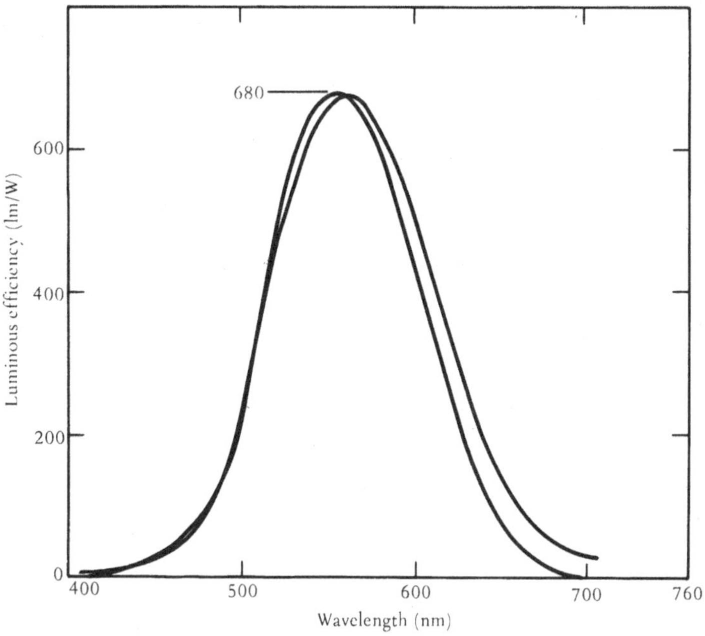

Figure 63
The luminous efficiency curve for the human eye (the phototopic curve for cone or daylight vision): the eye as a machine for converting radiant energy into sensation. The curve peaks at 680 lumens per watt. The colour line shows for comparison the luminous efficiency of the corrected selenium photocell type Megatron M

an acrylic dome (*Figure 9(a)*). In the second a layer of diffusing material is used.

3 It should have a series of ranges so that readings can be taken both inside and outside the museum.

SOME BASIC LIGHT UNITS

Some of the units related to luminous flux are not easy to grasp without study, and the situation is made more difficult because terms have only recently become standardised, so that the reader may find different ones in the older textbooks.

The stark minimum needed by the scientific conservator are: the *luminous intensity* of a light source, the much discussed *illuminance* of the light falling on an exhibit, and the *luminance* of the light source itself or of the exhibit regarded as a source of light. These are all rigorously defined terms, but it may help if we work round to them informally.

A light source gives out energy as a stream of photons, and the radiant flux is a measure of the power of these photons. But all the above three light units are concerned with luminous flux. The relation between these two measurements, radiant and luminous flux, has been given in the last section.

If we draw lines of luminous flux radiating from a point source, some of them passing through a hole, we can measure

Figure 64
A light beam from a point source emerging from a hole. Only the outermost rays (lines of flux) are shown. In three dimensions the angle of divergence is a solid angle, measured in steradians

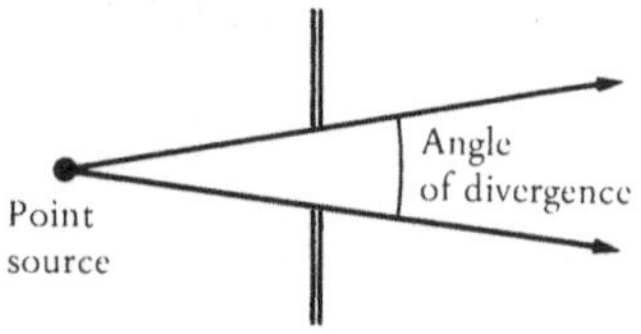

the angle of divergence of the light beam emerging from the hole (*Figure 64*). But our diagram is drawn on flat paper and the real situation is three-dimensional. We need the *solid angle* of the cone of light. Just as two-dimensional angles are measured in degrees or radians, there being 2π radians in one revolution round a point, solid angles are measured in *steradians*. The complete solid angle round a point in three dimensions measures 4π steradians. If our hole has an area of 1 m^2 and is 1 m away from the point source, the cone of light which emerges has a solid angle of one steradian.

We may now measure the luminous intensity of the light source by metaphorically counting the number of luminous flux lines in a unit solid angle of one steradian. (This imaginary operation is made possible by using a standard lamp to define the flux and comparing all other measurements to it.) Luminous flux per unit solid angle is therefore measured in lumens per steradian. Lumens per steradian are called *candelas* (cd). By following this system a little further, the luminous intensity in any direction of an extended source such as a fluorescent lamp can be calculated.

From here it is an easy step to the most important measurement for us, *illuminance*, for this is simply a measure of the lumens falling on a square metre, which are called *lux*.

The third unit, *luminance*, is not so simply grasped. It refers to the luminous flux given off by an area, and this area can either be part of a light source like a fluorescent tube or a patch of colour which is giving off light by reflection. Put two patches of colour side by side. Whatever their colour one will appear brighter than the other, and this is the patch of higher luminance. But luminance is not just a matter of comparison, it is a matter of measurement. For simplicity we shall ignore self-luminous sources (light sources) and consider only perfectly diffusing (matt) reflecting surfaces. Their luminance can be measured in two ways. Of attractive simplicity is the *apostilb* unit of measurement. The luminance of a matt surface in apostilb is simply the illuminance falling on it in lux divided by its reflectivity. Illuminate at 50 lux a matt surface which reflects 25% of the light which falls on it and its luminance is 0.25×50 apostilb. Unfortunately the more rigorous textbooks call this not luminance but luminous exitance or luminous emittance and state that true luminance should be measured in candelas per square metre (cd/m^2). If we want to convert from apostilb to cd/m^2 we have merely to divide by π. Thus an illuminance of 0.25×50 apostilb $= 0.25 \times 50/\pi$ cd/m^2.

An example. A 100-watt tungsten lamp emits about 1200 lumens more or less uniformly in all directions, and can be regarded as a point source. Its luminous intensity is then 1200/ 4 π candelas (lumens per steradian) or about 95 cd. Now at 1 m from the lamp one steradian covers 1 m^2 by definition. Therefore an exhibit 1 m from this reflector-less lamp will receive an

TABLE 20. Three Basic Light Units

Unit (abbreviation)	*Measures*	*Meaning*
Candela (cd)	Luminous intensity	Lumens emitted from a point source per unit solid angle (definition can be extended to large sources)
Lux (lx)	Illuminance	Lumens incident on a square metre
Apostilb (asb)	Luminance (strictly Luminous Exitance)	Lumens reflected from a square metre of matt surface, or lumens emitted from a square metre of diffusing source

Notes
1 lux = approx. 1/10th lm/ft^2 or foot-candle.
1 asb = 0.318 cd/m^2 = 0.0001 lamberts = 0.093 foot-lamberts.

illuminance of 95 lux. If this object has a matt surface with a reflectivity of 25% its luminance will be 0.25 × 95 = about 24 asb. Move the exhibit twice as far away, to 2 m, then by the inverse square law both *illuminance* and *luminance* must be divided by 4, to 24 lux and 6 asb respectively.

TABLE 21. An Example

Electrical power input to lamp		100 W
Light emission		about 1200 lm
(Luminous efficacy	= 1200/100	12 lm/W)
Luminous intensity	= 1200/4π	95 cd
Illuminance at 1 metre		95 lux
Luminance of a matt surface with 25% reflectance illuminated by 95 lux = 0.25 × 95 =		24 asb

These are in fact characteristic figures, but of course as soon as a reflector is put behind the lamp its 1200 lumens will all be directed forward, giving a much higher illuminance.

All these three units – candelas, lux and apostilb – can be measured with instruments, although they refer to light as seen by the eye. The term 'luminosity' is now reserved for something that cannot be instrumentally measured: the subjective perception of luminance in a given situation, as it is affected by surroundings, adaptation etc. Colloquially luminosity would be called brightness, but the term brightness is no longer allowed in technical lighting discussion.

VISUAL PERFORMANCE

It was said in Part 1 (p. 61) that neither visual acuity nor colour discrimination is likely to be at its maximum however the museum is lit, and that neither the visitor nor the artist would want this to be so because maximum acuity is only needed for finely detailed and demanding tasks such as restoration. The large body of work which has been done to measure visual performance under different conditions of luminance, size of object and contrast, is therefore barely relevant to museum exhibition.

Whether we can see a difference between two adjacent matt grey patches depends on the visual *contrast* between them, and this can be measured. If one of the patches has a luminance of L asb (luminance in apostilb = illuminance × reflectance), and the adjacent patch has a luminance of $L + dL$ asb, then the contrast between the two is dL/L. The greater the difference the higher the contrast, up to a maximum, which could be measured. We could turn this upside down and say that, if a contrast of dL/L can just be seen, the contrast *sensitivity* can be expressed as the reciprocal, L/dL. Thus in a situation where sensitivity is high, contrast need only be low for objects to be distinguishable.

Now we must imagine a whole set of test-patch pairs, some of high contrast, some of low, and a test situation where we record the lowest contrast visible at a particular light level. As the amount of light is increased there comes a point at which we see all the detail we will ever see, however bright the light. Our sensitivity at this point, measured by L/dL for the just visible difference, is then the maximum contrast sensitivity. Dim the lights again and find the test-patch pair that can just be seen at this lower level. The contrast sensitivity will be lower. We obtain the *Relative Contrast Sensitivity* (RCS)% at this point by dividing the contrast sensitivity by the maximum contrast sensitivity previously obtained.

The universal curve so obtained (*Figure 65*) expresses RCS on the basis that no further sensitivity is gained above a luminance of 10 000 cd/m² (10 000 π apostilb). That is to say, RCS = contrast sensitivity at a given luminance divided by contrast sensitivity at 10 000 cd/m². It will be seen that the curve falls gradually and smoothly from this maximum. For a painting of 20% average reflectance 10 000 cd/m² represents an illuminace of about 150 000 lux, so that, whether we illuminate at 50 or 500 lux we are far down to the left on the curve.

Figure 65
Universal Relative Contrast Sensitivity (RCS) curve showing how the sensitivity of the eye to small differences in contrast falls smoothly from a maximum at 10 000 cd/m² (about 30 000 asb). (Adapted from Henderson[236])

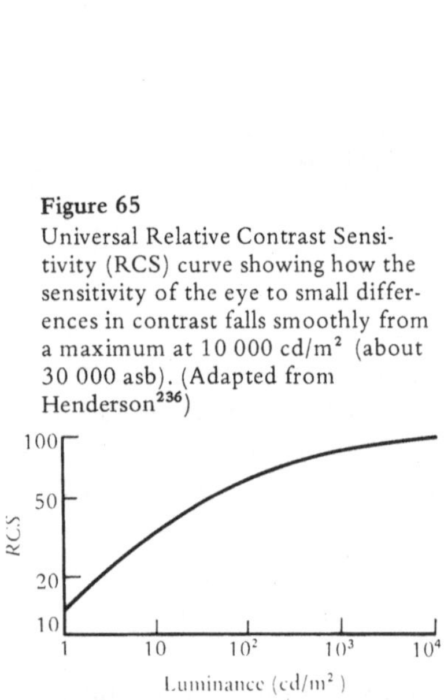

It is clear that this is very much a theoretical maximum, well beyond the illuminance required for even the most exacting task.

Brommelle[76] has discussed the trade-off, when the illuminance is raised, between the greater RCS so obtained and the greater damage caused by the light. He finds that under practical conditions the gain in RCS is likely to be more than offset by the increased rate of damage.

As for finding a formula which will actually guide our choice of illuminance, the curve of *Figure 65* leaves us as much in the dark as before.

Some of the factors affecting real lighting situations have been discussed. It may be worth ending this section with another.

It should be evident that, if an artist paints a dark picture or if a costume designer blends two near-matching tones, they are not creating things that thereby require more light for viewing. If an artist paints a 'high-key' portrait, this is not because he expects his picture to be especially dimly lit. An artist intuitively spaces out his tones to get the effect he has in mind under what he regards as ordinary lighting. Yet it has been maintained that Rembrandts ought to be especially brightly lit. If we must put this on a scale we are saying that RCS need not – should not – be so high for dark-toned exhibits as for light (though exceptions could be found where colours have faded or darkened). In other words the luminance scale becomes very elastic and loses usefulness, and we move yet further from being able to make use of visual performance data.

LUMINANCE AND SUBJECTIVE BRIGHTNESS

Having regard to what has just been said about the term brightness not being allowed, this subtitle should read *Luminance and Luminosity*. But it then really requires an unnecessary amount of thought to disentangle the meaning. The question at hand is, how many more lights do we have to switch on to make a room look twice as bright? Most people would realise that we would require more than twice the number of lights, but would stop at that stage. Physicists will recall a 'law' called Fechner's Law which was said to apply to all the stimuli we receive through our senses: sight, sound, touch, taste, smell. Fechner's law stated that the subjectively felt strength of the sensation was proportional, not to the strength of the stimulus but to the logarithm of its strength[237]. This means that, to receive twice the sensation you have to square the strength of the stimulus. Two pneumatic drills do not nearly double the sensation of the noise from one. According to Fechner you have to square the amount of noise energy reaching your ear. Although the decibel scale of sound (*see* p. 160) is based on Fechner's logarithmic law, this no longer appears to be quite correct[238]. However our interest is in light.

Applied to light, Fechner's law would imply that a room lit at 2500 lux looks only twice as bright as one at 50 lux ($2500 = 50^2$). But experiments have shown that the illuminance difference for brightness doubling is not so great as this, and is more likely to amount to something like 10 to 1. It may be stretching the experiment beyond what was intended but this could imply that, standing at the door adjoining two rooms, one lit at 500 lux would look twice as bright as one lit at 50 lux.

But it might be objected that to say one room looks twice as bright as another is pure guesswork. There appears nevertheless to be some kind of physiological scale of this kind which can be disclosed by experiment. Set up three boxes with identical objects in them in front of the experimental subject. Light the one on the left at 50 lux and the one on the right at 500 lux. Using a light meter we would light the middle box at the average between 50 and 500 = 275 lux. But Hopkinson's work suggests that without a meter, using our eyes alone, we would choose about 190 lux for the middle box*.

However, once again adaptation upsets the figures. If we want 50 lux to look satisfactory we must avoid just the situation described, where the observer sees two rooms lit at different levels. Without this direct comparison the subjective brightness scale loses its meaning. In a sense this must be our aim, for we must not give the visitor a chance to apply his brightness scale between the exhibition room lighting and some other overbright area.

THE BLUE WOOL STANDARDS

By using the Blue Wool standards it is possible to estimate the lightfastness of a dyed fabric or paint. This measurement does no more than grade the material on a scale from 1 (fugitive) to 8 (of good lightfastness); it cannot give us a very good idea of

*Hopkinson[239] suggested a relation between stimulus L (the luminance) and sensation M of $M = k\,L^{0\cdot 3}$.

how much exposure to light the material will stand in any situation.

The Blue Wool standards have been adopted as ISO (International Organisation for Standardisation) Recommendation R 105 and British Standard BS1006 (1961), so that sample cards are readily available. Each card contains 8 specially prepared blue dyeings on wool. They are so chosen that standard number 2 takes roughly twice as long to be perceptibly faded as standard 1, standard 3 roughly twice as long as standard 2, and so on through to standard 8.

To rate the lightfastness of our material we expose it together with a card of the Blue Wool standards, and from time to time check both our material and the standards for first signs of fading. This can most easily be done if one half of each patch of colour is covered with an opaque card throughout the test.

Attempts have been made to measure how much light exposure is required to fade the standards[240]. They have met with little success, since rates of fading are related to other factors besides the light, such as proportion of UV[241], humidity, etc.. The light may appear to be the same but the standards may be found to fade at quite a different rate, although they keep more or less in rank.

In the museum, however, we can limit our interest to an indoor situation where extremes of temperature and humidity are avoided and all the light comes through glass though without specifying UV-filtering. For this special situation Feller[242] has found that the blue wool standards can be very useful in grading material into three categories (*Table 22*). It should be noted that the lifetimes estimated in the table are for an average annual exposure of about 1½ million lux hours (1½ Mlx h). Under conditions controlled to 150 lux the annual exposure is about ½ Mlx h so that the figures could be multiplied by three. With no UV the multiplying factor would be higher still: six or more.

TABLE 22. Standards of Photochemical Stability for Materials in Conservation

Class	*Classification*	*Intended useful lifetime*	*Approximate equivalent standard of Photochemical stability*
C	Unstable or fugitive	Less than 20 years	BS1006 Class 3 or less
B	Intermediate	(20–100 years)	(3 to 6)
A	Excellent	Greater than 100 years	Greater than BS1006 Class 6

DAMAGE VERSUS WAVELENGTH

The potency of radiation for photochemical damage is related to its wavelength, the shorter the wavelength the more damaging the radiation. This follows from the relation between the wavelength and the energy of the photon (the quantum of light), where the energy of a mole of photons (6×10^{23} photons) is $119560/\lambda$ kilojoules (λ = wavelength in nm). But we cannot assume a direct numerical relation of this nature between damage and wavelength since each substance has its characteristic response to the colour of the radiation falling on it, absorbing some wavebands and reflecting others, sensitive to some but not to others.

However the possibility of arriving at some overall average for guidance in the museum has been tempting, and such a search is legitimate. In 1953 Harrison published a report[243] which was influential in museum circles. In it he proposed a 'probable relative damage' factor for wavelengths between 300 and 660 nm. The factor ranged from a maximum of 7.75 at 300 nm down to zero at 660 nm. This so-called Harrison damage factor was calculated from the damage caused to low-grade paper by light, and was proposed by him as an interim measure for calculating the relative effects of UV and of the different kinds of light sources on museum objects. To work out the relative amount of damage caused by different light sources all one had to do, after adjusting the spectral energy figures to equal illuminance, was to multiply the energy at each wavelength interval by the damage factor for that wavelength and sum the result. In this way, for example, the light from an overcast sky through glass was found to be five times as damaging as light from a tungsten lamp at the same illuminance.

Feller[244, 245] has pointed out that in the UV range the deterioration of rubber, the erosion of paint and the cross-linking of certain acrylic polymers follows a similar pattern to that for Harrison's low-grade paper, and that log (Harrison Damage) plotted against wavelength is close to a straight line with negative slope[246].

Let us then, *faut de mieux*, use this log relation until something better is found to replace it. But Harrison's calculation can be greatly simplified as follows.

We wish to compare light sources of different colour temperature, from which the UV has been removed where appropriate, as it should be, for their relative power to damage museum material. Divide the visible spectrum into five 50 nm bands from 400 to 650 nm and allot relative damage values

TABLE 23. Simplified Harrison Damage Factors

Waveband (nm)	D_λ *(rel. damage)*	V_λ *(lum. effic.)*
400–450	100	0.008
450–500	24	0.115
500–550	5.6	0.766
550–600	1.3	0.911
600–650	0.3	0.323

from the log scale (*see Table 23*, which also shows the appropriate luminous efficiency, V_λ, figures). Next read off the relative energy (E_λ) of the light source in question from its spectral energy distribution at the centre of each band, and multiply each of these five figures by its appropriate V_λ (*Table 24*). The sum of the five $E_\lambda V_\lambda$ products gives a relative measure of luminosity. But all light sources must be compared at the same illuminance, meaning here the same luminosity, i.e. the same $\Sigma E_\lambda V_\lambda$.

TABLE 24. Damage Factors of Daylight and Tungsten Compared

	425	*475*	*525*	*576*	*625nm*
E_λ for D6500 daylight	902	1154	1063	960	854
$\Sigma V_\lambda E_\lambda = 2104$					
E_λ for black body at 2850 °C	22.74	45.46	75.76	110.85	147.4
$\Sigma V_\lambda E_\lambda = 212$					
E_λ for 2850 °C normalised to $\Sigma V_\lambda E_\lambda = 2104$	226	451	752	1100	1463
$\Sigma E_\lambda D_\lambda$: D6500 = 125353; 2850 °C = 39504					
Damage ratio D6500/2850 °C = 125353/39504 = 3.2					

Table 24 compares standard daylight to tungsten light, and proposes that standard daylight (with the UV removed) can be expected to be just over three times as damaging as tungsten illumination. Whether or not the UV is included in tungsten makes negligible difference. One suspects that this result may be truer for moderately stable materials than for very fugitive dyes. This suspicion is based, for example, on some work by Kühn[7], but also on the extensive work of MacLaren[8].

In 1956 MacLaren published a diagram summarising the proportion of fading caused by the visible radiation in sunlight in about 100 modern dyes of all grades of lightfastness (*Figure 66*). MacLaren's illuminant was total sunlight. Sunlight through glass (which removes UV at 300–325 nm) will give relatively higher figures for fading by the visible portion of the spectrum. Harrison's and MacLaren's figures can be made to agree if we regard

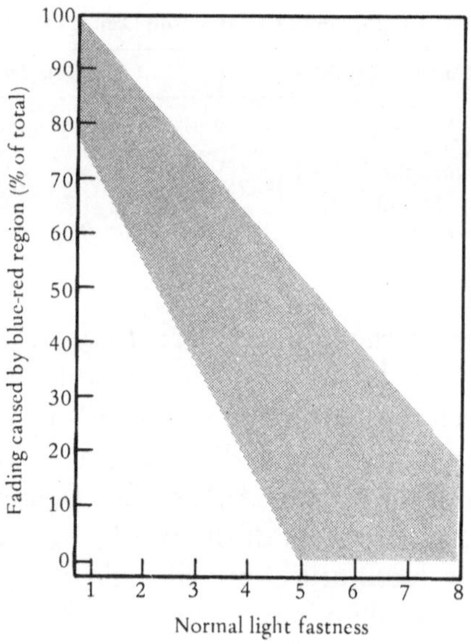

Figure 66
McLaren's diagram showing the relationship between fastness of a dye on the Blue Wool scale (x-axis) and the proportion of the fading caused by visible radiation (blue to red region) (y-axis). The diagram is for daylight. Only about 20 out of a total of 117 dyes tested by McLaren[8] gave points outside the shaded area. This modification of the diagram is due to Giles[6]

the fastness of the average museum material susceptible to light as grade 6 on the Blue Wool scale, a far from ridiculous assumption. In this case we can allot about a quarter of the damage to the visible radiation in sunlight.

But for more fugitive materials, which category includes many textile dyeings, the figure for visible radiation would be higher. On the other hand, colourless polymeric materials of good stability, especially modern synthetics, and the fastest dyes are probably affected only by UV. Thus it can be seen that no single figure can be given for damage versus wavelength.

HEAT RADIATED FROM LIGHT SOURCES

Electric lights convert an input of electrical energy into an output which is a combination of (a) visible radiation, (b) UV and (c) IR radiation, and (d) heat conducted and convected away from the light fitting and control gear.

The *luminous efficacy* (note: efficacy not efficiency) of a light source is defined by electrical engineers as the output of visible radiation in lumens divided by the input of electrical energy in watts: lumens per electrical watt (lm/W), *Table 25*. But this is not quite what we want for our present purpose.

A worry for all conservators is the extent to which different kinds of light source radiate heat and thereby warm the exhibits above general room temperature. Something on this has been written on p. 43. Our purpose here is to collect figures on the

TABLE 25. Conversion of Electrical Energy by Light Sources[236]

	(1) Visible	*(2) UV*	*(3) IR*	*(4) Remaining heat*	*(5) Efficacy (lm/elect.W)*
100 W tungsten incandescent lamp	5.75%	about 0.1%	75%	20%	12.6
Typical 80 W 'Daylight' fluorescent lamp	23%	about 1%	30%	46%	62

heat radiated with the light in various light sources. We cannot use efficacies (a) because the heat conducted and convected from the light source should not reach the exhibit except as general room heat, and this can be cancelled out by thermostatic control, and (b) because we want to bring daylight into the comparison.

It is at least worth considering that the daily cycle of radiant heating followed by darkness may be relevant to deterioration processes.

For this purpose we compare light sources by finding, not their luminous efficacies (lumens per electrical watt, lm/elect.W) but their lumens per *radiant* watt (lm/rad.W). *See Table 26.*

TABLE 26. Radiant Heat and Light (Lumens per Radiant Watt)

	lm/rad.W	*(lm/elect.W)*
Sunlight from which all UV and IR has been removed (theoretical)	220	–
Daylight through glass, about	130	–
High-efficacy fluorescent lamp	130	(70)
Low-efficacy (high fidelity) fluorescent lamp	85	(45)
Tungsten lamp with dichroic reflector ('Coolbeam')	40	(about 12.6)
Tungsten–halogen lamp	20–25	(15–22)
Ordinary tungsten lamp	16	(12.6)
Metal halide lamps under development	130 upwards	?

The 'Coolbeam' lamp is a tungsten lamp of normal colour temperature but with an integral reflector which has a dichroic coating designed to reflect only the visible light forwards while the IR is transmitted through the reflector to the back of the lamp. I have not been able to get the complete curve of radiation for this lamp, but heating experiments indicate that it might give a figure approaching that for a low-efficiency fluorescent lamp. In other words the 'Coolbeam' reflector spotlights are a most useful addition to lamps suitable for the museum.

The tungsten-halogen lamp run at its usual higher colour temperature than the normal tungsten lamp, though perhaps thus rendered 50% more efficient in lumens per radiant watt, does not approach the other light sources in coolness.

What do these figures mean in practical terms? If one lamp emits half as much lumens per radiant watt as a second lamp it will deliver twice the amount of heat at a given illuminance. But the rise in surface temperature of an exhibit will not be twice as much since the rate of cooling is greater the higher the temperature. A few figures for actual temperature rises are given on p. 43. Bearing this cooling factor in mind, the figures in *Table 26* can be regarded as heat ratings for light sources.

Lumens per electrical watt have been included in *Table 26* for the purpose of cost comparisons. Taking into account only the electricity consumed, and ignoring capital and replacement costs, the cost of electricity to the various light sources is inversely proportional to their efficacies. The most obvious comparison is between ordinary tungsten and high fidelity fluorescent. From the table the ratio can be seen to be 45/12.6 = about 3½.

It is relevant to the museum situation that people are satisfied with a 25% reduction in illuminance if high-efficacy (poor colour rendering) fluorescent lamps are replaced by low-efficacy (good colour rendering) lamps[247].

ACTIVATION ENERGY

Once we start examining the chemical processes of deterioration we become interested essentially in their rates. Basic to an understanding of rate processes, not only chemical but physical (e.g. diffusion), is the Arrhenius equation, which states that the rate of a reaction or process is proportional to a function involving two variables: temperature and activation energy. Before we express the Arrhenius equation we need to take a further look at the second of these variables, activation energy.

At the level of individual molecules, energy is acquired or relinquished in separate packets or quanta, so that the energy of a molecule is raised or lowered in jumps. If a molecule receives a quantum of energy large enough for it to react, this almost invariably occurs before a second quantum can be added to the first, so that we are considering single quanta. We can therefore say that, for a molecule to react, it must receive energy above a certain minimum in one quantum. This minimum is the activation energy.

We can picture each molecule as if it were a ball lying in a hollow. This is the 'ground state' of the molecule. It will stay there for ever unless pushed out. A push is the arrival of energy from elsewhere, a transfer of energy to the molecule. Surrounding the molecule are others of the same or different types, so that it has a number of possibilities for reaction. Each has its characteristic activation, high for an unlikely reaction, low for an easy reaction. Thus the basic necessity is that the energy which knocks it out of its ground state into an 'excited state' is at least as great as the activation energy for a particular reaction. Even then several other factors operate to prevent molecules from reacting, and there are many ways for it to lose its energy and to revert to its original ground state.

Since in the exhibition room all chemical reactions are deleterious, we must try to prevent them, and one fundamental way is to ensure that activation energy levels for all possible reactions are not exceeded. This means looking at the sources of energy available.

In the dark the only available energy, so long as high-energy radiation is at its normal low level, is in the form of heat quanta. As we have seen, however plentiful these are they will be impotent to cause chemical change unless a significant proportion of them have energies in excess of the activation energy of the easiest route to deterioration. But at any given temperature there is a spread of quanta over quite a range of energies, so that we cannot give a hard figure for the activation energy which must be exceeded for a material to be stable at, say, room temperature. In actual practice one would be unhappy if the minimum activation energy were much below 100 kilojoules per mole (kJ/mol)[248].*

The unit for the measurement of activation energy, kilocalories per mole, needs explanation. If we talk in terms of the energy changes of an individual molecule, these are of course very minute fractions of a calorie (cal). Chemists work in quantities one can handle, and the most convenient unit to them is the mole (mol), which is the molecular weight in grams. Thus 1 mole of water is a weight of 16 + 1 + 1 = 18 grams. The particular convenience of this is that, whatever the substance, one mole of it always contains the same number of molecules, actually about 6×10^{23}. Therefore when we speak of an activation energy of 100 kJ/mol we mean that the activation energy which must be given to each molecule for reaction is 100 000/(6×10^{23}) joules.

Now in contrast to heat quanta, which at a given temperature are spread over quite a range of energies, the energy of a photon of radiation is precisely defined by its wavelength, using Planck's formula. The energy of a photon of wavelength λ nm = 119 700/λ kJ/mol. This gives us a range of photon energies in the visible

*Cellulosic materials such as wood[249] and paper[250] have activation energies between 100 and 125 kJ/mol. Proteinaceous material is in the 125-170 kJ/mol range[251] Whereas pure cellulose fibres are stable in the dark, poor quality paper yellows and embrittles.

from about 150 kJ/mol in the far red to about 300 kJ/mol at the violet limit, and a more powerful set in the glass-penetrating UV of from 300 to 400 kJ/mol (*Table 27*). To get a usable supply of heat quanta in this UV range one would have to heat up to about 200 °C[252]. This emphasises the power of light compared to heat.

'Photolysis', that is the breaking of a molecule directly by the absorption of energy, certainly occurs with UV radiation of wavelength shorter than 300 nm, though it is probably unimportant with material of museum stability under radiation which is never in this range.

As with heat quanta, the presence of photons with the necessary energy for activation is not a sufficient condition for reaction – even less so, indeed. The photon must either be absorbed by the molecule itself or absorbed nearby and the energy transferred. The molecules of a colourless material, by reason of its colourlessness, absorb very few visible light photons but may absorb in the UV. A dye is coloured by virtue of the fact that its molecules absorb visible light photons of certain wavelengths only. But, whatever the case, even if the right energy is absorbed by the right molecule, it may very likely be lost before reaction can take place.

So, roughly speaking, at room temperature in the light there is a range of quantum energies available which, at the lower end, up to about 125 kJ/mol, will be predominantly heat quanta, and above this, up to about 350 kJ/mol, will be photons. That red light photons take over just about at the point where heat quanta become scarce is no coincidence. The eye builds up a picture of the world by receiving photons. If it had tried in the course of evolution to respond to photons beyond the red, in the IR (which would be an advantage to most animal species), it would render itself sensitive also to heat quanta of this energy, of which there is an adequate supply at blood heat (37 °C). This would blur the image. In the language of communication theory we could say that the light signal would be degraded by heat

TABLE 27. Wavelength and approx. Activation Energy

Wavelength	*300*	*400*	*500*	*600*	*700*	*760 (nm)*
Activation energy						
kcal/mol	95	72	57	48	41	38
kJ/mol	400	300	240	200	170	157
electron volts	4.1	3.1	2.5	2.1	1.8	1.6

Note
For activation energies in kcal/mol; kJ/mol; and electron volts divide the following by the wavelength in nm: 28 650; 119 700; 1240 respectively.

'noise'. Photographic film is limited in the same way, though its sensitivity can be extended to the very near IR if it is stored at low temperature. Infrared TV cannot get far beyond 1500 nm without cooling.

We have already recognised that museum material is at least pretty stable in the dark, with a few possible exceptions such as poor quality paper, and we can go a little further. Since our material is exhibited without immediate destruction by light we can also afford to forget about the weakest photons: those at the red end of the spectrum. The most fugitive material may be a little sensitive to yellow light, but generally speaking we can confine our attention in practice, firstly to the UV, and secondly to the shorter half of the visible spectrum. A reminder of MacLaren's generalisation (*Figure 66*) is appropriate: that every kind of organic substance, whether dye or medium or support, has its threshold wavelength. All wavelengths shorter than this threshold can cause damage if absorbed. The more stable a material the shorter this threshold wavelength will be. Thus fugitive material is destroyed by both visible and UV radiation, whereas fast material is likely to be susceptible only to UV.

The organic material in a museum is complex, consisting of a great many different kinds of molecule, mostly large. Furthermore it is usually the case that a reaction which is caused by UV or visible radiation seldom takes place in a single step – a chemical change to a new stable product caused directly by the absorption of radiation. The new chemical species formed photochemically is likely to be unstable enough to be affected by heat quanta, making further reaction dependent on temperature. And so on, until a stable situation is reached. The first and most important step, the photochemical reaction, is independent of temperature, but not necessarily the subsequent ones. The net effect is to make photochemical reactions in general rather temperature insensitive but not completely so.

Because of their complexity we can never hope to unravel most of the chemical routes to deterioration, which makes any valid generalisation important as a guide to conservation action. Thus, looking at the polymeric materials which make up the bulk of cellulosic or proteinaceous support material and of paint media and synthetic varnishes, we can say, firstly that under light they will deteriorate in a number of stages, nearly all leading in the direction of oxidation. These reactions will not be very sensitive to temperature, making temperature control in the museum more a matter of comfort for visitors rather than exhibits. Secondly, the most likely outcome is that a molecule splits off a fragment which evaporates or diffuses away. This kind of process leads to contraction and embrittlement. Molecules may also cross-link together, and though this is less damaging to their bulk properties, it is important in restoration materials which ought to remain soluble[253, 254].

The reciprocity rule (*see* p. 21) now becomes comprehensible. Because we get the same number of photons in each case, we can halve the time of exposure and double the amount of light to get the same degree of deterioration. However this needs watching. It has been stated that a photochemical reaction involves the absorption of one photon only, since reaction, if it occurs at all, occurs so fast that there is no time for a second photon to arrive at the same site. This rule breaks down under two conditions: (a) Very high light intensities, way beyond the museum range, where photons are so dense that an excited molecule may have its energy raised even higher by the absorption of a second photon. (b) Some excited species do *not* have short lives, and these are rather likely to be encountered in museum material. Once again, the conditions for two-photon absorption are present. Though this is so far completely hypothetical, the implications are important. Firstly, two visible light photons can confer energy by addition in excess of anything in the 'daylight-through-glass' UV, and therefore potent. Secondly, rate of damage on a pure two-photon basis becomes proportional, not to the light intensity itself but to its square, thus giving an added bonus to low light levels.

This section closes as it opened with the Arrhenius equation, relating temperature and activation energy to rate of reaction. It is usually expressed in two forms:

1 Reaction rate is proportional to the exponential,

$$\exp\left(\frac{-120E}{T}\right)$$

where E is the activation energy in kJ/mol
and T is the temperature in K (= °C + 273)

or alternatively

2 If the rate of reaction is measured at two temperatures T_1 and T_2,

$$\log_{10}\frac{\text{Rate}_1}{\text{Rate}_2} = 52\,E\,(1/T_2 - 1/T_1)$$

By putting the appropriate figures into the second form of this equation the old chemical adage that reaction rate doubles for a 10 °C rise in temperature can be seen to be true enough for the chemical laboratory (activation energies around 50 kJ/mol) but of little value in the museum. On the one hand cellulose only requires a rise from 20 °C to 25 °C for rate doubling (activation energy about 100 kJ/mol), so that cool storage can confer considerable benefit. On the other, the dark reactions which follow photochemical change might typically have activation energies around 10–15 kJ/mol, requiring a rise from 20 °C to over 70 °C for rate doubling, so that materials stable in the dark at 20 °C will remain unaffected by large changes in temperature while illuminated for exhibition.

THE PRIMARY PHOTOCHEMICAL REACTION

The oxidation of museum materials by UV radiation and light (photo-oxidation) involves a huge variety of chemical reactions. Each reaction path, and there may be dozens in a single material, consists of several stages. Few of these are known in detail.

The first three stages, however, have a certain uniformity.

1 Absorption of radiant energy. This is conventionally illustrated by the following equation:

$$R + h\nu \rightarrow R^*$$

where R is any organic molecule, in our case usually long-chain or polymeric; $h\nu$ is the energy of the single absorbed photon of frequency ν (or wavelength c/ν, where h is Planck's constant and c is the velocity of light). The asterisk * indicates that the absorption of one photon by R has rendered it electronically excited, that is to say one of the electrons in the molecule has been raised to a higher energy level.

2 The formation of a free radical, represented by $R^\bullet$, either at the site of absorption or elsewhere, even on another molecule. This is a matter on which ideas are changing rapidly. For a review of developments *see* Mills and Allen[255,256]. The chemical reader should be warned that until recently much work was concentrated on irradiation by UV of wavelength shorter than 300 nm. To the conservator such work is useless to the extent of being misleading. By contrast more recent research, concerning for example initiation by singlet oxygen, intermolecular transfer of energy, long-lived radicals, and two-photon

reactions, all stress mechanisms possible under near UV and visible radiation.

3 As soon as the free radical is formed there is a very high probability indeed that it will immediately combine with an oxygen molecule as follows:

$R^{\bullet} + O_2 \rightarrow RO_2^{\bullet}$ (a peroxy radical)

The attraction is so high between the two that this reaction is never likely in museum conditions to be rate-determining. (When a reaction proceeds by several stages, as with the types we are considering, some will go much easier than others, and the slowest becomes the rate-determining step, as in a traffic jam where the speed of progress is determined by one bottleneck.) No light is required for this reaction.

Two important points need emphasising. Firstly, in a slowly reacting organic surface up to a few hundred microns thick, rate of oxidation will not be limited by diffusion of oxygen. That is to say in the museum a varnish will not protect an underlying layer from oxygen. Secondly removal of oxygen from a museum case for conservation purposes will have little effect on deterioration unless the technology is good enough to exclude all but a few parts per million of oxygen. This conclusion[19] has recently been confirmed in practice[4]. Note that both the conclusions apply to museum conditions, but not necessarily to all others.

After this combination with oxygen the foremost photochemists in the world will be able to give no more than certain outlines of likely reactions, though some interesting stable products can be identified, in the gas chromatography of oil films for example[257].

In general an organic material exposed to light passes through three periods. In the first, often called the 'induction period', exposure to light appears to be having little effect. In fact light is being absorbed and reactions are occurring. But they are of a kind that does not appreciably affect the strength or colour of the material, and are often due to impurities or irregularities in polymer structure which take on the burden of photo-oxidation, thus temporarily protecting the polymeric structure from degradation. The length of the induction period is very variable, and very little work has been done on museum material. It may be absent altogether. Feller[258] found an induction period of 11 years on a well-illuminated wall before *n*-butyl and *iso*-amyl methacrylates started to form insoluble material.

In the second period the adventitious impurities which have been protecting the polymer have become exhausted, and so photo-oxidation of the polymer starts and accelerates. Finally, in the last stage, deterioration slows down because most of the damage that can be done has been done.

PLACING A COLOUR ON THE CIE CHROMATICITY CHART

From time to time we are faced in the conservation literature with a diagram of colours known as the Chromaticity Chart or Standard CIE Diagram[259]. This has already been briefly described in Part I (p. 55). We will now find how to place a colour on the diagram.

A colour can only be seen when light shines on it (unless it is self-luminous) and this diagram concerns seen colours. Therefore we must illuminate our test colour. Let us here take as our test colour some yellow-green grass in the foreground of a famous painting by Uccello, *St. George and the Dragon*, which hangs under daylight in the National Gallery, London (*Plate 16*). The reflectance at the point X is shown in *Figure 67*. From now on we need not handle the painting, nor even look at it. We handle only numbers.

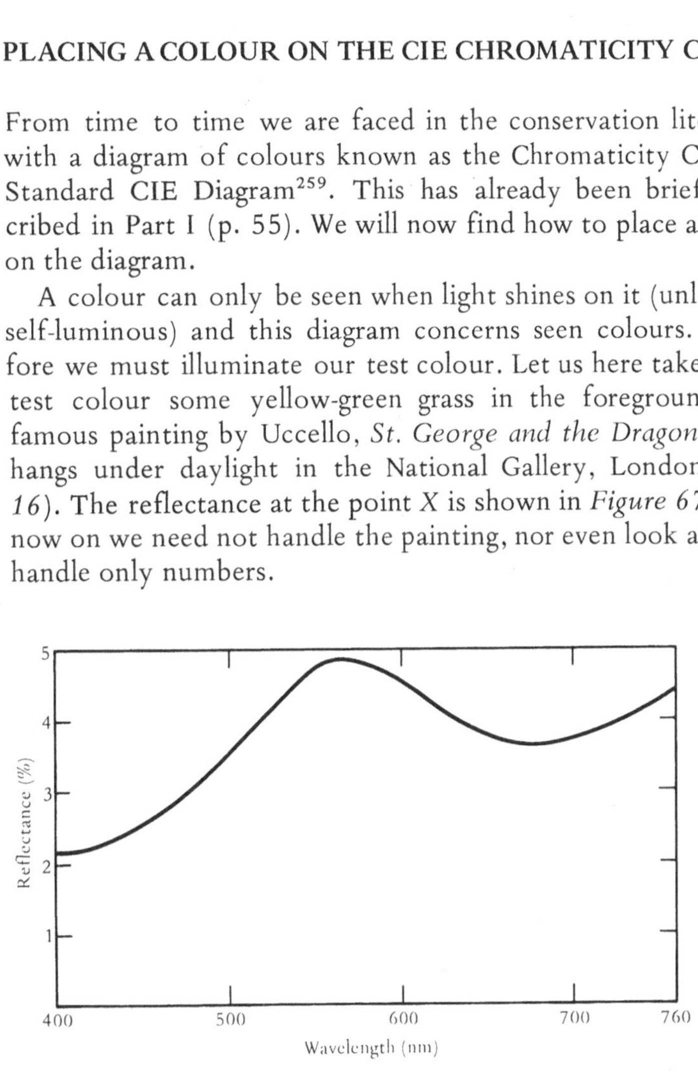

Figure 67
Reflectance curve of the grass on the Uccello at the spot marked X (diameter 4 mm) on *Plate 16*. Reflectance values for this curve will be found in column (3) of *Table 28*

Daylight is well represented by an average daylight spectral energy distribution known as D65[69] (*Figure 59(d)* and column (2) in *Table 28*). To illuminate our painting mathematically we have to multiply these two curves together, the reflectance curve and the D65 curve. We do this by multiplying each reading on one curve by the corresponding reading at the same wavelength on the other curve (*Table 28*, columns 2–4). We now have a curve for the light coming from the green grass on the Uccello, hanging in a daylit gallery (column 4). This is what we require.

TABLE 28. Placing a colour on the CIE Chromaticity Chart

Wave-length (nm)	*D65*	*Green refl.* (%)	*Refl.* × *D65* = *E*	*Colour-matching functions* $\bar{x}$	$\bar{y}$	$\bar{z}$	$\bar{x}E$	$\bar{y}E$	$\bar{z}E$
(1)	(2)	(3)	(4)	(5)	(6)	(7)	(8)	(9)	(10)
400	828	2.14	1772	0.0143	0.0004	0.0679	25	1	120
10	916	2.17	1988	0.0435	0.0012	0.2074	86	2	412
20	935	2.22	2076	0.1344	0.0040	0.6456	279	8	1340
30	868	2.28	1979	0.2839	0.0116	1.3856	562	23	2742
40	1049	2.38	2497	0.3483	0.0230	1.7471	870	57	4363
450	1171	2.52	2951	0.3362	0.0380	1.7721	992	112	5229
60	1178	2.69	3169	0.2908	0.0600	1.6692	922	190	5290
70	1149	2.86	3286	0.1954	0.0910	1.2876	642	299	4231
80	1159	3.08	3570	0.0956	0.1390	0.8130	341	496	2902
90	1088	3.30	3590	0.0320	0.2080	0.4652	115	747	1670
500	1094	3.52	3851	0.0049	0.3230	0.2720	19	1244	1047
10	1078	3.77	4064	0.0093	0.5030	0.1582	38	2044	643
20	1049	4.03	4227	0.0633	0.7100	0.0782	268	3001	331
30	1077	4.29	4620	0.1655	0.8620	0.0422	765	3982	195
40	1044	4.52	4719	0.2904	0.9540	0.0203	1370	4502	96
550	1040	4.70	4888	0.4334	0.9950	0.0087	2118	4864	43
60	1000	4.82	4820	0.5945	0.9950	0.0039	2865	4796	19
70	964	4.83	4656	0.7621	0.9520	0.0021	3548	4433	10
80	957	4.76	4555	0.9163	0.8700	0.0017	4174	3963	8
90	886	4.67	4138	1.0263	0.7570	0.0011	4247	3132	5
600	900	4.55	4095	1.0622	0.6310	0.0008	4350	2584	3
10	896	4.39	3933	1.0026	0.5030	0.0003	3943	1978	1
20	876	4.20	3679	0.8544	0.3810	0.0002	3143	1402	1
30	833	4.02	3349	0.6424	0.2650	0	2151	887	0
40	837	3.87	3239	0.4479	0.1750		1451	567	
650	800	3.76	3008	0.2835	0.1070		853	322	Z =
60	802	3.68	2951	0.1649	0.0610		487	180	30701
70	822	3.65	3000	0.0874	0.0320		262	96	
80	783	3.65	2858	0.0468	0.0170		134	49	
90	697	3.67	2558	0.0227	0.0082		58	21	
700	716	3.73	2651	0.0114	0.0041		30	11	
10	743	3.81	2831	0.0058	0.0021		16	6	
20	616	3.91	2409	0.0029	0.0010		7	2	
30	699	4.02	2810	0.0014	0.0005		4	1	
40	751	4.15	3117	0.0007	0.0003		2	1	
750	636	4.28	2722	0.0003	0.0001		1	0	
60	464	4.40	2042	0.0002	0.0001		0		
								Y =	
							X =	46003	
							41138		

$X+Y+Z = 117\ 842$ $\qquad x = \frac{X}{X+Y+Z} = 0.3491$ $\qquad y = \frac{Y}{X+Y+Z} = 0.3904$

Notes

Col. (2): Spectral power distribution for standard daylight D65.

Col. (3): Reflectance of the Uccello green, from the curve in *Figure 67*.

Cols. (5)–(7): The CIE 1931 colour-matching Functions (also called Distribution Coefficients).

Cols. (8)–(10): Each colour-matching Function is multiplied by *E* (col. 4) and each column is then summed to give *X*, *Y*, *Z*.

x and *y* are the co-ordinates on the CIE Chromaticity Chart (*Figure 69*).

Cols. (2), (5), (6) and (7) are from standard tables.

The data in this table are at 10 nm intervals. When using full colour-rendering calculations, 5 nm intervals are recommended. A computer becomes essential

Our eyes have three visual pigments for blue, green and red light. This has long been suspected but only recently verified (*Figure 68*). The method we are using is based on this assumption. The combined response of the three visual pigments gives us our sensation of colour. We now carry out the equivalent mathematic operation. The curves represented by the three columns of figures, $\bar{x}$, $\bar{y}$, and $\bar{z}$, (columns 5–7) can be thought as corresponding in principle to the sensitivities of our three visual pigments. Our test colour (column 4) has to be multiplied by each of the three columns in turn and the results in each column summed (columns 8–10). The sum of our test colour multiplied by the $\bar{x}$ column is called X, and similarly for Y and Z. These sums correspond roughly to the amount of red, green and blue light respectively in the colour.

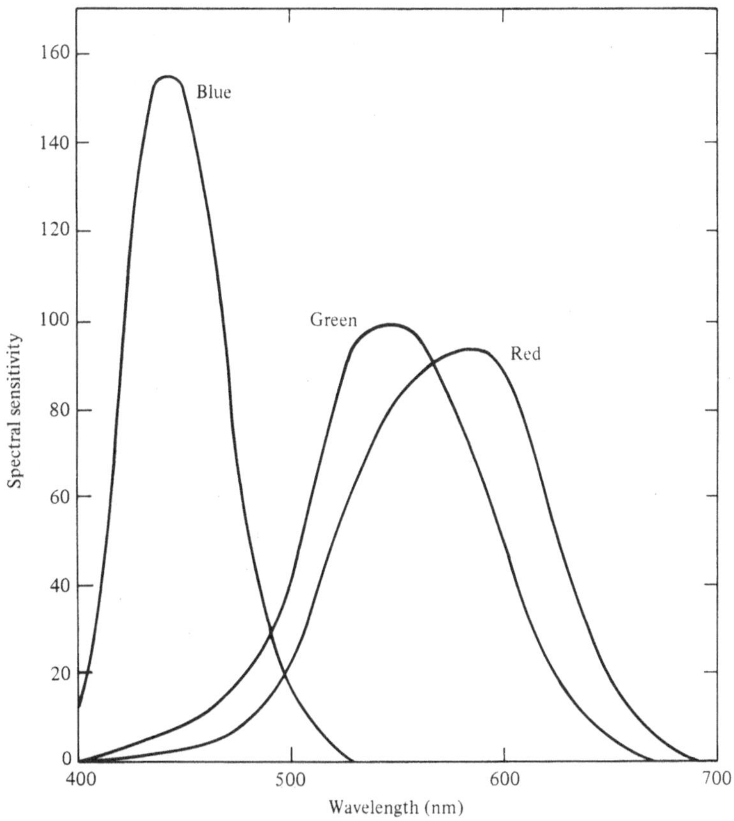

Figure 68
Cone sensitivity curves of the human eye[260]

We could now plot our colour on a three-dimensional 'colour space' using X, Y, Z axes. More usually X, Y, Z are reduced to x and y as follows:

$$x = \frac{X}{X + Y + Z} \quad \text{and } y = \frac{Y}{X + Y + Z}$$

x and y are the co-ordinates of our test colour on the Chromaticity Chart (*Figure 69*), often referred to as its chromaticity co-ordinates or simply chromaticity.

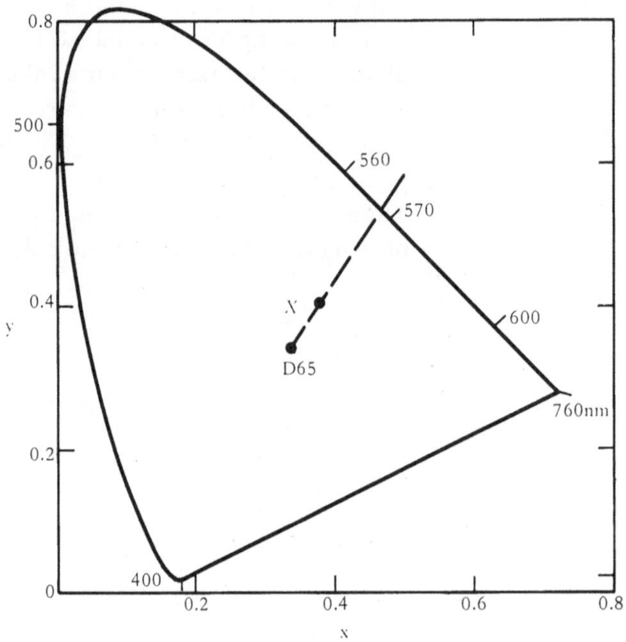

Figure 69
Chromaticity of grass on the Uccello at the point marked *X* on the painting (*Plate 16*). The chromaticity of the grass is also marked *X* on this chart, at $x = 0.349$, $y = 0.390$. The painting is viewed under daylight, which we have taken as D65, and this also is marked. A line from D65 through *X* cuts the boundary at about 567 nm, which is therefore said to be the dominant wavelength of this colour. It is close to the peak of the reference curve, *Figure 67* (1964 10° CIE diagram)

For reference in the next section the reader may note a mysterious fact, that the $\bar{y}$ column of figures (column 6) gives us precisely the Luminous Efficiency curve of the eye, already described on p. 177 (*Figure 63*). This was, of course, arranged that way for convenience. It means that Y becomes a measure of the luminance of the grass in the Uccello when illuminated by daylight of a known intensity.

Lastly, a useful short cut. To save us multiplying each test colour by the D65 curve, *Table 29* (*see* next section) gives $\bar{x}$, $\bar{y}$ and $\bar{z}$ already multiplied by this curve. Next time we do the calculation, we take the reflectance curve as it stands and straightway multiply it by the three columns in turn (columns 3–5 in *Table 29*). Then we proceed to sum the columns as before to get X, Y and Z, and from these x and y.

In many applications we are concerned with finding by measurement the small differences that the eye can distinguish between two colours. We may be following instrumentally the change in colour of a dye as it fades. Unfortunately the standard chromaticity chart of *Figure 69* is very deceptive in this respect. A change that can just be spotted by the eye, a 'minimum perceptible difference', is represented by a tiny shift in the blue corner of the chart, but by a shift ten times as big as the green apex. Also the shift depends strongly on direction.

Getting over this difficulty involves going back to colour space (three dimensions) and twisting the diagram into very peculiar shapes. However, such transformations are useful, particularly that known as CIELAB, where the three coordinates are labelled L^* for luminance, a^* and b^*. Colour shifts calculated in this system correspond very reasonably to what the eye sees.

THE COLOUR RENDERING CALCULATION

Although the colour rendering qualities of a light source are not connected with its power to damage, and therefore not directly with our concern for preservation, nevertheless the scientific conservator will frequently be asked whether a certain fluorescent lamp, UV filter or tinted glass will distort colours. Just as frequently someone will assert that distortion is occurring and the conservator will want to come to his own conclusion. Since subjective judgement is inadequate, the calculation here described is an indispensable aid.

It is not difficult, but, like the chromaticity calculation, it is tedious to carry out. Any kind of automatic calculating device, best of all a computer or a programmable calculator, is welcome.

There are two recognised methods: the colour-shift method advocated by the CIE[261] and the NPL–Crawford Spectral Band method[262, 263]. Both require us to compare mathematically the appearance of colours under our test illuminant (for example a certain type of fluorescent lamp) with their appearance under a reference illuminant of theoretically perfect colour rendering, chosen as being of the same Correlated Colour Temperature (CCT). The term CCT is used for the following reason. Strictly speaking an illuminant can only have a colour temperature assigned to it if its illumination matches that from a black body or Planckian radiator (p. 49). This means that the colour of the lamp can be represented by a point on the full curve in

Figure 22. But light sources with points not too far from this black body curve can be assigned a correlated colour temperature by using the convention in *Figure 70*. In both the colour rendering methods we have thus to find the chromaticity of our test lamp by placing it on the Chromaticity Chart as described in the previous section. We then find its CCT from *Figure 70*. Our reference illuminant, if the CCT is 5000 K or below, is then a black body radiator of that colour temperature. If the CCT is above 5000 K our reference illuminant is the standard daylight curve for that colour temperature.

The methods now diverge. The plan of the CIE method is to place a set of eight test colours illuminated by the test lamp in colour space, then to replot their positions when illuminated by a reference illuminant of the same CCT. In general, there will be shifts in the positions of the test colours. If these shifts are large then the colour rendering of the lamp under test will be poor. The CIE result is generally presented as a figure called the R_a which is derived from the average of the shifts of the eight test colours. A lamp with $R_a = 100$ has a perfect colour rendering. *Table 5* (p. 54) includes the R_a of some of the better fluorescent lamps. The value of the worst shift in the eight test colours is also given in the table as R_w. It sometimes happens that, though the average is acceptable, one or more test colours get very poor rendering.

The CIE method has won the day and become the standard. Nevertheless, in some ways the NPL–Crawford method is more suited to museum use. It has in fact been developed using, at one stage, museum personnel[265]. Also the computation is shorter and easer to understand.

To illustrate the use of the Crawford method we will take as an example a tinted glass. An exhibition room in a historic building is lit by side windows, and we wish to reduce the illuminance from the daylight by re-glazing with a tinted glass of about 30% transmittance (*Figure 71*) which we hope will be neutral enough not to distort colours. It is not the glass itself whose colour rendering quality we measure but that of the daylight which comes through it. We will therefore do the equivalent of shining standard D65 daylight through the glass by multiplying together the D65 spectral energy distribution and the transmittance of the glass (*Table 29*).

We now follow the procedure outlined in the section on p. 195 to place our test illuminant (D65 through the tinted glass) on the Chromaticity Chart. Using the form of this chart shown in *Figure 70* we find the CCT of this test illuminant: 6395 K. We may now expect to have to search for the spectral power

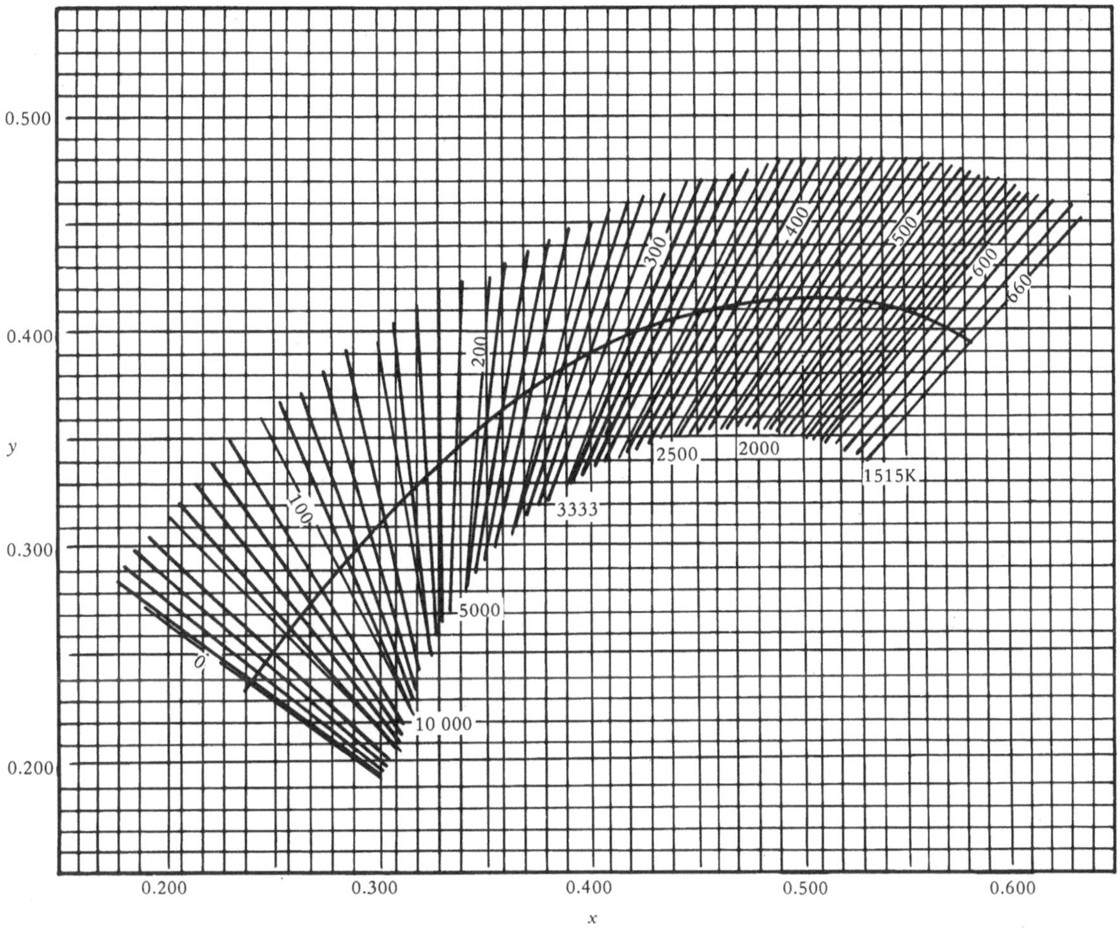

Figure 70
Finding the Correlated Colour Temperature (CCT) of a light source not on the Black Body Radiator Line. Example: a light source with (x,y) coordinates (0.420,0.440). The lines are labelled in Mireds, where M mireds = 10^6 /M K. Our light source is on the 280 mired line. Therefore its CCT is 280 mireds, or 3570 K[264]

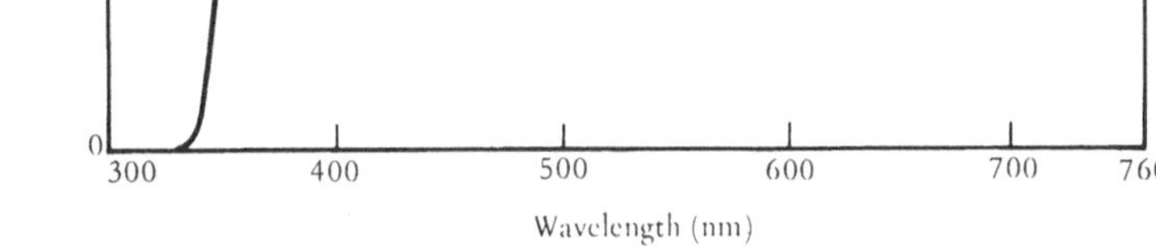

Figure 71
Transmission curve of a 'grey' glass whose colour-rendering quality is calculated in *Tables 29 and 31*. Transmission values of this curve are given in column 2 of *Table 29*. (It is interesting that this film actually increases the UV content by 45%. If it were to be used to reduce daylight to 200 lux, the UV exposure would be greater than if blinds had been used for this purpose)

distribution of the daylight reference illuminant at or nearest to 6395 K (such curves can be found, for example, in ref. 66). But there is a short cut which will be apparent when we have examined the basis of the Crawford method.

Crawford, in developing his spectral band method, used an apparatus by which he could subtract or add light at certain chosen wavebands and investigate the effects of these changes on viewers looking at colours and objects illuminated by the result. For this purpose he divided the visible spectrum into six

TABLE 29. The Crawford–NPL Colour-rendering Calculation

Wave-length (nm)	*Glass trans.* (%)	*Colour-matching functions weighted by D65*			*D65 through tinted glass*			*Norm-alised*	*Crawford Band Luminances*
		$\bar{x}E$	$\bar{y}E$	$\bar{z}E$	$\bar{x}E'$	$\bar{y}E'$	$\bar{z}E'$	$\bar{y}E'$	
(1)	(2)	(3)	(4)	(5)	(6)	(7)	(8)	(9)	(10)
400	58	0.112	0.003	0.532	6.4	0.1	30.8	0.006	
10	45	0.377	0.010	1.796	16.9	0.4	80.8	0.017	
20	32	1.189	0.035	5.711	38.0	1.1	182.7	0.044	
30	24	2.330	0.095	11.370	55.9	2.2	272.8	0.090	0.644
40	19.5	3.458	0.228	17.343	67.4	4.4	338.1	0.175	
450	19	3.724	0.421	19.627	70.7	7.9	372.9	0.315	
60	20	3.243	0.669	18.614	64.8	13.3	372.2	0.528	
70	21.5	2.124	0.989	13.998	45.6	21.2	300.9	0.839	
80	23.5	1.048	1.524	8.915	24.6	35.8	209.5	1.413	
90	25	0.330	2.142	4.791	8.2	53.5	119.7	2.112	10.96
500	26	0.051	3.343	2.815	1.3	86.9	73.1	3.431	
10	26	0.095	5.132	1.614	2.4	133.4	41.9	5.267	
20	25.5	0.628	7.041	0.775	16.0	179.5	19.7	7.087	23.23
30	25	1.687	8.785	0.430	42.1	219.6	10.7	8.669	
40	26	2.869	9.425	0.201	74.5	245.0	5.2	9.673	
550	27.5	4.267	9.796	0.086	117.3	269.3	2.3	10.634	
60	29	5.625	9.415	0.037	163.1	273.0	1.0	10.778	46.60
70	28	6.947	8.678	0.019	194.5	242.9	0.5	9.591	
80	25.5	8.304	7.885	0.015	211.7	201.0	0.3	7.937	
90	22.5	8.612	6.352	0.009	193.7	142.9	0.2	5.641	
600	22	9.046	5.374	0.007	199.0	118.2	0.1	4.667	12.53
10	22	8.499	4.264	0.003	186.9	93.8	0	3.703	
20	21.5	7.089	3.161	0.002	152.4	67.9		2.682	
30	21	5.062	2.088	0	106.3	43.8	Z =	1.730	
40	21	3.547	1.386		74.4	29.1	2435.4	1.148	
650	21.5	2.147	0.810		46.1	17.4		0.687	
60	24	1.252	0.463		30.0	11.1		0.438	
70	28.5	0.680	0.249		19.3	7.0		0.280	
80	34.5	0.347	0.126		11.9	4.3		0.171	
90	40.5	0.150	0.054		6.0	2.1		0.086	5.98
700	44.5	0.077	0.028		3.4	1.2		0.049	
10	45.5	0.041	0.015		1.8	0.6		0.026	
20	45	0.017	0.006		0.7	0.2		0.010	
30	44	0.009	0.003		0.3	0.1		0.005	
40	42.5	0.005	0.002		0.2	0		0.003	
750	40.5	0.002	0.001		0			0.001	
60	38	0.001	0			Y =			
					X =	2533.3		99.951	
					2259.1				

$X+Y+Z = 7227.8$ $x = \frac{X}{X+Y+Z} = 0.3126$ $y = \frac{Y}{X+Y+Z} = 0.3505$

Correlated Colour Temperature from (x,y) coordinates in table = 6395 K bands (dividing points 400, 455, 510, 540, 590, 620 and 760 nm) such that changes in the amount of light in any band were found to have an equal subjective effect. For this manipulation the light from the illuminant had to be split up into its spectrum. The additions or subtractions were performed with masks on the spectrum and the light was then optically recombined to form 'white' light. Crawford then found how much change could just be tolerated in any band before the *adapted* eye could notice a change in colour rendering.

On this basis one can now use Crawford's results to compare a test and a reference illuminant band by band. We still need to find our reference illuminant, but its complete spectrum is no

TABLE 30. Reference Sources for Colour Rendering

			Band limits (nm)			
Colour temperature (K)	400–455	455–510	510–540	540–590	590–620	620–760
Black body						
2500	0.148	4.59	14.80	43.96	21.95	14.34
2700	0.187	5.15	15.78	44.36	21.30	13.23
2854	0.216	5.56	16.43	44.55	20.72	12.52
2900	0.224	5.68	16.62	44.59	20.54	12.34
3000	0.242	5.93	17.00	44.66	20.20	12.00
3100	0.262	6.18	17.36	44.74	19.88	11.59
3200	0.282	6.40	17.68	44.76	19.58	11.26
3300	0.300	6.64	18.00	44.79	19.29	10.96
3400	0.320	6.86	18.30	44.80	19.02	10.70
3600	0.360	7.29	18.85	44.80	18.52	10.20
4000	0.435	8.04	19.78	44.70	17.67	9.39
Standard daylight						
5000	0.574	9.64	21.82	44.16	15.78	8.00
5500	0.652	10.24	22.34	44.04	15.22	7.52
6000	0.724	10.77	22.76	43.87	14.78	7.13
6500	0.791	11.22	23.11	43.68	14.38	6.82
7000	0.852	11.65	23.39	43.45	14.08	6.56
7500	0.906	12.02	23.64	43.25	13.83	6.36
8000	0.956	12.34	23.84	43.05	13.60	6.18
8500	1.001	12.59	23.92	43.11	13.36	6.02
9000	1.040	12.90	24.18	42.67	13.24	5.91

Notes
Luminances in the six Crawford bands (normalised to total luminance = 100). For black body radiators up to 4000 K and for Standard Daylight from 5000 K. (Calculated from Wyszecki and Stiles[66] *Tables pp. 8 and 692*)

◀ *Notes on Table 29 (opposite)*
The transmission of the tinted glass whose curve is shown in *Figure 71* is given in col. (2). First the Correlated Colour Temperature of D65 through this glass must be found by placing it on the Chromaticity Chart, just as in *Table 28*. Multiplying cols. (3), (4) and (5) by col. (2) we find cols (6), (7) and (8), whose sums give X, Y and Z. X, Y and Z give x and y. So far the calculation follows *Table 28*, but using the short cut of combining D65 with the standard Colour Matching Functions (1931). The Correlated Colour Temperature is found (x, y) on the enlarged Chromaticity Chart of *Figure 70*. It is 6395 K. We now normalise the luminance col. (7), that is to say we multiply each item in it by 100/2533.3 so that the new col. (9) sums to 100 (actually 99.951). Finally, in col. (10) we split the luminance into the six Crawford bands and proceed to *Table 30*. The calculation for a light source, such as a fluorescent lamp is slightly shorter. Turn to *Table 28* and delete cols. (2) and (3). Write in col. (4) the Spectral Power Distribution of the lamp, and complete the table to find (x,y) for the lamp and hence its Correlated Colour Temperature from *Figure 70*. Finally normalise the luminance and split into the six Crawford bands, as in cols. (9) and (10) above.

longer required, merely the six bands, which for our example of 6395 K can be found by interpolation from *Table 30*.

We must first of all adjust our test illuminant so that it has the same illuminance as the reference illuminant found from *Table 30*. Clearly if the test illuminant is less strong there will be a deficiency of light in all of its six bands, and vice versa. We have already noted that the column headed y in *Table 28* (col. 6) represents the luminous efficiency of the human eye. Multiply this by the energy distribution of the light source and we get the actual response of the eye to each wavelength (*Table 29*, col. 7, headed $\bar{y}E'$). Sum this column and we get the illuminance Y, the strength of the light through the glass as the eye sees it. By using the appropriate factor we could actually turn this into a lumen value, but there is no need to do this for the present calculation. Y for our test illuminant is 2533.3 (sum of column 7). But it has been arranged that all the reference illuminants in *Table 30* have $Y = 100$. Therefore each $\bar{y}E'$ value must be multiplied by 100/2533.3, giving the figures in column 9 for 'normalised' $\bar{y}E'$. Checking these, we find that they now add up to 99.95, which, though not 100, is as near as we need. The 'normalised' $\bar{y}E'$ column is now grouped into the six Crawford bands to give the final column 10 in the table. The readings are at 10 nm intervals, but the Crawford bands are split to the nearest 5 nm. Therefore proceed for the first band as follows, and the others likewise. Band 400–455 nm is the sum of half the 400 nm reading plus the readings for 410, 420, 430, 440 and 450 nm.

We now compare test with reference illuminant band by band by finding test/ref % (*see Table 31*). The deviations from 100% appear next, plus or minus. We then average neighbouring bands and do the same operation on them, for the double and deviations.

For the final stage Crawford found that we can tolerate a 10% deviation in any single band, but only a 5% deviation in double bands. Thus anything over 10%, ignoring sign, in the deviations for single bands is put in the last column (excesses), and anything over 5% for double bands. The final figure is the sum of all these excesses. The higher the figure for excesses the worse the colour rendering.

A Class A illuminant is not allowed any excesses at all. Colour rendering for Class A illuminants can be regarded as completely satisfactory. Class B are allowed up to 32 excesses, and Class C up to 64. The discrimination of the method is fine enough to put both these classes in the category of good colour rendering. All below this, however, must be classed as poor, or low fidelity light sources (compare *Table 5*, p. 54). Some fluorescent lamps give excesses of up to 300.

Our tinted glass cannot be put in the top class because it has acquired 25 excesses, though it should not be considered to give poor colour rendering. Nevertheless there are grey classes in Class A, so they become preferable, other things being equal.

Incidentally it should now be clear that we are not necessarily comparing the colour rendering of the light that gets through the tinted glass with that through a completely colourless glass or even through ordinary glass. We are comparing it to daylight at the same colour temperature as the light that gets through the tinted glass. In general tinted glasses do two things to the light: they distort its colour rendering to a degree which may or may not be significant, and they change its colour temperature. This latter effect, by definition, has no effect on colour rendering. A very good example is the 'bronze' glass popular in new buildings for reducing solar heating. This appears quite strongly coloured, and yet the calculation shows that it introduces no significant distortion, but *does* lower the colour temperature considerably. An example on the author's files, Pilkington Antisun Bronze 6 mm (1974), gave no Crawford excesses, but lowered D65 by 940° to 5560 K.

TABLE 31. The Crawford–NPL Colour Rendering Calculation – Final Stage

	Band luminances		*Ratios*	*Deviations from 100%*		
Crawford bands (nm)	*Test source: Tinted glass through D65*	*Ref source: Daylight at 6395 K*	*Test Ref* (%)	*Single bands*	*Double bands*	*Excesses over tolerance*
400–455	0.644	0.777	83	–17		7
					–9	4
455–510	10.96	11.13	98	–2		0
					0	0
510–540	23.23	23.04	101	1		0
					4	0
540–590	46.60	43.72	107	7		0
					–3	0
590–620	12.53	14.46	87	–13		3
					–13	8
620–760	5.98	6.89	87	–13		3
					Total excesses	= 25 *(Class B)*

Notes

Band luminances for the Test Source are taken from col. (10) of *Table 29*. The Reference Source must have a colour temperature of 6395 K to match the Test Source. *Table 30* gives band luminances for Reference Sources at 6000 and 6500 K. Band luminances for 6395 K are found from these by interpolation. Deviations for double bands are found by averaging the ratios of the bands above and below. Thus the third double band is the average of bands 3 and 4. (101+107)/2 = 104. 100–104 = 4. For the single-band excesses anything over 10% is recorded, and for the double bands anything over 5%. Signs are ignored. The Test source is evaluated as Class B on the following scheme: class A, 0 excesses; class B, 1–32 excesses; class C, up to 64, etc. The divisions between classes being at 32, 64, 128, 256 . . . excesses

COLOUR RENDERING AND THE BLACK BODY CONVENTION

Although there could be no reasonable objection to using the phases of daylight as reference illuminants in the range down to 5000 K, many curators feel uneasy about the use of an entirely theoretical black body series as reference illuminants for colour temperatures below 5000 K. The fact that this means, in effect, regarding tungsten or tungsten–iodine as the reference illuminant for colour temperatures around 3000 K makes matters no better, it is felt.

Quite enough has been said in the previous pages about the adaptive powers of the eye for the reader to realise that we cannot judge matters by transferring coloured objects quickly from one illumination to another. We must judge only in the adapted state, and this means interposing a lapse of time during which the objects are not visible, but our memory of colours makes this process inaccurate.

Some experimental attempts have been made to get over this difficulty. The most relevant for our purpose[266] used an arrangement in which subjects were trained in the use of the Munsell Colour Atlas (p. 59). With practice selected subjects were able to estimate hue, value, and chroma to within one point without reference to the colour atlas itself and purely from internal reference as it were. They were next confronted with colour chips, having adapted themselves to the illuminant, and asked to specify them in Munsell terms. They did this under both daylight and tungsten illumination. Certain results appeared to be significant, but were at least as dependent on the colour of the surroundings and seemed to show little overall pattern. However a tendency for purples to become redder under tungsten seemed consistent.

All we can say at present is that the expected effect (that reds should look brighter under tungsten and blues darker), if indeed significant, must be small. No strong grounds exist for discarding the black body convention.

CHOOSING A FLUORESCENT LAMP

Lamps of moderate to good colour rendering fall into three categories (*see Table 5*, p. 54). Capital cost of the old halophosphate lamps is about half that of the other two, but halophosphate lamps of good colour rendering cost more to run because of their low efficacy. Top colour rendering is to be found among certain of the new polyphosphates.

Triphosphor lamps have been a matter for indecision in the museum world. There is considerable pressure to use them because of energy economy. Thornton proposed in 1971[267–9] a source which emits only at three wavelengths: 450, 540 and 610 nm, corresponding approximately to the three peak responses of the human visual system. Narrow-band phosphors emitting at these three wavelengths were mixed in different proportions to make lamps over a range of colour temperatures. All the major manufacturers now produce triphosphors. Their CIE colour rendering index, R_a, can be good, but because of the big gaps in their spectra there is usually at least one test colour that gets a bad rating (R_w index down to 40). For the same reason they get a bad Crawford class. In summary there will be certain museum situations not demanding the best colour rendering where they will be the choice[51].

Of the colour temperature categories (6500, 5000, 4000 and 3000 K) 6500 K is too cold for normal exhibition use but may find favour in the restoration studio. Most lamps in this category have good colour rendering. In the 5000 K category came a lamp with the highest colour rendering in any of the author's tests: the Matsushita FL 40S-N-EDL, with $R_a = 98$ and $R_w = 96$. But the most acceptable choice for rooms combining daylight and artificial light is 4000 K. For 50 lux one usually considers tungsten or 3000 K. The author has found only one 3000 K lamp in the top colour-rendering category: the Philips 93.

Compact fluorescent lamps are available for use in tungsten fittings (e.g. Philips SL and Osram/Wotan Compacta). They have longer life and can be more economical. Their colour rendering is in the triphosphor category.

Recommended specification for best colour rendering:

R_a about 90 or better,
Worst R about 80 or better,
Crawford Class A, B or C.

DIMMING

Lighting control involves dimming. Both tungsten and fluorescent lamps can be dimmed. Present-day dimmers use phase control circuits which alter the phase between current and voltage and hence the power delivered. They waste very little energy.

Dimming or under-running a tungsten lamp also decreases its colour temperature (making the light redder) and increases its life. For example, dropping the illuminance to half its value will increase the life of the lamp 20-fold but might drop the colour temperature by 200 K, which is near the acceptable limit.

With suitable equipment, fluorescent lamps can be dimmed to 10% of the rated luminance without any significant change in colour.

Humidity
Part Two

Humidity

THE STANDARD HYGROMETRIC (PSYCHROMETRIC) CHART

When air is heated or cooled without any evaporation or condensation of moisture, its Moisture Content on a weight/weight basis remains unchanged. In Part I we measured moisture in grams of moisture per cubic metre of air because a cubic metre of air is easily visualised. But air expands as it is heated, so that the grams of moisture in a cubic metre of air slightly fall. In the usual working range the inaccuracy thus introduced is not serious, and is likely to be less than 2%. However we now change to one of the standard forms of the Hygrometric Chart (*Figure 72*), in which moisture content is measured in kilograms of moisture per kilogram of dry air. It is also possible to read off from this chart the wet bulb temperature, the specific volume (the reciprocal of the air density) and the specific enthalpy for any state point on the chart. The latter term, *enthalpy*, will be explained below. The chart is based on a sea-level barometric pressure of 1013.25 mbar. High altitude will affect these relations, so that an adjustment of around 2 or 3% may have to be made in RH for museums at one or two thousand metres elevation if this is measured by a wet-and-dry bulb instrument[270].

Figure 72 *(opposite)*
The Standard Hygrometric (Psychometric) Chart, relating dry-bulb and wet-bulb temperatures, RH (Percentage Saturation), Moisture Content, Specific Enthalpy and Specific Volume. *Figures 72, 73, 75, 83* and *84* are reproduced with the kind permission of the Chartered Institute of Building Services (22 Balham High Road, London SW12 9BS) from whom pads of charts for permanent records may be obtained

A change in air temperature where there is no gain or loss of moisture is termed a 'sensible' heat change, since no latent heat of evaporation or condensation is involved, merely a change in temperature, which one can sense. Where a change in moisture content is also involved, we must also consider the change in 'latent' heat.

In Part I (pp. 77–81) we considered two basic directions of movement on the chart. The first was a sensible heat change, as just explained, where movement must be along a line of constant moisture content. The use of a simple room heater-

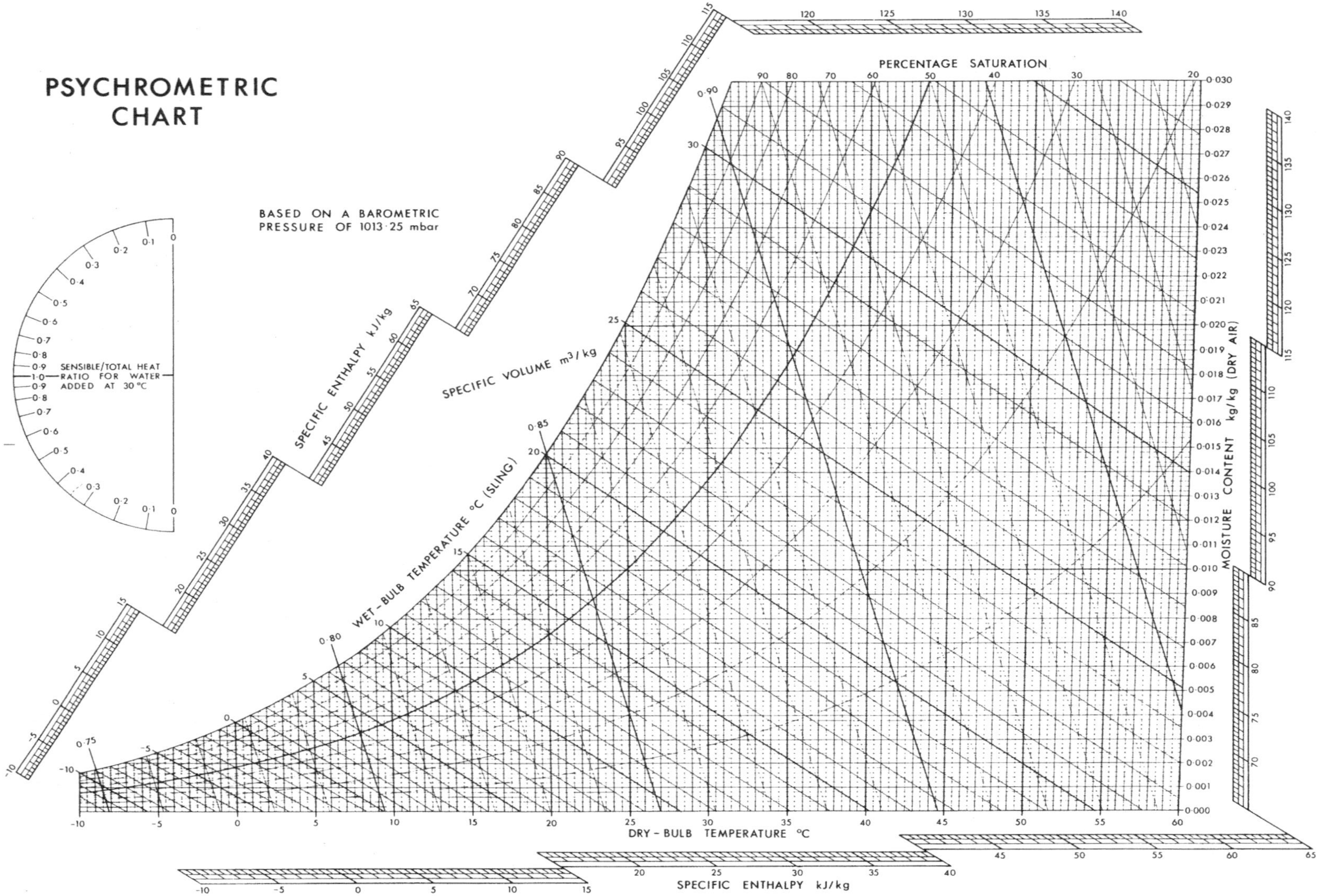
PSYCHROMETRIC
CHART
BASED ON A BAROMETRIC
PRESSURE OF 1013·25 mbar
SENSIBLE/TOTAL HEAT
RATIO FOR WATER
ADDED AT 30 °C
SPECIFIC ENTHALPY kJ/kg
SPECIFIC VOLUME m³/kg
WET-BULB TEMPERATURE °C (SLING)
PERCENTAGE SATURATION
MOISTURE CONTENT kg/kg (DRY AIR)
DRY-BULB TEMPERATURE °C
SPECIFIC ENTHALPY kJ/kg

radiator is a perfectly adequate example of a simple sensible heat change, though the purist would have us consider the effect that heating the air in the room would have in driving moisture out of wood, etc.. We could equally well consider cooling, provided that no condensation takes place.

The second direction of movement was made in Part I, for purposes of simplified explanation, up or down the chart: addition or subtraction of moisture without temperature change. But in reality this is a difficult change to perform, since increasing moisture content by evaporating water involves subtracting latent heat from somewhere, as happens, for example, with an evaporative humidifier. The air passes through the wet sponge and water from the sponge evaporates into the air stream by drawing the necessary latent heat from it, thus cooling it.

In fact the natural direction of change is an *adiabatic* change: one which involves no exchange of energy between the air and anything external to it, such as a heater or cooling coil. The example just given of humidification by evaporation (without using a heater) is just such an adiabatic change.

In such a case the air and the water vapour picked up from the sponge have to use their own store of energy, and they do this by converting some sensible heat into latent heat: there is no net gain or loss of energy.

Now we cannot very easily measure the total energy of a volume of gas (some of which energy is locked in molecular and atomic levels), and there is no need to do so. All we need is a scale which takes into account the energy which is exchangeable in air-conditioning operations: sensible heat and latent heat. The scale appropriate for this is energy measured as *enthalpy*. For further simplicity the enthalpy of dry air is given zero value at 0 °C, just as we commonly measure height above mean sea level.

On the chart will be seen scales of 'specific enthalpy', meaning enthalpy for a specific or standard quantity of air, here 1 kg. The enthalpy is measured in the standard energy units of kilojoules. There are scales on either side, enclosing the chart, so that a line of constant enthalpy at, say, 50 kJ/kg can be drawn by laying a ruler to cut 50 on both scales (*Figure 73*).

A curious and most useful coincidence will then be noticed: lines of constant enthalpy are near enough parallel to lines of constant wet-bulb temperature. Therefore, to a perfectly reasonable approximation we can use lines of constant wet bulb temperature in place of lines of constant enthalpy. Movement along a line of constant wet bulb involves almost no energy

Figure 73 *(opposite)*
The green line has been superimposed at a Specific Enthalpy of 50 kJ/kg (kilojoules per kilogram). Note that it is approximately parallel to the lines of constant wet-bulb temperature

PSYCHROMETRIC CHART

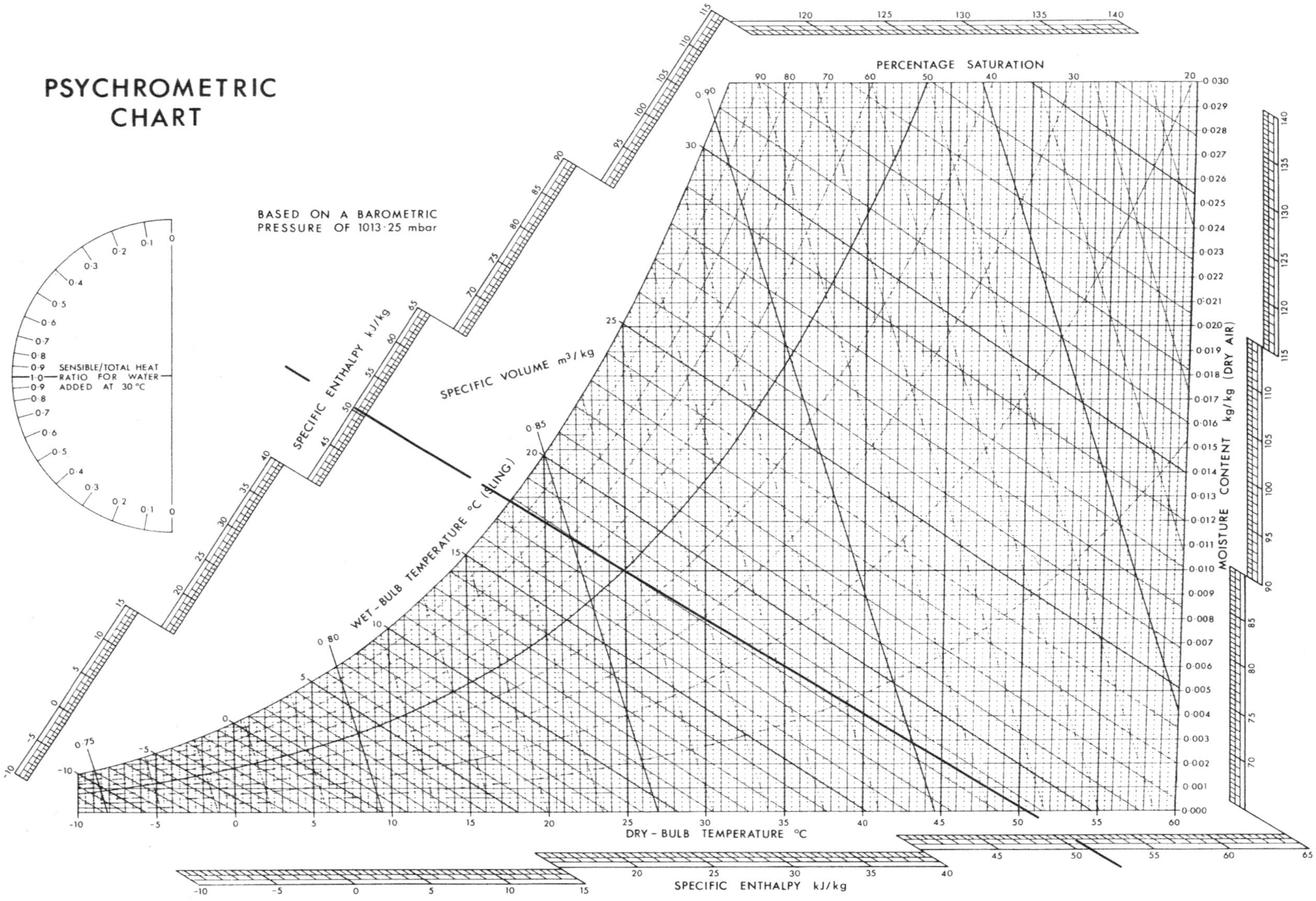

(enthalpy) change, and this is important because energy changes cost money.

It may help to remember that, although energy and enthalpy use different scales, an enthalpy gain or loss strictly equals an energy gain or loss.

THE CLASSICAL AIR-CONDITIONING OPERATION

The classical system achieves control of temperature and RH with an elegant simplicity. With one simple thermometer the fresh air is brought to saturation at a Set Point on the hygrometric chart. A second thermometer in the room or duct supplying the room then re-heats this air so that room conditions of temperature and RH are achieved. From time to time through the year the Set Point requires manual adjustment.

WINTER

The operation is most easily explained by following it through for cold winter air. Such air, even if saturated, carries insufficient moisture for ordinary room conditions. To force this air to absorb moisture it must first be heated. This is pre-heat (line OA in *Figure 74*).

After pre-heat the air passes through a humidifying device. If there is no attempt to heat or cool the humidifying water the path of the air up the hygrometric chart will be not straight up, but along a line of constant enthalpy, since energy is neither added nor subtracted. This line slopes to the left, indicating cooling due to the evaporation of water. A physical change without energy change is called 'adiabatic'. Conveniently lines of constant enthalpy are approximately parallel to line of constant wet bulb so that the wet-bulb temperature at A and B is the same.

The air leaves the humidifier saturated or nearly so (point B). Consequently its wet- and dry-bulb temperatures are the same. So an ordinary control thermometer (it does not have to be a wet bulb) after the humidifier will register whether the wet-bulb temperature is high or low. To adjust to the correct set point B, this thermometer controls the pre-heat.

But at this stage (B) the air is saturated and also too cold. However the set point has been so arranged that addition of heat alone – the re-heat stage – will bring about the correct conditions in the room (C). Re-heat is controlled by a second thermometer in the room.

This is not quite the whole story: the air in *Figure 74*, having

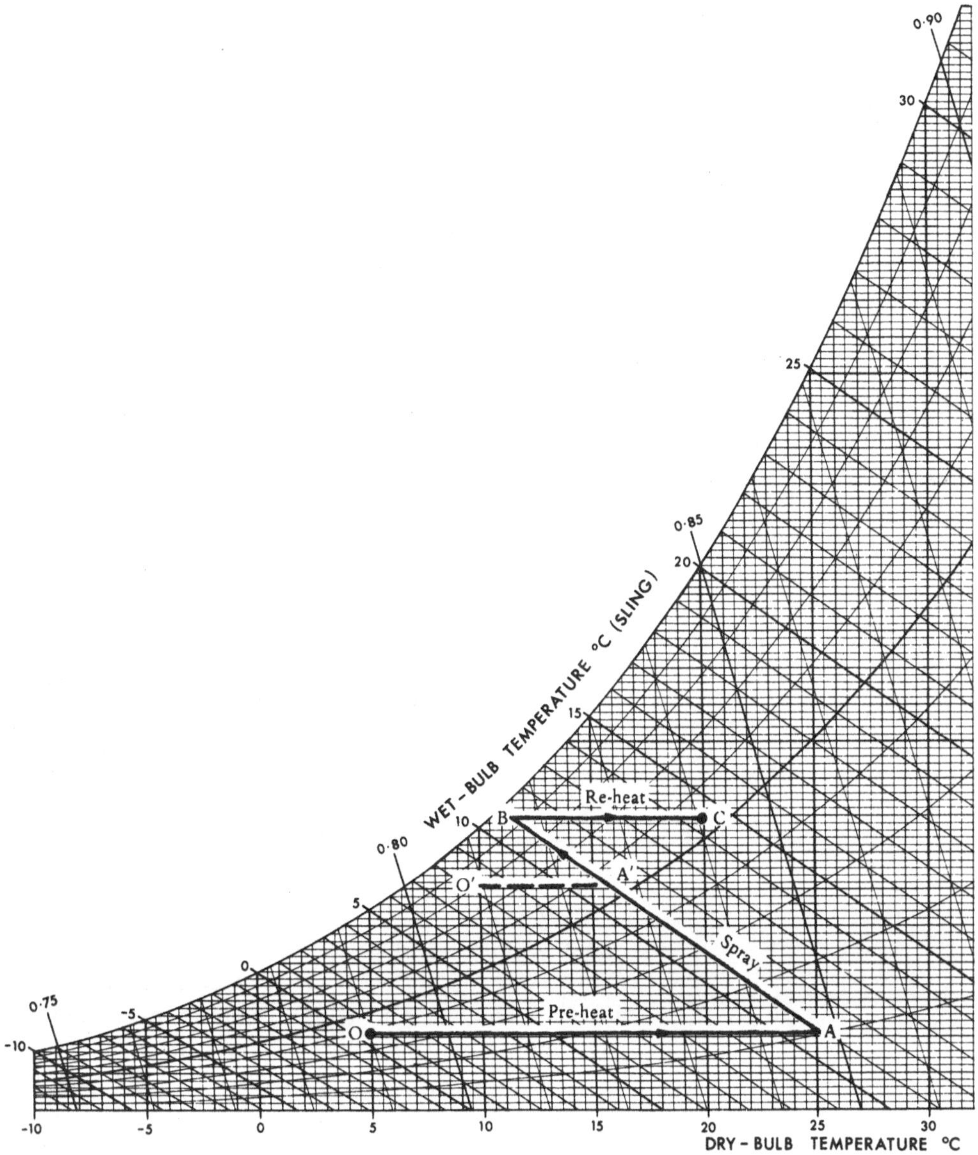

Figure 74
A basic air-conditioning operation: winter heating and humidifying. All the incoming air is first pre-heated (*OA*) and saturated adiabatically (*AB*). It is then ducted to sets of rooms, each of which has controlled re-heat (*BC*) to warm the air further and to reduce its RH to the correct level. The system is controlled by two dry-bulb electrical thermometers. The first, situated after the spray (at *B*), reads in effect the wet-bulb temperature at this point, since wet- and dry-bulb temperatures are equal in saturated air. If this is too low it calls for more pre-heat. The second is a zone thermostat controlling the amount of re-heat for a set of rooms. For simplicity, recirculation and certain control refinements have been left out of account. There is no RH sensor in this very basic scheme. If the outside conditions were *O'* rather than *O*, the path followed would be *O'A'BC*

passed through the room, is thrown away, which is extravagant. In practice most of it is recirculated. So a complete diagram would show a little fresh air mixing with recirculated air. But there is no further control difficulty in taking this into account.

It should be noted that, although the air traces a Z-shaped path on the hygrometric chart, at no point is energy added only to be later subtracted, which would be a nett waste: energy is increased from O to A, held constant from A to B, and increased again from B to C.

Recalling that the dew point of any packet of air is measured by lowering its temperature to the point – the dew point – where water starts to condense from it, one can see that the set point is the dew point of the air supplying the room. Hence this system of air conditioning is appropriately called Dew-point Control.

SUMMER

Dew-point control can also operate under summer conditions. No pre-heat will be needed if the temperature of the outside air is above that required for the room supply. Indeed because of various heat sources in the building, including lights, the temperature of the outside air may fall to 12 °C before pre-heat is required. However, we must now consider what will happen if we pass warm outside air directly through a spray-chamber to saturate it in an attempt to arrive at the dew point of the room supply air. As before, the air will travel up a line of constant wet bulb, but it will now arrive at a dew point some way above that required. For example warm air at 50% RH, 25 °C will travel up the 18 °C wet-bulb line, arriving at a dew point near 18 °C. With all the other heat sources in the room (sun, lights and people) we may find the room air back near 25 °C but with its RH raised to between 60 and 70%.

To offset all these heat gains we need a set point (dew point) of about 8 or 9 °C. This can only be achieved by cooling the spray water, or a chiller following the spray, under the control of the set point.

Figure 75 shows the path of fresh air from outside (O) to saturation (A), followed by cooling to the set point (B) (though these two changes would be concurrent). Air at B then mixes with recirculated air to give correct room conditions at C. Note that though passing through a water spray, the humidity of the air is reduced because of the coldness of the water.

Looking at the enthalpy changes that take place under warm conditions, we can trace the following:

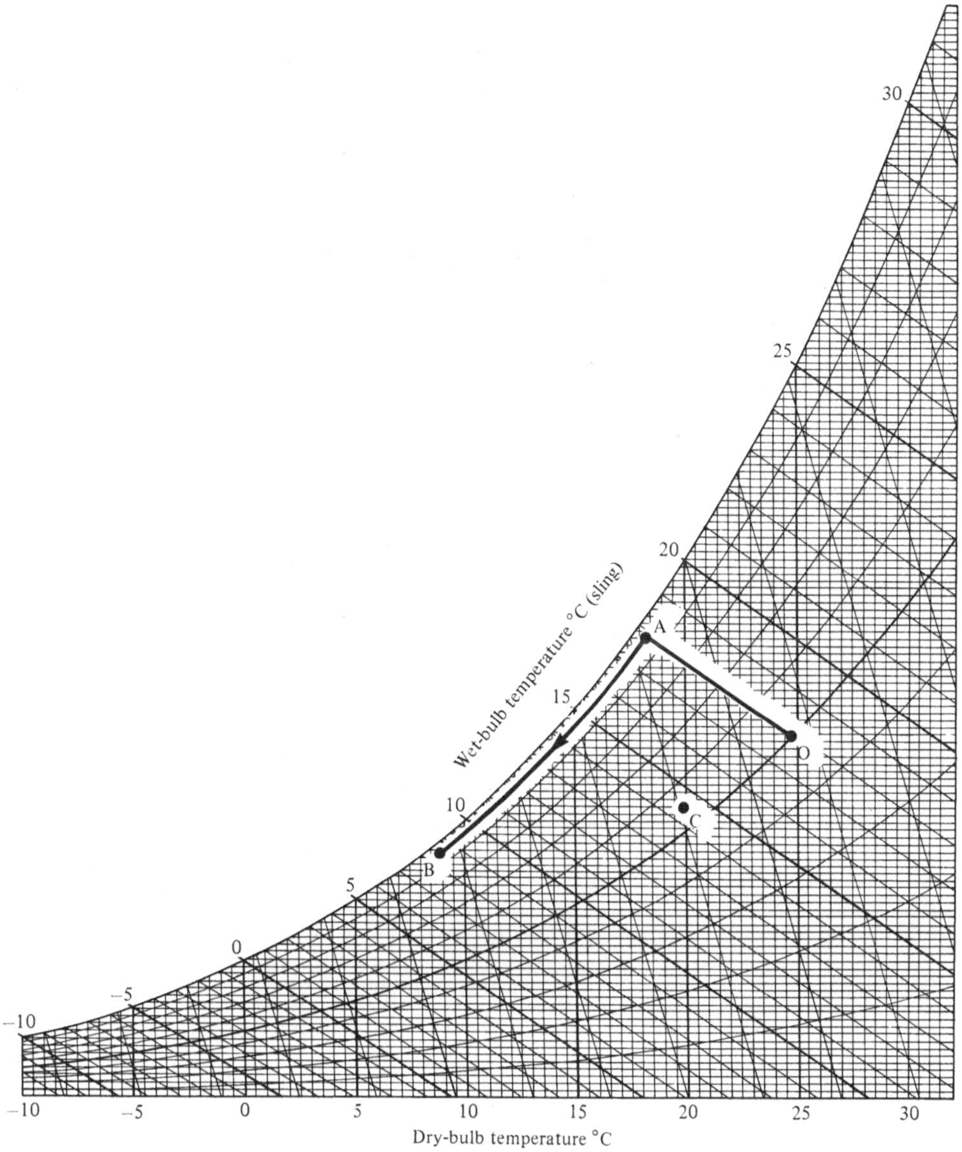

Figure 75
Cooling outdoor summer air (O) to point B so that it can mix with air that has been heated in the room (and in its passage through the system) to give a mixture at the correct conditions to re-enter the room (C)

(1) Fresh air is cooled to the set point (energy extracted from air).
(2) Air re-heated to bring it to room supply conditions (energy added to air).
(3) Heat is gained from sun, lights and people (further energy added).

(3) is not under the control of the air-conditioning system, but (2) is and represents wasted energy. So the classical system could be improved. For museums it could be improved in a second respect also.

The system was evolved to keep controls simple and actuated by temperature sensors of proved reliability. It will deliver constant RH to rooms with similar load, but this is not possible for rooms with different characteristics of gain and loss of heat and moisture.

A MUSEUM AIR-CONDITIONING SYSTEM

It is only recently that humidity-control sensors have become a practical proposition. Even now only a few show evidence of long-term reliability. But the use of triple redundancy (*see* page 221) will overcome this problem, and the case with which controls can be made to interact electronically sets aside the premium on simplicity of the classic scheme.

The following control scheme follows suggestions made by Reading specifically for museums and galleries[271].

Referring to *Figure 76*, we first locate the Room Specification, R, at 20 °C, 8 g/kg (corresponding to about 55% RH). The enthalpy corresponding to R is E. To minimise energy waste this

Figure 76
Enthalpy considerations

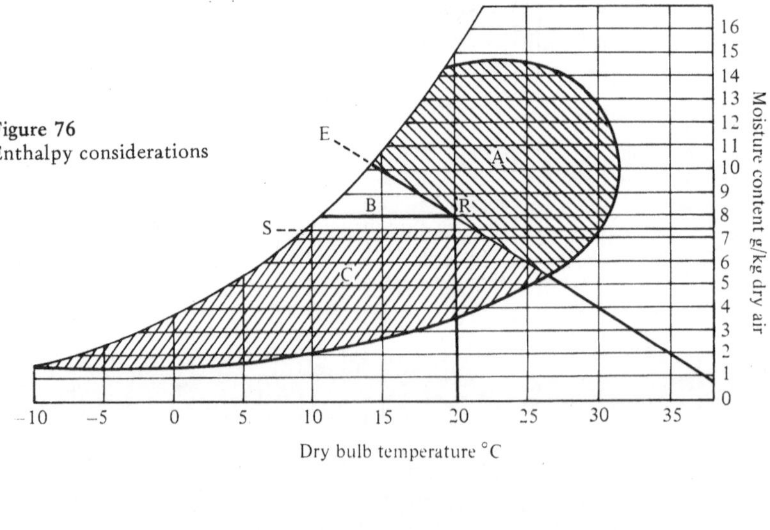

line of constant enthalpy E should not be crossed. The set point for the fresh air (which joins recirculated air before entering the room) is just under saturation at S. We add a shaded zone enclosing all normal weather conditions throughout the year. We then subdivide this zone by relating enthalpy and moisture content to R.

First, in area A, above the room enthalpy line E, the outside air has more energy than the room air, so it is best to use as little of it as possible, since cooling is expensive.

Area B covers air with lower energy but higher moisture content than the room. This excess moisture will have to be removed by chilling. Since the room air will certainly pick up more energy in recirculating it is most economical for area B to use full fresh air. Air in area C needs both moisture and energy.

CONTROL

The flow chart for the control system is outlined in *Figure 77*. The enthalpy sensor determines whether the energy of the out-

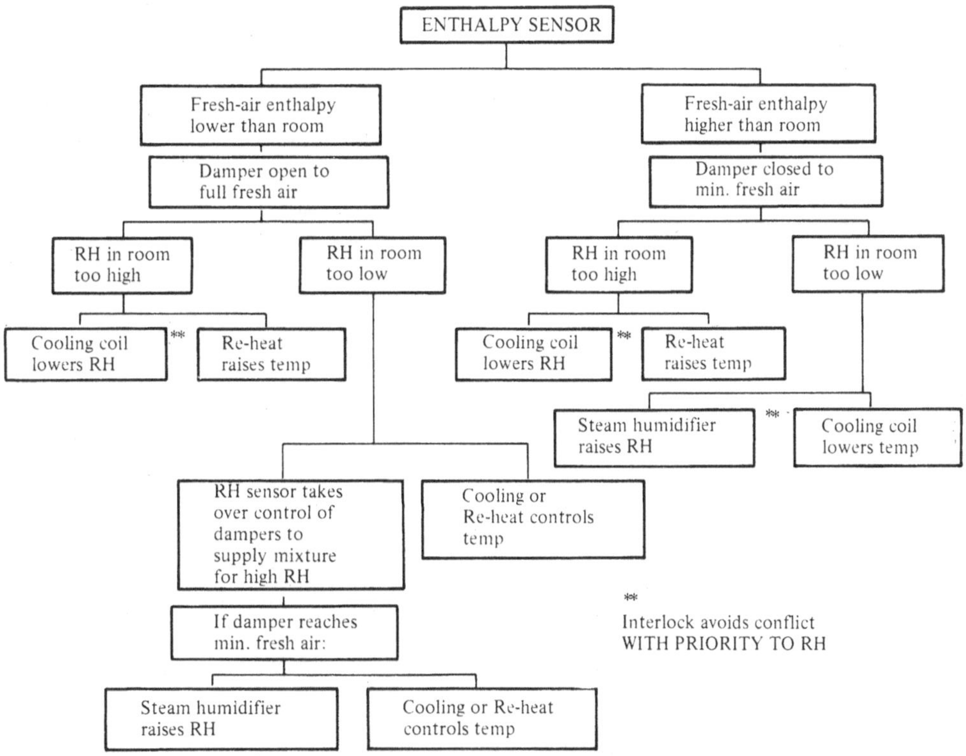

Figure 77
Flow chart for a museum air conditioning system. (Adapted from ref. 271).

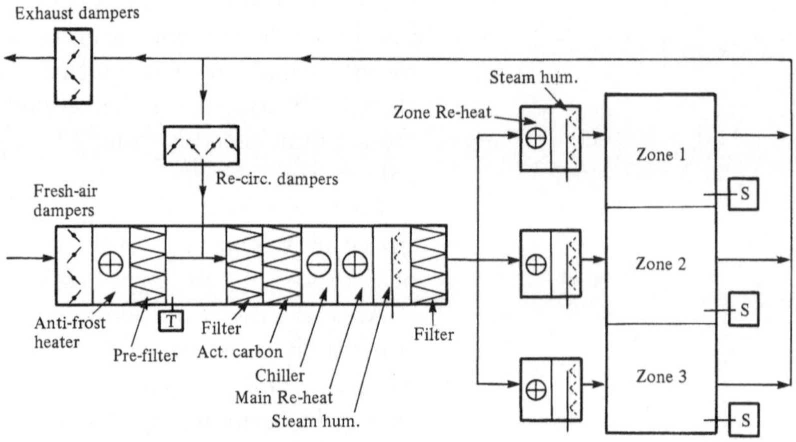

Figure 78
A museum air-conditioning system. Fresh air passes through dampers which work together with the exhaust and recirculation dampers to produce a controlled mixture of fresh and recirculated air. It then passes through a small heater for cold conditions and a pre-filter which need not be of high specification. Fresh and recirculated air now join to pass through the main filters for particulate and gaseous pollution. The mixture is then heated or cooled as appropriate and humidified. In each zone (a small set of similar rooms) re-heat and steam humidification allow final adjustment. S = RH and temperature sensor. These are all connected to a priority selector. This determines which of the zone RH and temperature sensors controls the main treatment plant, on the following principle:
The zone RH sensor which is demanding the lowest RH is in control of the main steam humidifier.
The zone temperature sensor which is demanding the lowest temperature is in control of the main re-heat.
For other zones the small zone units make up the difference.
T = temperature sensor to bring on anti-frost heater. At night the setting is for full recirculation. Zones using toxic vapours, such as laboratories and varnishing areas, should have no recirculation. Many other arrangements are possible. (Adapted from ref. 271)

side air is lower than that required in the room. If so, dampers are set to maximise fresh air. But this setting can be overridden by the priority room RH sensor (the RH sensor in the room requiring least humidification) if the room RH is too low. This priority RH sensor takes control and reduces the fresh air ratio until the desired RH is attained. If, after closing to minimum fresh air, the RH is still too low, then the main steam humidifier is brought into action.

For other situations, *Figure* 77 can be followed.

By ensuring primacy of RH control this system confers another advantage. In classical control failure of one component can rapidly lead to dangerous RH conditions. Under the Reading scheme the decision can be left to the triple-redundancy RH sensors as to whether the plant should be shut down. Before this happens, however, there is a great deal of flexibility in ways of maintaining RH. For example, the zone humidifiers would be able to cope with a failure in the main humidifier if the load were not too high. In other situations, temperature constancy will automatically be set aside in favour of RH maintenance.

If this fails and the RH goes outside boundaries (suggested at 40 and 70% RH) then the plant is automatically shut down. Even then a further decision could be made on the following lines. Shut down all heating and humidification but:

(1) Either keep fans running with fresh air and keep zone open to public,
(2) or run full fresh air but close zone to public,
(3) or shut down fans and close zone.

A fuller discussion will be found in ref. 271.

HEATING AND COOLING LOADS

Examination of *Figure 76*, which is characteristic of one part of the world, London, would seem to show that in any one year, more heating than cooling is required. This is also the general feeling: heating is essential in winter but cooling in summer is a luxury.

Analysis shows, however, that, in such a climate, the total energy available in a museum is sufficient to balance heating requirements for a large part of the year[272]. The break-even point is 11 or 12 °C. Fresh air below 11 °C requires supplementary heat. But since the average annual daytime temperature is 12 °C in London, heating and cooling requirements are pretty evenly balanced.

SENSORS

There are two tasks for the room temperature and RH sensors: the first is to control the system, the second to monitor it. Remembering that RH sensors tend to drift, the simplest way is to have one sensor pair (temperature and RH) for control and one for monitoring. This redundancy gives us a chance of spotting malfunction.

A better way, however, is to employ a triple redundancy arrangement. Each room has not two but three RH sensors electronically wired together so that, so long as they all agree, the signal returned is their average. But if one disagrees with the other two its reading is ignored and the average of the other two is returned. At the same time a warning light appears. If all three sensors disagree or give a null signal the system closes down with a further warning.

This triple redundancy arrangement raises reliability to the point where the three RH sensors can be given the jobs of both control and monitoring. The cost of three sensors and associated logic circuitry in place of two will be offset by the increase in reliability.

EXTERNAL DESIGN CONDITIONS

In estimating the size of installation required to give constant indoor conditions throughout the year a knowledge of the climate is essential. For example temperatures below −10 °C are too rare in England to justify the extra heating necessary to deal with them, but they are to be expected over a large part of Europe and North America.

TABLE 32. Some External Design Conditions

City	*Winter* (°C)	*Summer* (°C)	%RH
Helsinki	−27	27	47
Ottawa	−27	33	43
Moscow	−26	31	40
Vienna	−15	31	40
Ankara	−14	36	25
New York	−12	36	37
Washington	−10	36	41
Madrid	−4	36	29
Tokyo	−2	33	58
London	−1	28	42
Canberra	−1	36	25
Rome	−1	36	33
Athens	−1	37	26
New Delhi	4	40	
Hong Kong	7	33	68
Rio de Janeiro	13	34	53

Thus, each country or climate area will have agreed Design Conditions based on expected climatic extremes, cool seasons and hot seasons. A common criterion is that winter External Design Conditions are conditions which are exceeded in an average winter for no more than 2.5% of the winter period. For southern England a winter Design Condition of −1 °C saturated implies that the temperature in an average winter is only expected to drop below −1 °C for 2.5% of the winter-heating period. Even if it does this is likely to be for a period short enough for the thermal insulation of the building to cope. The RH at this temperature makes very little difference.

Summer conditions are developed in a similar manner. Some Design Conditions are shown in *Table 32*. They form quite a useful index for assessing and comparing climate.

Of course a stringent specification may demand special Design Conditions. For instance, ref. 272 suggests a winter External Design Temperature of −5 °C rather than −1 °C and for summer 29 °C and 50% RH rather than 28 °C and 42% for museums and galleries in the London region.

DIMENSIONAL CHANGES CAUSED BY RH VARIATION

The key substance here is wood, because of its complex response to RH change, though it is not intended to go into great detail here on wood/moisture relationships.

The process of seasoning timber is a process of allowing water to evaporate from the wood in a controlled manner until its moisture content has reached a level suitable to the use to be made of the wood. This evaporation causes shrinkage, and we

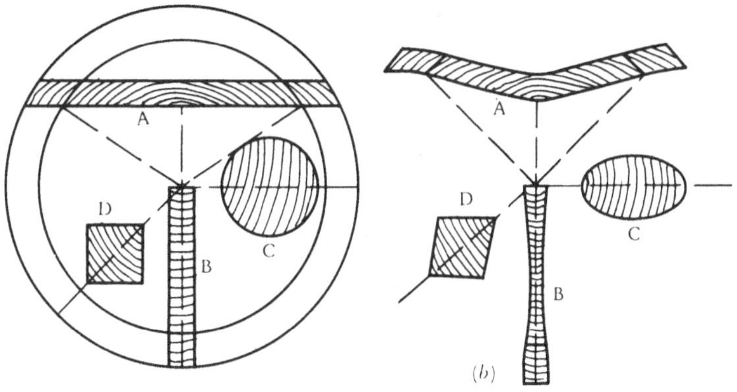

Figure 79
Anisotropic drying of timber after Tiemann[85]. *(a)* the cross-section of a tree trunk with the inner circle indicating the extent of the heart wood. Because radial contraction is less than tangential, pieces A, B, C and D dry as shown in *(b)*

want all the shrinkage to take place before the wood is fabricated. Thereafter it will, of course, swell and shrink by small amounts in response to changing RH but should, on average, remain the same size if the seasoning has been carried out correctly.

Although the moisture contents of different woods vary to some extent they are similar enough for us to give an average figure suitable for use in a particular climate and situation (*Table 33*).

In response to a change in moisture content the dimensions of a piece of wood change much more across the grain than along it. Even the cross-grain change is not uniform, being greater tangential to the trunk than radial. The result of these anisotropic changes will be apparent from a well-known diagram (*Figure 79*). The radial direction is from the centre of the trunk to the edge. Tangential is at right angles to this. Thus the most stable planks are radially cut.

Textbooks usually quote dimensional changes from the green unseasoned state to the dry state and relate these to moisture content. We are interested in the more modest changes of seasoned wood in response to RH changes in the middle of the range. These are roughly linear and the contrast between radial and tangential shrinkage is not so great. They do not change with age[274]. Some examples are given in *Table 34*.

By comparison dimensional changes caused by changes in temperature are small being typically 0.002% for a 5 °C change along the grain and 0.02% across it. This is why RH control is so much more important than temperature control.

The changes caused by RH variation shown in *Table 34* may seem small. However if a poplar panel 1 m wide across the grain and 0.5 cm thick were to have a completely moisture impervious coat on one side so that all shrinkage initially took place on the other, uncoated side, a 10% RH drop would cause it to bow so that, resting face up on a table, its centre would be about 10 cm above the table. Though a paint layer constitutes something of a moisture barrier and warps of this nature actually occur, the amount of warp for a 10% change would never be so

TABLE 33. Wood and Climate

Climate	*Examples*	*Moisture content*
Desert	Inland north Africa, Arabia, Dry south-west U.S.A.	6%
Cool temperate (rain all seasons, severe winter)	North-eastern U.S.A., Inland Europe, Central Iran	8%
Steppe	Hyderabad, Pakistan, Inland south Africa	8–10%
Warm temperate:		
Rain all seasons, cool summer	U.K., France	10%
Rain all seasons, hot summer with dry season (winter or summer)	South-eastern U.S.A., Spain, Italy, North-eastern Iran, North-central India	10–12%
Tropical:		
With dry season	South-central India	14%
Rain all seasons	South coastal India, Malaysia, Indonesia	15–18%

Notes
Generally recommended moisture contents for modern furniture and interior decorative wood.

NB. These figures have been compiled from a variety of sources which do not always agree, and so they can only be regarded as approximate. They are *not* made for museums. Since winter heating levels have risen greatly in the present century, the moisture contents quoted for cold climates are too low for antiquities.

TABLE 34. Dimensional Change in Wood
Per cent dimensional change for a 10% RH change in the middle of the range. (Figures are for 50 to 60% RH.)

	Tangential (%)	*Radial* (%)
Mahogany (*Khaya* spp)	0.45	0.30
Chestnut (*Castanea sativa*)	0.50	0.35
Walnut (*Juglans regia*)	0.50	0.40
Scots pine (*Pinus sylvestris*)	0.50	0.40
Poplar (*Populus canadensis*)	0.60	0.30
Oak (*Quencus robur*)	0.70	0.40
Birch (*Betula pubescens*)	0.85	0.65
Beech (*Fagus sylvatica*)	0.90	0.60

Notes
(Expansion along the grain for a 10% RH change might be about 0.03%).
These figures must be read as very approximate since dimensional change depends on the manner in which the wood has been dried, and also on where in the tree the wood is taken from. They are estimated from shrinkage from the green condition[85] taking into account the average moisture/RH relation (Stevens[276]).

great as this because moisture immediately moves in the wood to reduce differences[84].

The dimensional changes of canvas paintings are less predictable. In the first place multi-ply twisted thread may contract on wetting due to the extra twisting which results (*see also* p. 81), and in the second place there is usually a nett shrinkage when the material is first wet while free to contract. There is evidence that even century-old canvasses may be susceptible to this initial shrinkage, with potentially disastrous results. Nevertheless, in spite of these two contractile effects, we can expect a painter's canvas on a stretcher, whether sized, painted or raw, to expand when the RH goes up and vice versa, just like other water-absorbent materials. The amount of expansion is very variable, but appears to be rather less than the cross-grain expansion of wood.

As with the poplar panel we may consider a stretched canvas 1 m wide. If a 10% RH rise causes an expansion of 0.1% this will make the canvas sag about 2 cm in the middle if placed on a table, so that a really measurable change is involved. But here again the sag could be less in the practical situation if the paint layer and ground exercised some restraint on the canvas. Macleod reports an expansion of about half this on a canvas oil painting per 10% RH rise, while disclaiming attainment of complete equilibrium[275].

Paper comes in many forms, some modestly water-repellent. But the expansion characteristics of lightly-sized plain rag paper cannot by its nature be very different to the cross-grain expansion of the wood it came from.

DIMENSIONAL CHANGE IN PAPER

Machine-made paper expands more across the machine direction. Typical figures for a drawing paper for a 10% RH change in the middle of the range are 0.30% across and 0.05% along[277]. For handmade paper which expands evenly, we could expect a figure between these two.

DIMENSIONAL CHANGE IN IVORY

Lafontaine and Wood report dimensional changes around 0.3 to 0.4% for a 10% RH change[278]. By contrast, a very large temperature change (30 °C) caused less than 0.2% dimensional change.

DIMENSIONAL CHANGE IN PAINT

No figures could be found for dimensional change under normal RH, though this is clearly much less than for wood across the grain, but it is interesting to note that modern white lead paints (pigment volume concentration 45%) expand by 1.5–3.75% on immersion in water[279]. The figure will be less for more highly cross-linked mature paint.

DIMENSIONAL CHANGE IN CANVAS (SIZED, WITH GROUND)

Cornelius[87] and Berger[88] show how variable and inconsistent these figures can be, probably because canvas and glue oppose each other to a varying extent. His figures reduce to something of the order 0.05–0.1% expansion for cotton or linen (10% RH increase).

OUTDOOR CLIMATE AND RESPONSE OF OBJECTS INDOORS

We can continue a little further a topic broached in Part I, *Climate inside and outside the museum* (p. 87). Tables of RH throughout the year are readily available for most parts of the world[103]. But we cannot directly relate the behaviour of, say, wood indoors to the climate outside, even where there is no indoor heating. An example is shown in *Figure 80*. Here the wood was merely sheltered, and so followed the outside RH with a lag which would depend on its size, type and surface treatment, and also on whether it was extensively cracked and worm-tunnelled. The effect of the lag is to reduce the extremes of change somewhat.

There is no difficulty in principle in calculating the amount

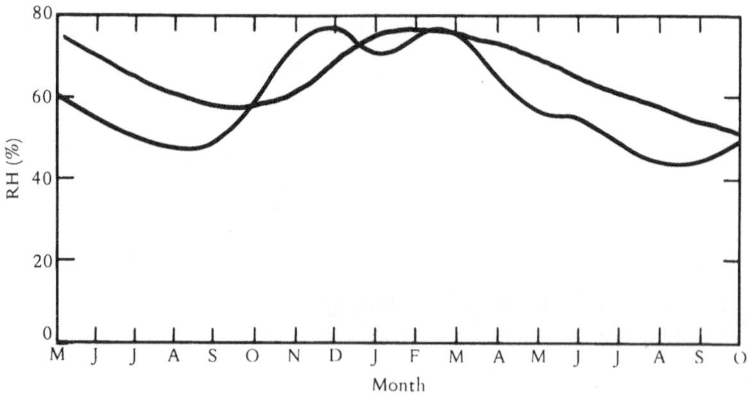

Figure 80
Lag behind atmospheric RH of wood in a sheltered location in Thessalonika, Greece. The record spans a year and a half, starting in May. Black curve, RH of the air: colour curve, moisture content of the wood in terms of equivalent RH (for relation between equilibrium moisture content of wood and RH *see Figure 86*). (Adapted from Tsoumis[280].)

of lag suffered by a given wooden object. If we equilibrate a piece of wood at one RH, say 40% RH and then transfer it to an atmosphere at, say, 90% RH, the rate at which it absorbs water will follow a curve of exponential decay, as can be verified, for example by examining some curves published by Stevens[281]. Such changes can be characterised by a half-time (*see also* p. 232), the time taken to reach the half-way condition, which in the above example is a moisture content corresponding to ½(90+40) = 65% RH. The half-time of Steven's unpainted 5-cm panels (of both beech and Scots pine) was about 15 days. We are generally more interested in dimensional than in moisture change, but the lag between cause (moisture change) and effect (dimensional change) is small for an unconstrained piece of wood of this size. For larger pieces we have to remember that the interior constrains the exterior until the new moisture content is more or less equalised throughout the wood.

Knowing the half time we can equally state how long it would take to reach any other stage of the change. We might consider that 90% of the theoretical change is a good measure of the greatest change likely to be suffered. In the above example, $T_{0.9}$, time to reach 90% of the change, means the time to reach a dimensional change corresponding to

$$40 + 0.9\ (90 - 40) = 85\%.$$

If $T_{½}$ = the half time,

$$T_{0.9} = \frac{\log_{10} ½}{\log_{10} 0.9}\ T_{½} = 6.6\ T_{½}$$

The 15-day half-time of a bare 5-cm panel implies that 90% of the total change will thus occur in about 100 days. But it is worth noting that Buck's figures for four 'small' painted panels[135] gave an average figure corresponding to a half-time of only 5 days. In this paper he asked how effective a hot wax treatment might be in stabilising a panel painting. The result was a half-time extended by about 6 or 10 times – 30 to 50 days. But the formula above can be used to show that, even then, 70 or 80% of the change will have occurred in a heating season of 100 days.

A single sheet of paper such as a mounted print would respond so quickly to RH change that it could be used to measure hourly variation. We have just seen that small untreated panel paintings might have half-times of the order of one week, and so would respond closely to monthly RH variations. The calculated response of a 5 cm (2 in) panel painting, with a half-time

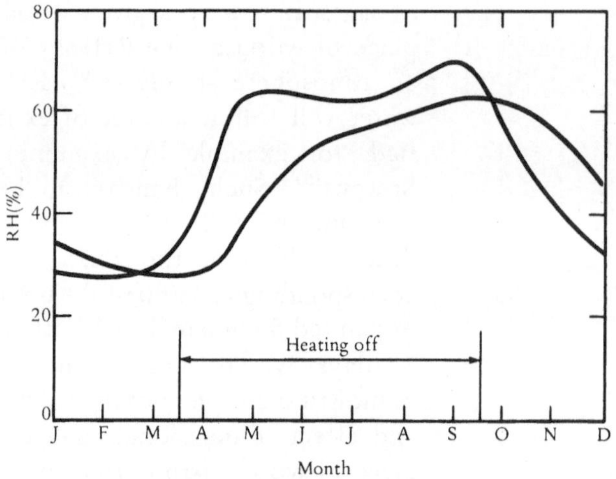

Figure 81
Lag behind atmospheric RH of a hypothetical 5 cm painted panel with a half-time of one month. Black line, expected indoor RH in London through the year, with a heating period from mid-September to mid-May at 20 °C; colour line, moisture content of panel in terms of equivalent RH

of a month (longer than Steven's 15 days because of the paint on one side) is shown in *Figure 81*. Though the panel might never reach equilibrium with its surroundings, its cross-grain dimension could change by over 3% in the course of the year.

Further than this we cannot generalise. Constrained and bulky wood probably moves discontinuously in response to RH change, perhaps giving way disastrously by suddenly cracking.

DOES CONSTANT RH KEEP DIMENSIONS UNCHANGED AT ALL TEMPERATURES?

One of the fundamental axioms on which we have built our conservation rules, one of the rocks in the foundation, is that, if we keep the RH of the air constant, no harm will come to moisture-absorbent materials through swelling or shrinkage however the temperature changes, within reasonable limits. We have been saying, in effect, that within reasonable temperature limits a constant RH implies a constant equilibrium moisture content (EMC).

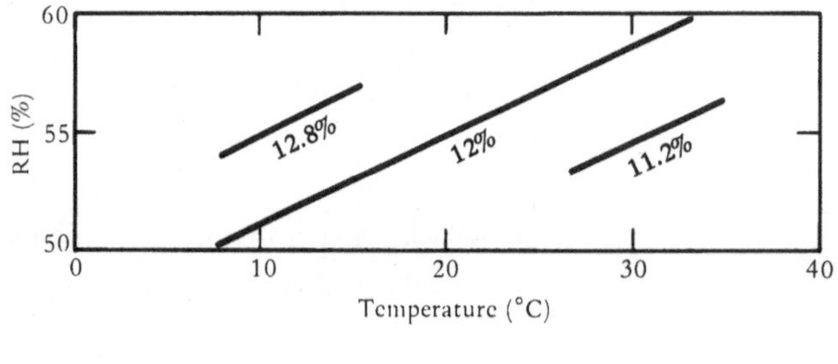

Figure 82
Centre colour line: RH required to maintain an equilibrium moisture content of 12% in wood under varying temperature

An average sort of wood has an EMC of 12% at 55% RH, 20 °C. The diagrams (*Figures 82 and 83*) show that, if we wish to keep the EMC absolutely unchanged under changing temperature, we have to adjust the RH to some extent, not keep it constant. Thus, for the large fall from 20 to 10 °C we ought to drop the RH from 55 to 51%. But most air-conditioning plants are set to run within ±5% tolerance limits so that it is beyond their capacity to make such fine adjustments.

All in all, the rule works out in practice and shows that we need not, indeed cannot, be highly accurate in our climate control.

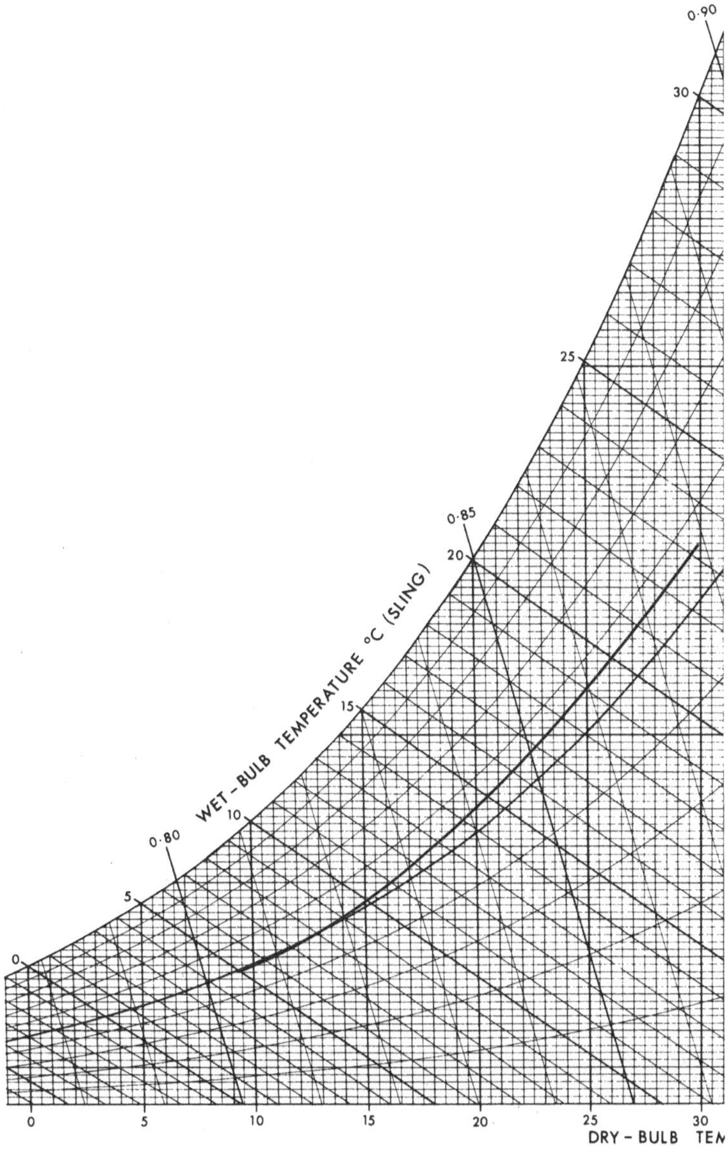

Figure 83
The information in *Figure 82* is here plotted on the Hygrometric Chart

EFFECT OF PEOPLE ON RH AND TEMPERATURE

In a crowded exhibition room people affect the climate by giving off both heat and moisture. The poorer the ventilation the greater will this effect be. But a heat input lowers the RH while a moisture input raises it. Will the RH rise or fall?

The sensible heat ouput of a standing person is about equal to that of a 100 W light bulb. Jones[282] gives a figure of 105 W at 20 °C, reducing to 72 W at 26 °C.

Moisture output per day in normal comfortable conditions (no feeling of sweating) is around 1000 ml from both skin and lungs, but can rise to ten times this figure during work in the tropics[283]. However the Standard Hygrometric Chart (*Figure 84*) allows us a shortcut on the calculations by using a figure for 'latent heat' in place of moisture loss. Thus, we can state that our standing man at 20 °C gives off, in addition to 105 W of 'sensible' heat, 45 W of latent heat[282], since this is the heat required to evaporate about 1500 ml of water per day at 20 °C (latent heat of water at 20 °C = 585 cal).

We can now use the protractor on the left of the hygrometric chart, which is marked for various ratios of sensible to total heat from an air-conditioning plant. We regard each person as a miniature air-conditioning unit. His sensible heat we have found to be 105 W and his latent heat 45 W, giving a ratio of 105/105 +45) = 0.7.

Whichever way we work the figures out, either by using the protractor on the Hygrometric Chart or by calculating the actual addition of moisture to the air we will find that the immediate effect of a mass of human beings in an insulated ventilation-free room is to raise the temperature and lower the RH (*Figures 84* and *85*).

Ventilation will reduce the effect but it will not alter its direction because the same proportion of heat and moisture enters the air whether it is replaced or not.

But this is not the whole story. The rest depends on ventilation and insulation. If there is not much ventilation and the wall insulation is poor, moisture will remain but heat will be lost. When the exhibition closes or when the church service ends the RH rises and, if the walls are cold, condensation may occur (*see also* p. 119).

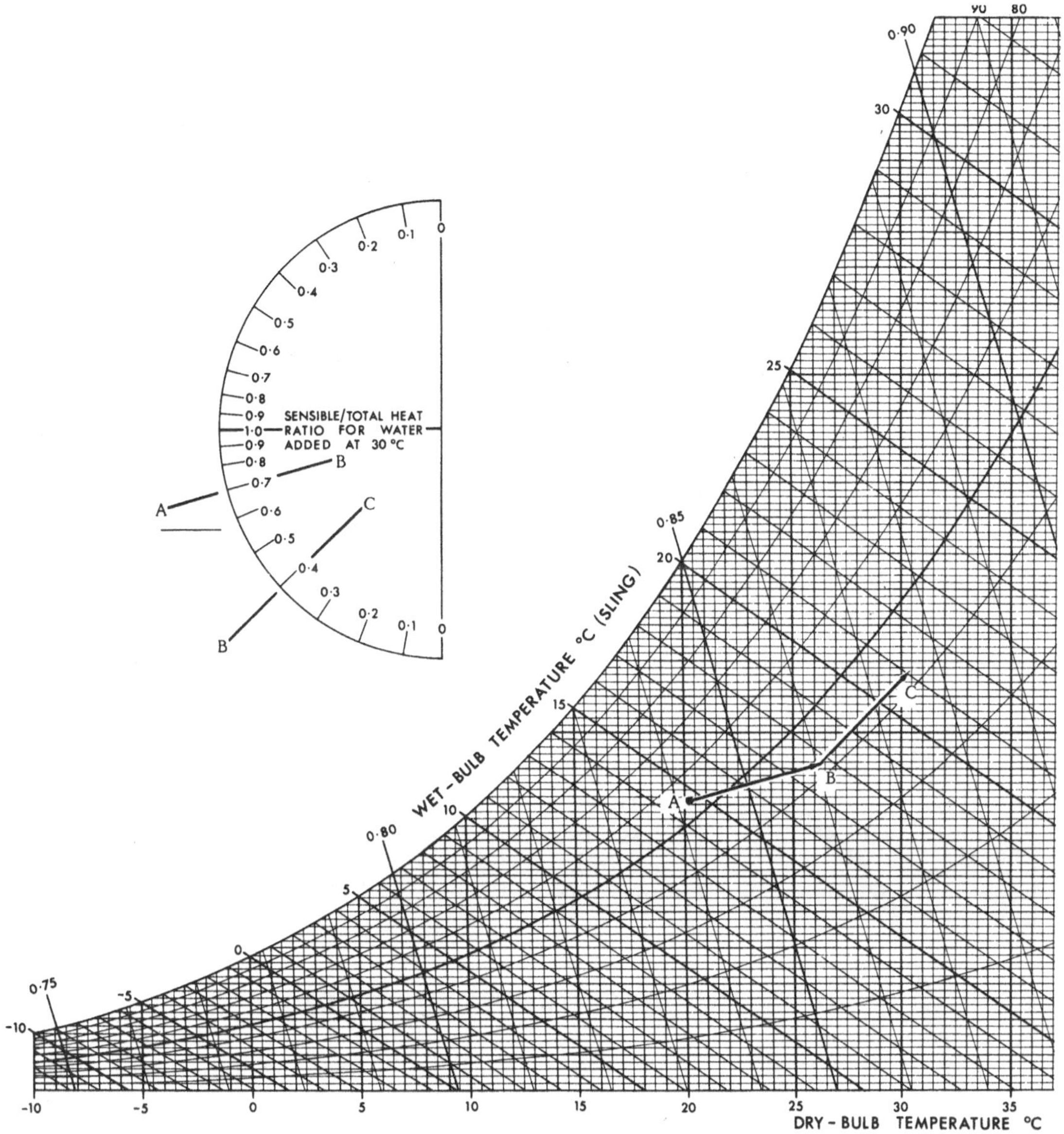

Figure 84
Calculating the effect of people on RH in an exhibition room. The arrow *AB* represents the direction of movement assuming no heat losses through the walls of the room. The direction of the arrow is obtained by using the protractor on the left (see text), though it was put there for calculating the effects of an air-conditioning installation

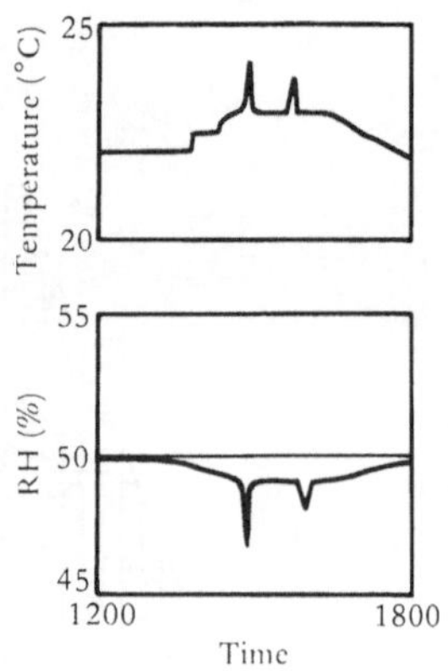

Figure 85
Effect of people on RH. 16 people visited the laboratory in the afternoon and were shown round in two groups. Each group saw, among other things, instruments for RH measurement and so gathered quite closely around the thermohygrograph from which this record was taken. The ceiling here is 2.5 m high, but the effective room volume cannot be given because areas interconnect. However both the general rise in temperature/fall in RH and the effect of clustering twice round the instrument can be seen

USE OF THE AIR MOISTURE-CONTENT SCALE

Throughout this book the emphasis is on relative humidity. But the air moisture content (MC) often provides a useful clue as to whether there are hidden sources of moisture such as damp walls. The air MC can be read off the hygrometric chart knowing the temperature and RH.

For a room ventilated with outside air the MC inside and outside should be about equal. If the room air MC is consistently higher than the outside air MC then an internal source of moisture can be suspected.

Figure 86
Equilibrium moisture contents (EMC) for some moisture-absorbing materials. Obviously the EMC varies to some extent depending on type and treatment[288]

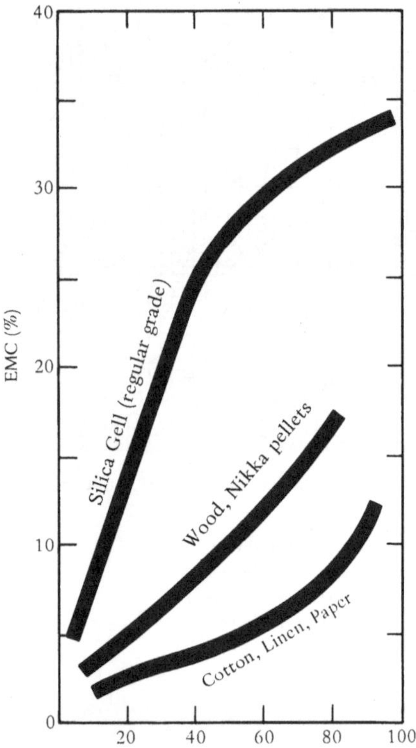

THE CLOSED AND BUFFERED MUSEUM CASE

There is now a body of literature on this subject. Since the theory is not obvious the literature tends to divide into two parts: theory and practice. For general theory and testing see refs. 123, 124 and 130. For experimental work on buffers see 288 and 291. For practice in the museum see 125 and 284–287.

HYGROMETRIC HALF-TIME

Progress is being made in providing 'alternative technology' RH stabilisation for museum show-cases, that is to say in providing a cheap non-mechanical method for stabilising the humidity of the air in a case through one whole season. The season might be a cold winter in the northern countries or a monsoon in the tropics. As Padfield has said, 'The show-case could be the greatest single aid to conservation'[289].

One radical answer is to seal the museum case hermetically: to make it airtight. However, whatever its advantages this turns out not to be cheap. In fact it is not at all easy to build a show-case that is airtight and yet can be opened from time to time by the curators. Of course if a case can be opened it will be opened, and probably not re-sealed properly. A truly airtight case would have to be permanently closed and constructed of glass and metal since wood and plastic are far too permeable. Having got this far we might as well replace the air in the case with an inert gas such as helium or nitrogen. All this has been done with care for the U.S. Declaration of Independence in Washington[290]. Nevertheless, although absence of oxygen is of general benefit, there are enough exceptions for this not to be advisable as a general rule.

We are therefore left with the leaky museum show-case and the problem of providing either a reservoir of water for arid winters or a controlled desiccator for humid monsoons. No elegant solution has yet presented itself, but as a step towards such a solution it might help to define the problem a little more closely.

There is hope that one day the mass of roaring machinery which is today's air-conditioning installation will shrink into a corner and even be rendered unnecessary in all but the most refined situations. The concept of 'hygrometric half-time' has been suggested in the hope that it will be useful in working towards this goal[123].

An unhermetically sealed case can exchange air and moisture with its surroundings by several mechanisms: air convection through cracks (aided by temperature differences within the case), water diffusion through the air in cracks, water diffusion through the materials of the case, pumping due to diurnal temperature fluctuations or to air pressure changes[130]. Padfield has established that pumping is negligible in normal circumstances[112]. Convection and diffusion are summarised in ref. 130.

A poorly constructed case or a case with sliding doors running on separate tracks can suffer an air change every couple of hours*, and it would obviously be hopeless to stabilise it. Cases much better than this have been disappointing[284]. A well constructed case with no cracks or warping and a snugly fitting door might leak at the rate of one air change per day. Ramer describes very carefully constructed cases with air change rates of one per 2 or 3 days[286]. At this level of control the limitation might well be diffusion of water vapour through the material of the case. For example diffusion through a sealed case made of 6 mm polymethyl methacrylate amounts to about an air change per 10 days[130].

*One air change does not mean that all the air in the case has somehow neatly slipped through the cracks to be replaced entirely by room air. If the rate of air flow through the cracks is V cubic metres/day and the case volume is C cubic metres, then the rate of flow can also be expressed as V/C air changes per day

Setting aside a goal of absolute stability, would it be possible for an orthodox case containing a reasonable amount of buffer to survive at an acceptable RH through a whole season of adverse room RH and then recuperate in the mild season? If so, we could achieve a virtually maintenance-free system. Adverse conditions are, of course, in the temperate zones, low RH during winter heating and, in the monsoon countries, high RH during the rainy season.

The first point is that such a 'natural' system will cycle through the year around an average RH which will be the *same* as the average room RH. In temperate zones with winter heating but with no humidification this average annual RH is a great deal too low. Estimated values (allowing for some summer cooling as well as winter heating in Washington and Ottawa) are: Ottawa 33%, Washington 40%, London 46%. In the inland monsoon areas, such as central India, average annual RH without any climate control is good (New Delhi 50%), but in the humid tropics it is hopelessly high (Singapore 76%).

However some humidification in winter, but not necessarily to 55% or to a precise ±5%, would bring European and North American museums to a reasonable average level. But in the humid tropics dehumidification and ventilation are the only answers, and the closed case is not a good idea since stagnant air promotes mould growth.

The next stage is to find out whether the system is indeed feasible. A theoretical examination[123] shows that, if we keep the room RH constant and start with a different case RH, then measure the difference between case and room RH, this difference will decay exponentially with time. Better known examples of exponential decay are the fall in temperature of a cup of hot liquid and the decay of radioactivity. Every radioactive isotope can be characterised by a half-life, which is the time taken for its activity to halve. It is no good trying to measure the full lifetime of an exponential decay process because it becomes slower and slower, and eventually infinitely slow. The half-time of the process, on the other hand, can be measured starting at any time during the process, and merely measuring the time taken for the variable we are concerned with (radioactivity, temperature above room temperature, RH difference between case and room) to drop to half its starting value. This rather mysterious property, which allows us to start measuring at any convenient time, is explained by the mathematics of exponential decay (*Figure 87*).

So our case with its buffer content can be characterised by a half-time, being the time taken for the RH difference between

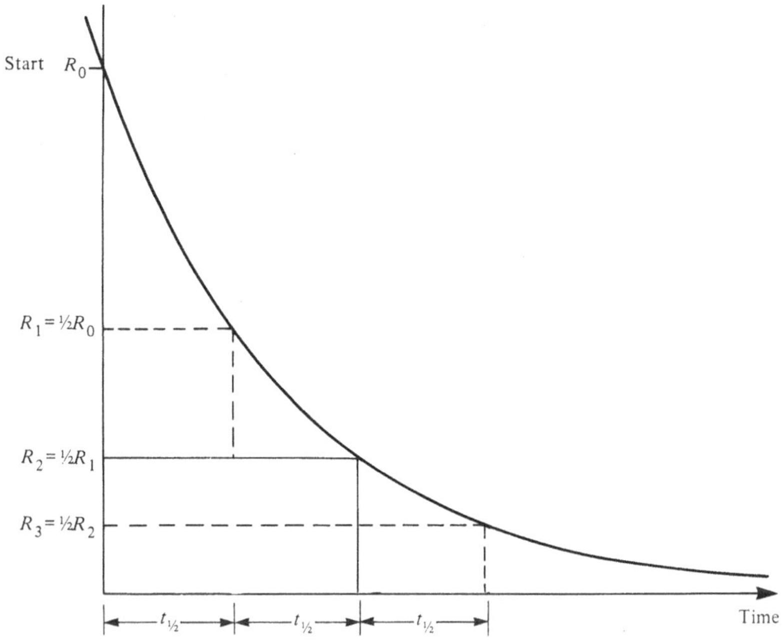

Figure 87
Exponential decay and hygrometric half-time. This curve shows three attempts to measure the half-time of a property R, say RH in a closed case, which declines exponentially. Each measurement gives approximately the same value for the half-time, $t_{1/2}$. We start timing when the RH is R_0 and note how long it takes to fall to half this value, $R_1 = \frac{1}{2}R_0$. This gives our first measurement of the half-time. Repeating the process gives us further measurements

case and room to fall to half its starting value. For example, a case conditioned to 55% RH is placed in a room at 35% RH. The time taken to reach ½(55+35) = 45% is found to be 67 days. Wishing to check this we start where we ended, at 45% RH, and must now measure the time taken to reach ½(45+35) = 40% RH. We find that this is again about 67 days.

The actual relation between the half-time of the case plus contents, its leakiness, and the amount of buffer, works out to be quite simple:

$$T = \frac{4MB}{N} \text{ days *}$$

where T = 'hygroscopic half-time' in days
N = number of air changes per day
B = quantity of buffer in kg/m^3 of case volume
M = 'specific moisture reservoir' of buffer, being the moisture in grams picked up by 1 kg of buffer on a rise of 1% RH.

Although all three of the variables, M, B and N can quite readily be calculated, the formula shows us without any calculation that we can double the half-time in three ways. We can

*This is a working formula, strictly only applicable at constant temperature. The dimensionally correct SI formula, where B is in kg buffer per *kg* air in case, and M is also in kg/kg, becomes: $T = 4760MB/N$.

double the quantity of buffer, or we can reduce the leakage rate by half, or (though there is less scope here) we can double the efficiency of the buffer, measured as Specific Moisture Reservoir. This last factor, *M*, now needs to be looked at.

MATERIALS USEFUL AS BUFFERS

A good buffer should be able to release or absorb a large quantity of water following a small change in RH. This implies a steep EMC curve rather than a high moisture content (*Figure 86*). The buffer should also respond quickly. Silica gel (regular grade) has a steep gradient at low RH. Its preparation as granules ensures a speedy response. At median to high RH, Nikka pellets and wood both have steeper curves than silica gel. Nikka pellets are particularly suitable because of their speed of response and lack of hysteresis (*see* below)[291].

M, the Specific Moisture Reservoir, can be obtained from the gradients of the curves. For all buffers it varies with RH (*Table 35*).

Inevitably there is a bit of hysteresis in these curves. This implies that, although we get the curves of *Figure 86* when we increase the RH and plot the EMC, if at any stage we reduce the RH we cannot follow quite the same curve downwards, but a flatter one. The buffer clings to its water and so gives up a little less than the curve would indicate[288]. The result is an *M* for regular-grade silica gel which is smaller when it faces drier air than when it absorbs from damp air. Some allowance has been made for this in composing the practical values in *Table 35*.

TABLE 35. M, Specific Moisture Reservoir, (g kg^{-1} % RH^{-1})

RH range:	*Low (30–50%)*	*Mid (50–60%)*	*High (60–80%)*
Silica gel (regular)	4	2	1
Nikka pellets	1½	2	3
Wood	2	2	2½
Cotton	1	1	1½

Note
These figures are approximate and can be changed (usually lowered) by treatment.

To get some idea of what the formula $T = 4\ MB/N$ implies, suppose we take a well made case with a leakage rate of one air change per day, and we require a half-time of 150 days. Then $MB = 150/4 = 37.5$. For either Nikka pellets or regular silica gel we can make $M = 2$. Then to summarise:

Half-time T = 150 days
N = one air change per day
Specific Moisture Reservoir M = 2

$$B = \frac{TN}{4M} = \frac{150}{4 \times 2} = 18.75 \text{ or approx. } 20 \text{ kg/m}^3$$

Hence the recommendation for a well made case of:
20 kg silica gel per cubic metre of case volume.

Making use of the exponential relationship we can simulate on a computer the behaviour of a case with known half-time in any annual climatic cycle. This has been done for a case with half-life = 150 days in New Delhi in *Figure 88*. It can be seen that a very satisfactory stability of ±4% RH should be achieveable without any mechanical device whatsoever.

PENETRATION OF OXYGEN AND WATER VAPOUR THROUGH PLASTIC FILMS

Attempts have been made to reduce the oxidation of both metals and organic material by placing them in sealed cases. Fabrication in acrylic sheet is easier than in glass. Will this keep out oxygen?

Polythene sheet is now widely used for storage and transport of museum objects, very often because it is supposed to be a

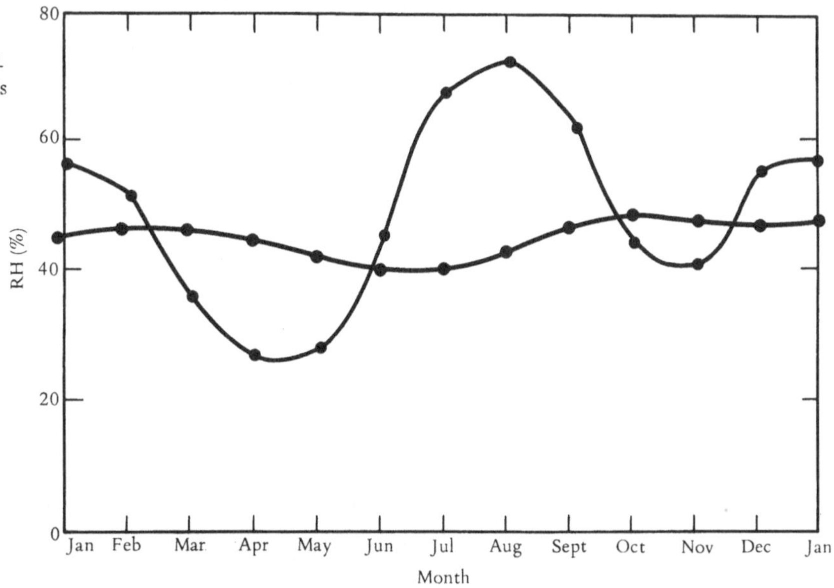

Figure 88
Computer simulation of the RH behaviour of a case with known half-time in an annual climatic cycle. This example shows a case with a half-time of 150 days responding to the annual climatic cycle of New Delhi

barrier to water vapour. How effective is it?

A knowledge of the permeability of plastic materials will enable us to answer these questions.

The permeability of a material to a gas is a measure of the rate at which the gas passes through a standard area and thickness of the material when there is a standard pressure difference between one side of the sheet and the other. The pressure difference refers to the partial pressure of the gas in question and not to any other gases present.

For a sheet of the material of area A and permeability P we can assume that

$$Q = PA \frac{dp}{dh} T$$

where Q = the quantity passing through in time T
dp/dh = the pressure gradient across the sheet.

For a thin film

$$\frac{dp}{dh} = \text{approx.} \frac{p_o - p}{h}$$

where p = the pressure on one side, say in a container,
p_o = the outside pressure, and
h = the thickness of the film.

So $$Q = \text{approx.}\ PA \frac{(p_o - p)}{h} T \qquad (1)$$

TABLE 36. Permeabilities of some Polymers[292]

	To oxygen 10^{-14} m^2 s^{-1} atm^{-1}	*To water vapour* 10^{-10} kg m^{-1} s^{-1} atm^{-1}
Polyvinylidene chloride	0.05–0.6	0.003–0.6
Nylon	2	0.4–10
Polyvinyl chloride (rigid)	4–7	1.5
Polystyrene	45–130	6
Polymethyl methacrylate	7.5	–
Cellulose acetate	30–60	9–33
Polytetrafluorethylene	100	0.2
Polyethylene (low density)	220	0.5–1.25

Note
Quantity Q is in m^3 STP for oxygen and kg for water vapour
Unit of pressure: atmospheres (atm)

The first question at the head of this section concerned an acrylic sheet, which usually means polymethyl methacrylate (e.g. Perspex, Plexiglas), to be used as an oxygen barrier.

We can imagine a metal case with one Perspex window of area 1 m^2 and thickness 0.5 cm (0.005 m), in which the air has been replaced by nitrogen. It contains unstable iron, so that any oxygen which does manage to diffuse through the window will react quite quickly with the iron and the full partial pressure difference between oxygen inside and outside the case will remain at 1/5 atmosphere. We can put our question into numerical form by asking how long it will take under these conditions to form 1 g of rust, Fe_2O_3, which is equivalent to fixing 48/160 = 0.3 g oxygen.

Since one molecular weight of a gas in grams occupies 22.4 l at normal temperature and pressure, 0.3 g oxygen occupies 0.3 X 22.4/32 l, or 0.3 X 0.0224/32 m³, so that in equation (1)

$$Q = 0.3 \times 0.0224/32 \text{ ml.}$$
$$A = 1 \text{ m}^2$$
$$h = 0.005 \text{ m}$$

from the table, $P = 7.5 \times 10^{-14} \text{ m}^2 \text{ s}^{-1} \text{ atm}^{-1}$

and $(p_o - p) = 1/5$ atm

From equation (1),

$$T = \frac{Qh}{PA(p_o - p)} = \frac{0.3 \times 0.0224 \times 0.005 \times 5}{32 \times 7.5 \times 10^{-14}} \text{ seconds}$$

T = about 800 days.

One gram of rust will be formed in 800 days, assuming that the case remains perfectly sealed during the period. We might regard this as a temporary rather than a permanent solution.

The second question, about the passage of water vapour through polythene sheet, requires us to develop equation (1) to give us a half-life of the kind described in the previous section. Suppose, for example, we had an object which was wrapped in polythene at 50% RH and the room RH then went up to 100% and stayed there. How long would it take for the RH within the wrapping to reach the half-way stage of 75% RH?

We re-write equation (1) in the form

$$\delta Q = PA\frac{p_o - p}{h}\delta T \qquad (2)$$

meaning that in a small period of time δ T sec a quantity δ Q kg of water vapour has passed through the film. But in the case as water vapour passes into the container the water vapour pressure difference falls, so that it is necessary to integrate. For convenience we first convert to concentration units:

If δQ kg of water vapour diffuse into a volume inside the polythene wrapping of V m³, the increase in the concentration of water vapour $\delta C = \delta Q/V$ kg/m³ (or g/l)

If the concentration of water vapour at any time within the wrapping equals C g/l, or 22.4 C g in 22.4 litres, then its partial pressure is 22.4 × C/18 atmospheres, since 18 is the molecular weight of water and the molecular weight in grams occupies 22.4 litres.

Therefore

$$p = \frac{22.4 \times C}{18} = 1.24\,C$$

and equation (2) can be rearranged to become:

$$\frac{\delta C}{C_o - C} = 1.24\,\frac{PA}{hV}\,\delta T$$

On integration this equation yields

$$\ln \frac{(C_i - C_o)}{(C - C_o)} = 1.24\,\frac{PA}{hV}\,T$$

where C_o, C_i and C respectively are the concentrations of water vapour outside the case, inside the case at the beginning, and inside the case at time T.

At the half time,

$$\ln 2 = 1.24\,\frac{PA}{hV}\,T_{1/2}$$

so that the half time $T_{1/2} = \dfrac{0.693\,hV}{1.24\,PA}$ (3)

To find the half-time of an actual example, we know from *Table 32* that P equals 0.065–0.17 for low-density polyethylene. Let us take the better of these two values, 0.065.

All that remains is to find the thickness of the polyethylene, h, the volume it wraps, V, and its surface area, A.

For a 125-micron thick polyethylene wrapped to dimensions 50 × 50 × 10 cm,

h = 0.000125 m
V = 0.025 m^3
A = 0.7 m^2

From equation (3)

$$T_{1/2} = \frac{0.693 \times 0.000125 \times 0.025}{1.24 \times 0.5 \times 10^{-10} \times 0.7} \text{ seconds}$$

$T_{1/2}$ = about 14 hours

This implies that the RH inside a polyethylene wrapping of the dimensions listed above, *containing no buffer*, which starts at 50% RH, will be 75% in less than 14 hours if the outside RH rises to 100%. Polyethylene is thus not very impressive barrier to water vapour. To improve the situation, for instance in wrapping an object for transport, we might use a polyvinylidene film such as Saran or a metallised plastic film.

Air Pollution
Part Two

Air Pollution

PLOTTING THE SIZE DISTRIBUTION OF PARTICULATES

Because our choice of air filter will depend on the sizes of particles floating in the air around our museums we need to know something about the distribution of their sizes. It might seem a simple matter to produce an ordinary histogram (or frequency distribution) with particle diameter along the x-axis and the proportion of particles in each diameter range represented by a series of vertical bars. The trouble is that we have to deal with a huge range of particle sizes from 0.01 to about 20 microns, a range of 2000 to 1. At the bottom end of the scale there is a great deal of difference between trying to catch on a filter a particle of 0.01 microns diameter and one of 0.1 microns, but there is no difference at all in this respect between particles of 9.91 and 10 microns, though the interval on a linear scale is the same in both cases. So a linear scale will not do, and a short diversionary explanation becomes necessary.

Being interested in the possibility of recording colour over a whole painting the National Gallery Scientific Department studied scanning methods which would break the image of the painting into small elements called, in the language of image analysis, *pixels*, and store colour information on each of these pixels successively. Since computer storage capacity is always limited, though it may appear to be vast, we needed information on the areas of paintings in the collection. If, for example, we find that a pixel size of 2 × 2 mm will do, and we have a storage capacity for up to ¼ million pixels the maximum area of a painting which we can handle will be 1 m^2 (500 × 500 2 mm squares). Will this cover a reasonable proportion of the collection?

A random sample of 100 paintings was selected by recording the area of every 20th entry in the National Gallery *Illustrated General Catalogue* (1973). A normal histogram of this sample (*Figure 89*) showed that about ¾ of the collection are estimated to be below 1 m^2 in area. But though this histogram answers our

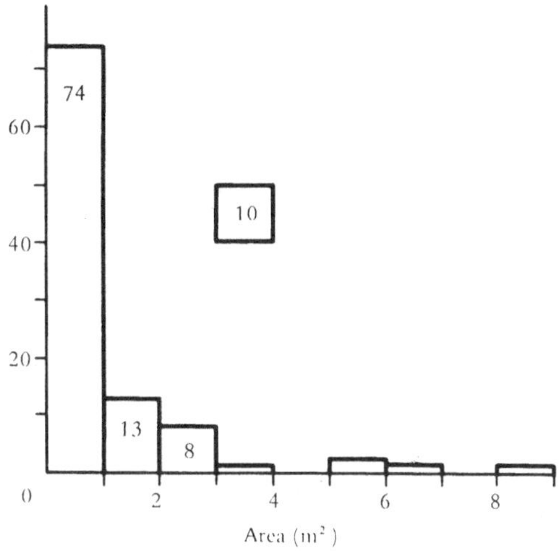

Figure 89
Histogram of National Gallery paintings. The areas of 100 random paintings from the collection were recorded by taking every 20th entry from the National Gallery *Illustrated General Catalogue* (1973). The resulting histogram shows, for example, that paintings of area between 2 and 3 m^2 form 8% of the collection, on this estimate

immediate question it gives us very little information about size distribution since most of the paintings are in the left-hand column.

Incidentally we should note carefully that the number of paintings in any size range is given by the *area* of the column, and this is the convention for *all* histograms. Since here the width of each column is the same, the height is directly proportional to the area, and so the number of paintings can be read directly off the y-axis. But as soon as we have a reasonable number of readings we shall usually want to turn a histogram into a curve – many of the curves used in physics and chemistry are in fact equivalent to histograms – and since the bars disappear in this process we can no longer label the y-axis in this way. Our development of the paintings histogram will make this clear.

If we attempt to extent the information about the pictures below 1 m^2 by subdividing our scale we shall find ourselves with a very long x-axis and the paintings still crowded up on the left.

The answer is to plot the distribution on a $\log_{10}$ scale (*Figure 90*). We now have the pictures divided up in a fairly symmetrical fashion and we can see that the average lies between 0.1 and 1 m^2, probably close to 0.4 m^2 (the mid-point on the log scale between 0.1 and 1 is 0.3162). Note that the number of pictures represented by each bar is given by the area of the bar, as before, though we are free, as in all histograms to put different scales on the x and y axes. Here, for example, 1 cm on the x-axis represents one unit, but 1 cm on the y-axis represents 5 units, hence the number of paintings represented by the 3rd bar is 1×28. The one unit on the x-axis for this bar is the difference between two logs, multiplied by 2 for convenience: 2 (log 0.3162

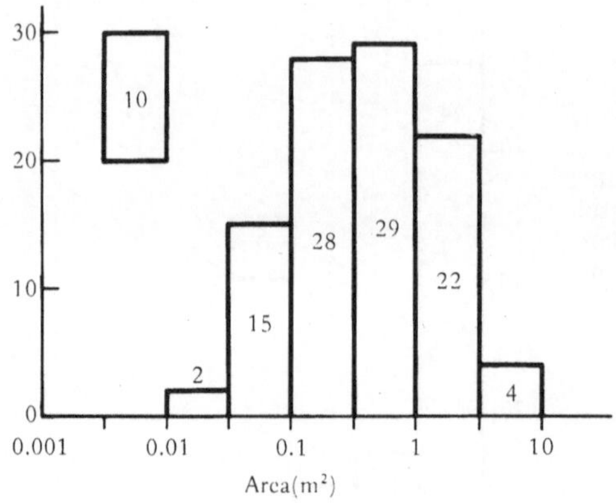

Figure 90
Histogram of National Gallery paintings on a log scale

– log 0.1) = 1. This holds for every bar. It also holds for every log histogram, though the scaling factor, here 2, is a matter of choice.

If we want to go further we can replace the bar histogram by a curve drawn through the mid-tops of the bars (*Figure 91*).

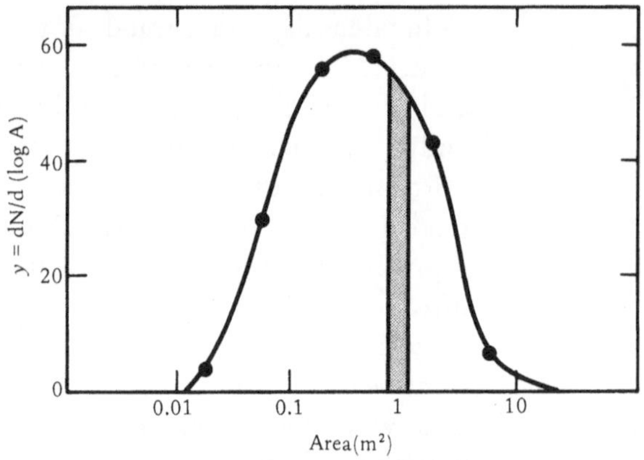

Figure 91
Curve derived from the histogram of *Figure 87* by joining the mid-tops of the bars. The proportion of paintings with areas between 0.75 and 1 m² is represented by the shaded region (*see* text)

This gives us more freedom, but at the cost of abandoning all hope of the y-axis giving us populations directly, since we are no longer relying on fixed-width bars.

The fact that the y-axis label is now $dN/d(\log A)$, or the rate of change of population with log area, does not greatly complicate the use of this graph if we write the following equivalence:

$$y = dN/d(\log A),$$

or for finite intervals

$$\Delta N = y\Delta(\log A).$$

This means that the number of paintings, ΔN, of area between A_1 and A_2

$$= y(\log A_2 - \log A_1)$$

Figure 92
Cumulative distribution of National Gallery paintings. By finding the point on the x-axis corresponding to 50% on the curve we can estimate that half the paintings will have areas less than 0.4 m^2

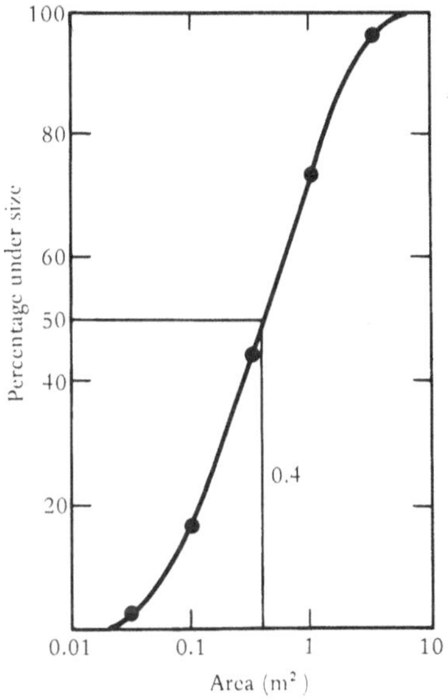

We can put this another way. We wish to find the number of paintings ΔN between A_1 and A_2. This is the area of the column drawn between A_1 and A_2. The width of the column is ($\log A_2 - \log A_1$) or $\Delta\log A$. The height of the column, y, being area/width = $\Delta N/\Delta\log A$, which for the very narrow columns of a continuous curve becomes $dN/d\log A$.

As an example, the proportion of paintings in the National Gallery collection with areas between 0.75 and 1 m^2 is represented by the shaded region in *Figure 91*: y at the mid-point = approx. 53, and ($\log A_2 - \log A_1$) = ($\log 1 - \log 0.75$) = 0.125. This gives us an estimate of $53 \times 0.125 = 6.6$ paintings in 100, or 6.6%.

Particulate distributions are today most commonly represented in this fashion, following the work of Junge[142]. The only differences are that we usually divide particulates into groups by diameter rather than area, and it is more useful to consider the mass of particulates in a given size range rather than their number.

In the hope that the reader will have become familiar with the painting distribution just described it will be helpful to use this same example to explain the other common way of representing particulate distributions: the cumulative distribution (*Figure 92*). Here the x-axis is the same as before – the areas of the paintings in m^2 on a log scale. The y-axis ranges from 0 to 100%. Its use can be illustrated as follows. From our first histogram (*Figure 89*) we discovered that 74% of the paintings are expected to be under 1 m^2 in area, and sure enough the curve in *Figure 92* gives 74% against 1 m^2. We can immediately read off from this cumulative distribution against 50% that half the collection will have an area less than 0.4 m^2 – information which we obtained rather less precisely from the histogram in *Figure 90*.

The particulates in the air – the actual suspended dirt particles – fall into three distinct size ranges (*Figures 93 and 94*)[293].

Starting at the large end, the upper size limit is governed by the maximum size of particle still in the air at some distance from its source. This is obviously a variable quantity: in a sand-

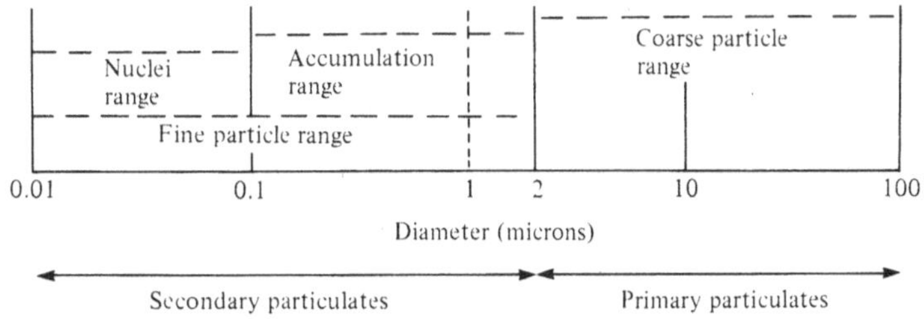

Figure 93
Classification of particulates

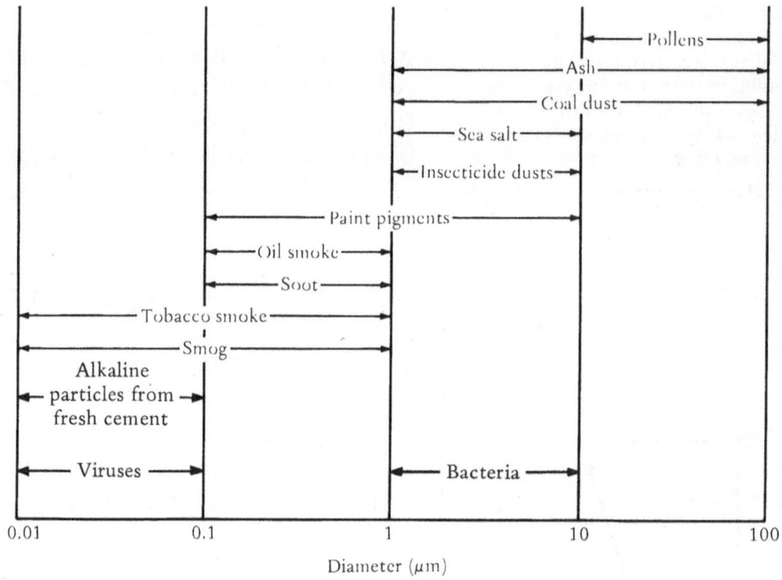

Figure 94
Size ranges of some particulates

storm grains well over 1 mm (1000 microns) in diameter can be carried large distances. In still air everything over about 15 microns settles, but smaller particles remain indefinitely suspended until they hit and stick to something. Our log particle diameter scale can be sure of covering everything that could penetrate well inside a building (apart from internally produced material like fibres and scales from the skin) if it stops at 100 microns, though there will usually be nothing much above 20 microns[294].

The size range of the largest particulates, from about 1 to 20+ microns, is called the *Coarse Particulate* range, because it has been formed either by direct mechanical action (e.g. dust formation, grinding processes, fine sand) or shot directly into the air as particles (e.g. ash, pollen, sodium chloride from the sea). Rain rapidly scavenges coarse particulates but has little effect on particulates below 2 microns.

But not all particulates are thrown into the air directly. Some are formed in the air itself. So we now turn our attention to the smallest particulates, in the *Nuclei* range, comprising all particles below 0.1 microns diameter. A more or less complete coverage is obtained with a scale that begins at 0.01 microns. One typical man-made source of these ultrafine particles is the automobile (there is a wider range from the combustion of coal), but nature produces them too from aromatic plant evaporates which polymerise in the air on dry sunny days to form a haze of Nuclei.

Whatever the source, the Nuclei are unstable and therefore

Figure 95
Distribution by mass of Nuclei produced from car exhausts near Los Angeles.
(Adapted from Whitby[297])

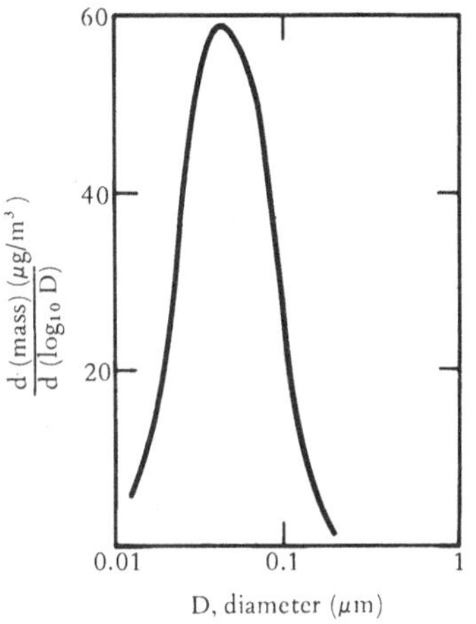

transient, with a half-life of no more than 12 hours[295]. Indeed to bag a fair sample of man-made transient Nuclei from a heavy traffic route the sampler should be less than a couple of kilometres away.

By coagulation the Nuclei transfer themselves to the Accumulation range[296]. This, our third and middle range, comprises diameters roughly between 0.1 and 2 microns. We shall now be able to see that this Accumulation range is the most important for museums to deal with and therefore dictates the type of air filter to use.

First, the black tarry matter in smog started life as gases formed in the higher-temperature combustion processes (cars and oil-fired power plants). Photochemical reactions yielded particulates in the Nuclei range, which presently coagulated into the Accumulation range.

Secondly, this range includes sulphates and nitrates, the end oxidation products of SO_2 and NO_2, together with ammonium.

Thirdly, it can be calculated that most of the surface in an average collection of particulates is on those in the Accumulation range[293]. All particles adsorb gases and vapours. Therefore, most of the adsorbed pollutant gases will be on particles in the Accumulation range.

For these three reasons it is most important to choose a filter which will stop particulates in this range, 0.1 to 2 microns in diameter.

Figures 95–100 are distributions of actual particulates in the air. Since, as we have seen, the Nuclei start to coagulate into the Accumulation range as soon as they are formed we cannot get a sample of air containing pure Nuclei under natural conditions. *Figure 95*, however, was obtained rather ingeniously from two particulate counts. In the first the wind was blowing directly

Figure 96
Distribution consisting mainly of aged particulates from car exhausts in the accumulation range (large peak), but with some Coarse particulates (small peak). (Adapted from Sverdrup[298])

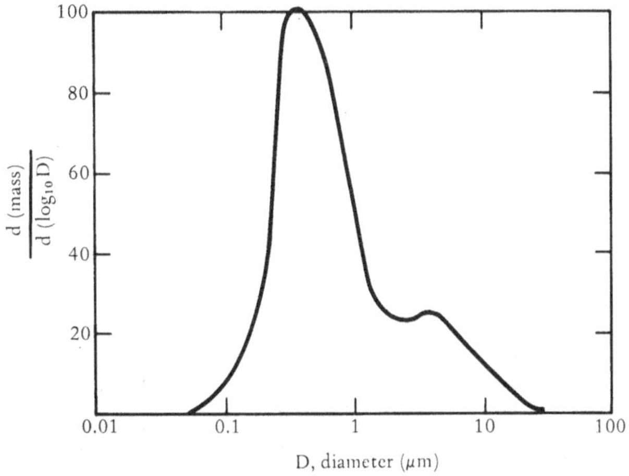

towards the sampler from a heavy traffic route during the rush-hour near Los Angeles. It was therefore loaded with Nuclei from automobile exhausts, though some had already started to age. In the second the wind was blowing from the opposite direction, still carrying predominantly automobile exhausts, but from some distance so that they had aged towards the Accumulation range and also become diluted. Subtraction of the second count from the first gave the curve of rather pure man-made Nuclei shown in *Figure 95*.

To make sure of a sample where the distribution of particles had aged, falling into the Accumulation range of about 0.1 to 2 microns, the sampler was taken out to the desert 200 km NE of Los Angeles where local emission of larger particles was very low and cars were not present. Nevertheless, when the wind was in the right direction a fairly large sample of aged Fine particulates carried from Los Angeles was collected and is shown in *Figure 96*. The small hump between 1 and 10 microns shows that some Coarse particulates were also present.

Figure 97
A typical bimodal distribution of particulates, the right-hand peak consisting of Coarse and the left-hand of Fine particulates. City Maintenance Yard, Denver, U.S.A., 1974. (Adapted from Willeke[299])

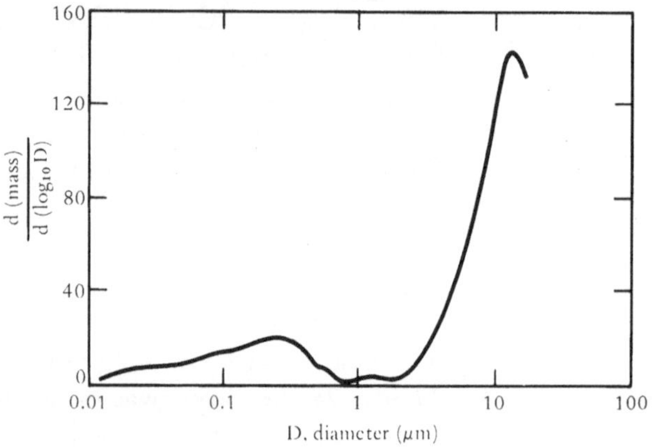

Figure 98
Distribution of total particulates in Nagoya, Japan, 1974. (Adapted from Kadowaki[296])

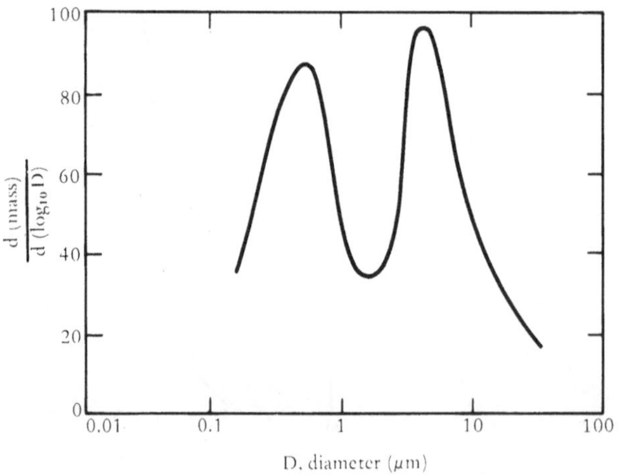

In general particulate counts have this bi-modal distribution, meaning that there are two peaks. The peak with mass maximum around 10 microns shows the Coarse particulates. The second peak, near 0.1 microns represents Nuclei and Accumulation particulates. Other of these bi-modal distributions are shown in *Figures 97 and 98.*

The particulate distribution shown in *Figure 97* also appears in two other forms in *Figures 99 and 100. Figure 99* merely

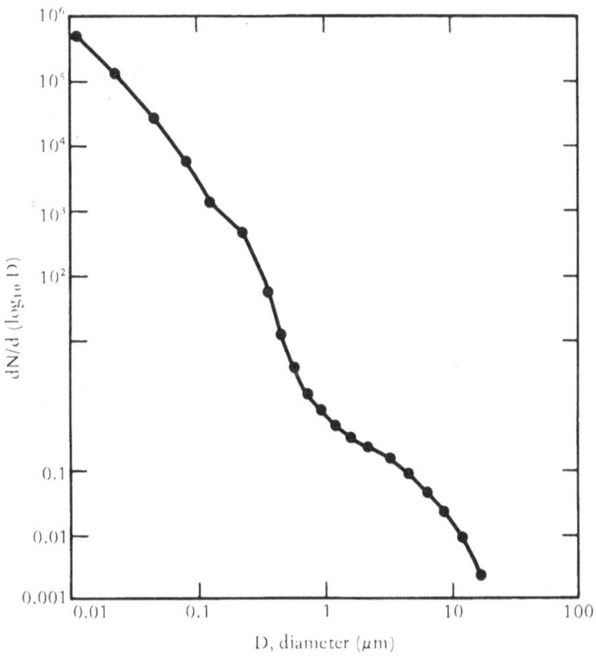

Figure 99
The particulates of *Figure 97* plotted as a number distribution

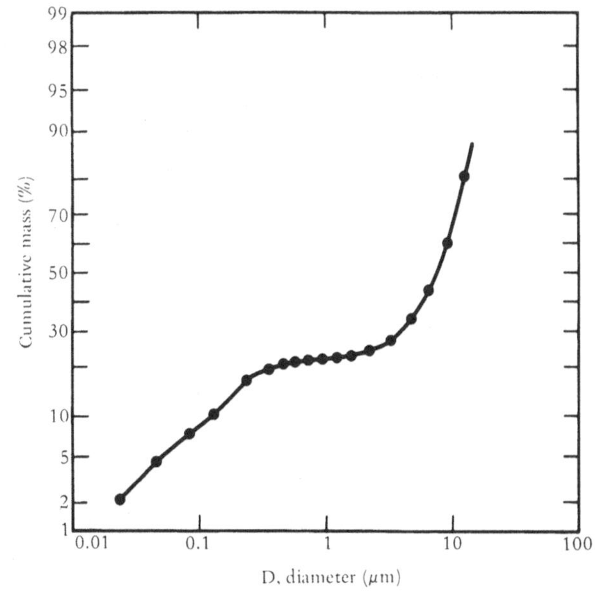

Figure 100
The particulates of *Figure 97* plotted as a cumulative distribution by mass

shows that there are vastly greater numbers of small particles than large, so that the *y*-axis has also to be put on a log scale to accomodate them all. But numerous though they are, the small particles contribute very little to the mass. We can get a good comparison of the mass of particulates above and below 1 micron by comparing the areas (which are proportional to mass) under the peaks in the bi-modal distributions (*Figures 97 and 98*). But a much better examination of mass distribution is obtained from the cumulative mass curve of *Figure 100*. Here we can read off directly, for example, that just over 20% of the mass is made of particles with diameters less than 1 micron. In all cases the mass below 1 micron comprises quite a large part of the total. Indeed in a selection of samples in London, Waller[300] found in all cases that the mass under 1 micron diameter comprised just over 50% of the total (*see also* Junge[203] p. 139 and Lee[301], p. 615).

CHOICE OF PARTICULATE FILTER

We are basically interested in reducing as far as is practicable the mass of dirt which reaches our antiquities, whatever its particle size. But filters have a characteristic lower limit so that below a certain particle size their efficiency deteriorates. We have just seen that this lower limit must certainly be well below 1 micron, otherwise the secondary particulates are going to get through. Therefore what we really want is a test dust corresponding in size distribution to the Secondary particulates.

This is what is provided by the Eurovent 4/5 Efficiency Test for the EEC area and the American ASHRAE 52/76. The main filters of a museum air-conditioning system should have an efficiency of about 85% (*see* p. 134).

We do not exaggerate the efficiency we require. Where doors are always swinging open and the public brings its own contamination it would involve mis-spent money and energy to pump the air through 'absolute' filters appropriate to operating theatres and radioactive laboratories. Also we must take into account the benefits of recirculation. Commonly 80–90% of the air in a conditioning system is recirculated, and thus a high proportion of the air passes through the filters more than once.

Recirculation has been taken into account in the curves in *Figure 101*. From this we can see that, as an example, with 10% fresh air a filter with a measured efficiency of 60% should take out about 97% by weight of the particulates in the measured range. To be on the safe side, let us take the Eurovent 4/5

Figure 101
Theoretical filter efficiency taking recirculation into account. (Adapted from Beal[302])

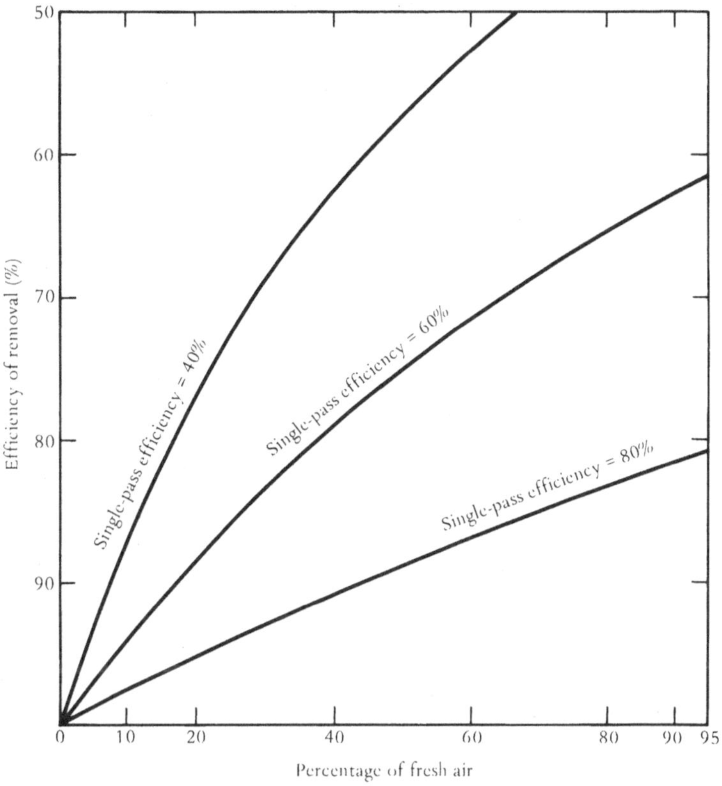

test dust to represent the whole of the particulate mass, not just the secondaries. Then a Eurovent 4/5 efficiency of 60% becomes at least 97% in practice for average urban suspended dirt.

This has been borne out by measurement. The filters in the first air-conditioned rooms of the National Gallery (Vokes 'Standard Kompak') had a Eurovent 4/5 efficiency of no better than 30%, yet measurements carried out by the Medical Research Council in 1959 showed efficiencies of over 90% in all cases, except where filters had been changed and some dust knocked into the ducts (a hazard of poor maintenance).

Since that time filter efficiency has improved, and it is now becoming the practice to specify at least 80% efficiency on Eurovent 4/5 or equivalent (as was done for the recent extensions to the National and Tate Galleries).

EFFICIENCY OF ACTIVATED CARBON FILTERS

Activated carbon is basically carbon prepared in granular form with a very high surface area for adsorption of gases. A good quality grade will have a surface area of 700 m^2/g. This can be raised to 1100 m^2/g but one should be wary that the result may be a granule of poor cohesiveness, leading to troubles with powdering. The granules are held in a mesh frame which is presented to the air flow obliquely as with particulate filters (see *Figure 58*). Activated carbon filters cannot be regenerated.

One test[303] on a 700 m^2/g activated carbon, gave:

gas	*removal efficiency at one pass*
nitrogen dioxide	90%
sulphur dioxide	95%

It should be remembered that recirculation will raise efficiency considerably. All forms of activated carbon are completely effective against ozone, and satisfactory with hydrogen sulphide.

ROOM AIR CLEANERS

Up to now we have assumed a ducted system, but room units for dealing with both particulates and gaseous pollutants can be effective. Such a unit would consist of a filter or filters mounted in a box with a fan to draw the room air through the filter.

Assume a fan passing a volume equal to the room volume through the filters in the unit 10 times an hour (10 air changes per hour). Assume also that the filter is at least 95% efficient for sulphur dioxide, as an example.

Tests show that if the room ventilation is completely shut off the concentration can be brought down to quite a low level (say 5–10 $\mu g/m^3$) in an hour.

In the more usual situation where there is some ventilation the reduction in pollutant can be rendered approximately as follows:

$$\text{Reduction in pollutant} = \frac{\text{Room ventilation rate}}{\text{Filter fan rate}}$$

For example, if the room ventilation rate is one air change per hour and air is drawn through the filters at 10 air changes per hour, pollution will be reduced to about 1/10th[304].

MEASURING CONCENTRATIONS OF POLLUTANTS IN MUSEUMS

These measurements are not lightly undertaken by a nonspecialist. Concentrations are, or should be, low and therefore the most sensitive equipment is needed. It is expensive and delicate. Alternatively, if averages over weeks or months are acceptable, rather than on-the-spot readings, absorption systems can be used, which depend on bubbling the air through, or passing it over, an absorption medium. A good example is the West-Gaeke method for sulphur dioxide. The absorption medium is an aqueous solution of sodium or potassium tetrachloromercurate, which forms a stable complex with sulphur dioxide. The sulphur dioxide is finally colorimetrically determined at 548 nm after adding pararosaniline. But these absorption methods require chemical manipulation.

Where possible an air-pollution laboratory should be asked to carry out a survey.

An example of on-the-spot analysis of each of the four main pollutants is given here. Sensitivities are suitable to museum purposes.

Sulphur dioxide: flame photometric detection. The air is fed into a hydrogen flame and light emitted by electronically excited sulphur species is detected with a photomultiplier.

Nitrogen dioxide: chemiluminescence. Light is emitted when nitric oxide (NO) and ozone react together. Therefore, nitrogen dioxide must first be catalytically reduced to nitric oxide, then mixed with excess ozone to produce the chemiluminescence. But since there may already be nitric oxide in the air the reaction is carried out both with and without catalytic conversion, and the nitrogen dioxide determined by difference.

Ozone: chemiluminescence. In this case the air is mixed with excess ethylene. The chemiluminescence produced when ethylene and ozone react is measured by photomultiplier. Another method involves direct measurement of ozone by UV.

Particulates: Particulate concentration is usually determined by weighing the solids collected on a filter or impactor from a known volume of air. The simplest devices measure total weight. But it is of interest to know the distribution by size (diameter). This can be done using a cascade or multistage impactor, where the air in its passage through the apparatus is caused to change direction so that it progressively deposits smaller and smaller sizes on collector discs.

THE FATE OF SULPHUR DIOXIDE IN THE ATMOSPHERE

We have already noted that about half the SO_2 in the atmosphere is man-made, half naturally produced[171]. The man-made SO_2 all comes from the burning of fuels, whereas the bulk of natural SO_2 starts life as the decay products hydrogen sulphide and dimethyl sulphide, which are oxidised in the atmosphere to SO_2.

Unfortunately in an industrial zone like Europe or North America this ratio man-made/naturally-produced is very different, only about a tenth of the SO_2 in the air being of natural origin. To deal in round figures, the worldwide natural background might be 5 or less $\mu g/m^3$ whereas we would not be surprised to find 50 $\mu g/m^3$ in any part of Europe, and higher in the industrial areas.

As the products of combustion leave the chimney and drift downwind some of the sulphur dioxide may impinge on an alkaline particle of grit and be neutralised. This falls out of the smoke plume, while the bulk continues on its way. The next hazard for SO_2 molecules is a water droplet. All water droplets have been nucleated on a particle of some kind, and such particles, having been formed by combustion, are quite likely to contain traces of iron, manganese, vanadium or other metals which have the ability to catalyse the further oxidation of SO_2. The SO_2 dissolves in the droplet as sulphurous acid, H_2SO_3, mostly in the form of the ions H^+ and HSO_3^-. Oxidation converts this to sulphuric acid. UV radiation can perform this operation without the help of a catalyst, but does not appear to be necessary. Estimates for the conversion of SO_2 to H_2SO_4 in airborne droplets vary from 5–25%[305–307]. Most of the oxidation to sulphuric acid takes place in the first hour of travel in the smoke plume[308]. Sulphuric acid, being hygroscopic, the droplet which contains it will grow in size, increasing its chance of precipitation. If it remains suspended it will continue to absorb and convert SO_2 so that its acidity will rise to perhaps pH4, a common figure for rain in the industrial world. By the law of mass action this accumulation of acid would soon bring further reaction to an end were it not for the presence of natural ammonia, another very soluble chemical, in the air. By neutralising the acidity the ammonia allows further conversion of SO_2 to occur. We now have SO_2, H_2SO_4 and $(NH_4)_2SO_4$ (as well as some $(NH_4)HSO_4$) in the air, the SO_2 being mainly gaseous, the others in solution in droplets. But most of the sulphur will still be unchanged SO_2. Rain will precipitate the droplet material in preference to the gas, but all will be adsorbed to some extent on surfaces with which the wind comes in contact. Thus SO_2 passes into the atmosphere and passes out again. Its residence time will vary greatly according to conditions. This might be as short as half a day[306,307] or as long as a fortnight. With a wind speed of 20 km/h even the shorter residence time implies a spreading of SO_2 for more than 200 km.

Thus as we proceed downwind from the source of pollution, the sulphur in the air is (a) diluted and (b) increasingly converted to acid. One result of this is that the rain in rural countries or areas adjacent to industrial zones, such as Scandinavia and the farming states of the U.S.A. is increasingly acid even when the SO_2 is low enough for lichens to grow freely.

The longer the sulphur resides in the atmosphere the more likely it is to end up in the form of ammonium sulphate. In fact it has been suggested that the widespread natural background of airborne sulphur, amounting to about 5 $\mu g/m^3$, is mainly in this form, and as such contributes to the formation of haze[309]. This is a good illustration of the difference between ordinary laboratory chemistry and its aerial form. Mix even highly dilute solutions of ammonia and sulphuric acid and they will have reacted with each other after one stir. Yet in the atmosphere, though both are present, their concentrations are so low that they coexist without ever fully reacting with each other.

A study of statuary in towns will confirm that the greatest erosion is on rainwashed surfaces. The gypsum ($CaSO_4.2H_2O$) resulting from the attack is carried away by the rain or by seepage through the stone, to be deposited elsewhere, very often on lower sheltered parts, mixed with soot. Up to 5% of industrial airborn sulphur is in the form of sulphuric acid[307,309]. Most of this will be washed out with rain so that rainwater will be the main source of sulphuric acid, but some will reach the stone in droplet form at high humidities, causing attack also in sheltered places. Direct gaseous attack may also be possible[310]. Sulphur-metabolising bacteria have been found in decayed stone, but have not been directly implicated in deterioration[311].

What does this mean in the museum? (a) Most of the sulphur will still be in the form of SO_2 in the town museum. A survey in the National Gallery, London showed that only 2–3% entered as H_2SO_4. (b) SO_2 is readily converted to H_2SO_4 on surfaces such as paper, textiles and stone, which usually contain traces of iron or other catalyst[308]. Droplets and particles with H_2SO_4 will be potent centres for metal corrosion. Fast corrosion characteristically begins at points. (c) Ammonium sulphate can also be expected, and may manifest itself as bloom on certain varnishes. Ammonium sulphate is a deliquescent salt which absorbs water at RH above 80%, but forms dry crystals below

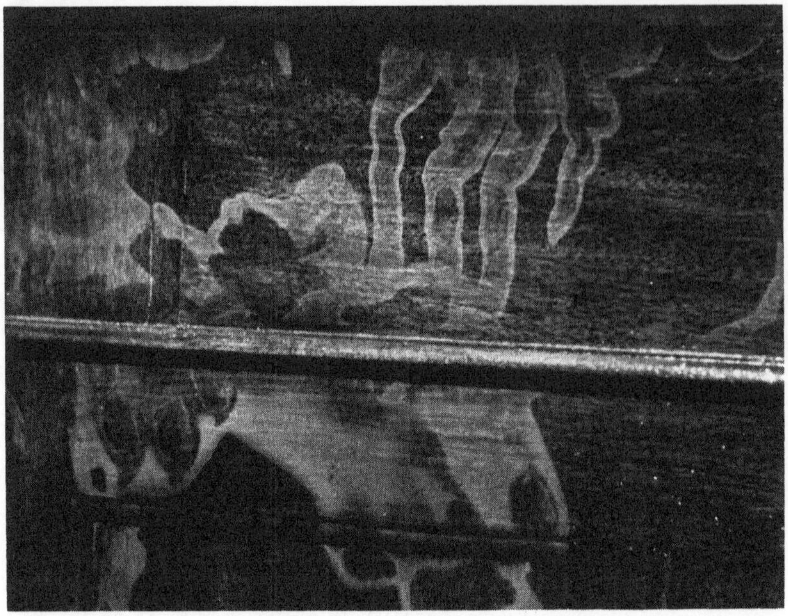

Figure 102
A large desk in a London office was disassembled and repolished (French polish, shellac). When dry it was reassembled. At this stage copious deposits of bloom formed where the fresh polish was touched, leaving a record of the movements of the polishers' hands during reassembly. Note particularly the upper right deposit of four streaks portraying the movement of the four fingers of the left hand skidding over the surface

this humidity. Fine particles of ammonium sulphate, probably less than 1 micron in diameter, if present in the atmosphere, will be deposited on attractive surfaces, though remaining invisible. Particularly favourable appear to be shellac and natural resin varnishes after a week or two of oxidation. Any episode of RH above 80% will allow the crystals to take up moisture, form into droplets on the surface, and, as the RH falls, to recrystallise into much larger crystals, 50 microns or more in diameter, though a surface covered with 5 micron crystals will exhibit the characteristic appearance of bloom. Human sweat encourages the deposition of SO_2 [312] and also of bloom formation (*Figure 102*) tempting one to suppose that the ammonia is supplied by the sweat.

THE FORMATION OF OZONE

Ozone is created in two processes, one natural and one artificial. Natural ozone is formed in the upper atmosphere (maximum concentration at about 30 km) by the action of UV radiation on oxygen:

(i) Oxygen is split into two highly reactive oxygen atoms.

$$O_2 \rightarrow O + O$$

(ii) Some of these react with further oxygen to produce ozone.

$$O + O_2 \rightarrow O_3$$

In this reaction the excess energy of the oxygen atom must be carried off by collision with another unspecific molecule in the air, otherwise the ozone would fly apart as soon as formed.

The presence of this ozone over our heads is of vital importance to life because it filters out the dangerous UV radiation of wavelength shorter than 300 nm. Some of it diffuses down to ground level, accounting for a natural ozone background of from 40–80 $\mu g/m^3$.

Whether or not this natural background of ozone contributes significantly to the deterioration of antiquities is still an unanswered question[210]. My own opinion is that it probably does. It is unquestionably highly reactive to a wide variety of organic materials (p. 150). Therefore it is certainly of concern to us that the activities of man can considerably increase the concentration of ozone – to around 500 $\mu g/m^3$ – not only in Los Angeles, where the reaction was first studied, but all over the industrial world.

Characteristic weather conditions necessary for its formation are anticyclonic: clear skies and low winds. For a build-up of pollutants a temperature inversion can put a lid, as it were, over an industrial area or town. A temperature inversion means that the usual conditions, where the air gets cooler with height, do not occur: the lid consists of a layer of warmer air. Thus pollutant gases cannot rise and disperse by their relative warmth, and so their concentration rises.

The series of chemical reactions which result in man-made ozone start with nitrogen oxides from automobile exhausts. Nitrogen oxides occur in considerable variety in the exhausts from all combustion processes, but nitric oxide, NO, is predominant in car exhausts. Typically nine molecules of NO are here produced for one of NO_2, the next commonest oxide[216].

Now a series of complicated and interlocking reactions occur (an excellant review is given by Durbin and Hecht[313]). The reaction that sets the whole process going is the breaking up of nitrogen dioxide, NO_2, by sunlight to form NO + O. The oxygen atom created by this reaction combines with normal oxygen to produce ozone, just as it does in the stratosphere, and this

Figure 103
A flow chart of the changes involved in the formation of ozone from the nitrogen oxides from automobile exhausts

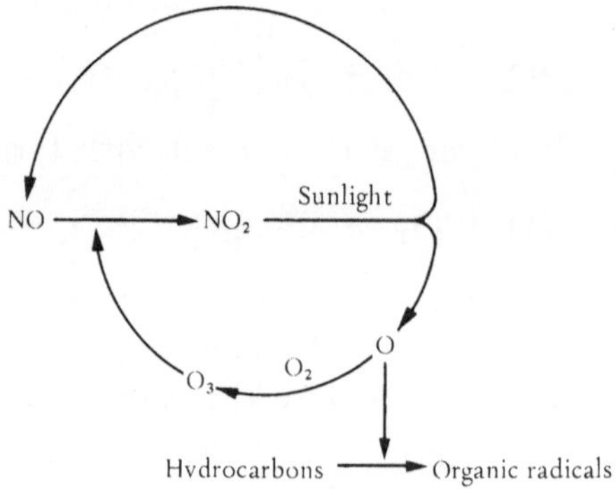

ozone then attacks NO*. *Figure 103* shows the changes involved better than the usual list of chemical reactions (adapted from Durbin and Hecht[313]).

Once the cycle has started up with the breaking of the first NO_2 molecules, the ozone formed by the combination of an oxygen molecule and an active oxygen atom oxidises NO to NO_2. Some oxygen atoms also attack unburnt hydrocarbon material in the exhaust producing reactive organic radicals. Not shown in this diagram are many further reactions which follow the production of organic radicals. The notorious end-product PAN (peroxy acyl nitrate) is formed by a series of reactions involving both NO and NO_2 with organic radicals.

The diagram tells us little, however, of the relative strengths of the reactants. Observation shows the following pattern:

1 Traffic increases in the morning before the sun is high, causing NO and hydrocarbons to build up in the atmosphere.

2 Sunlight, as it gets stronger, sets the cycle going. Ozone remains low so long as it is being used up in attack on NO and hydrocarbons.

3 The final stage occurs after NO_2 has taken over from NO as the predominant nitrogen oxide. The NO supply is reduced as the gases diffuse over the city. Now the concentration of NO_2 in turn falls as it takes part in the formation of ozone and PAN. Since ozone has less NO to attack its concentration rises.

Thus the chief final pollutants are ozone and PAN, which characteristically build up to a maximum shortly after midday

*Chemists will have to forget one well-known laboratory reaction: the spontaneous combination of the colourless nitrogen oxide, NO, with the oxygen in the air to form brown fumes of nitrogen dioxide, NO_2. At the very low concentrations we are here concerned with, this reaction becomes of no importance[212].

Figure 104
Build-up of ozone (colour curve) and PAN (black curve) during an average day in Delft (wind speed 2–4 m/s[1]). (Adapted from Nieboer[314])

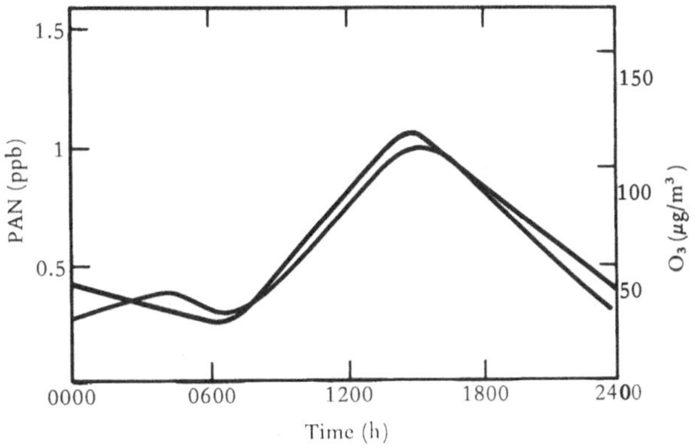

(*Figure 104*), and NO_2 whose concentration is highest in the evening. PAN is troublesome to all living things, both animal and plant, but probably of minor importance to antiquities. Ozone, as we have seen, is a potent antiquities poison.

Because sunlight is stronger in summer, there is created an annual cycle for ozone, man contributing the extra during the summer. Ironically, the better the regulations against producing smoke, the sunnier life in the city becomes, and therefore the more ozone is produced.

Ozone is so reactive that it was supposed until recently to have a lifetime of only a few hours in the open air. Cox and others[315] have shown, however, that, like SO_2, it can be transported over distances up to 1000 km from its place of origin, implying a lifetime in the air of one or two days. Thus, for example, with the westerly winds associated with a 'high' (anticyclone) over N.W. Europe, ozone produced in France can be transported to the remotest parts of W. Ireland and be measured there at 3 times the background level.

The possibility of ozone acting as an accelerator of photo-oxidation has been raised byBriner[209]. One could conclude from a paper by him that each molecule of ozone reacting with the surface of an exhibit causes around 100 000 oxygen molecules also to react with it. If this is the case, concentrations of ozone much lower than the natural background level could influence the deterioration of all organic material in antiquities. This is at present an open question, but it is obviously of prime importance for future scientific research on preservation.

Fortunately the end of the story is not so black. Few studies of the indoor concentration of ozone have been made, but two are reassuring, two less so. Derwent[219] carried out measurements in the National Gallery when the the concentration outside in Trafalgar Square was 80 $\mu g/m^3$ and found vanishingly low concentrations (less than 0.5 $\mu g/m^3$) in the exhibition rooms, whether air-conditioned or not. Mueller and others[218] suggest from their experiments that the half-life of ozone indoors (in a typical bedroom) is about six minutes, because of its close proximity to surfaces with which it reacts. But Davies *et al.* found ozone at the Sainsbury Centre (Norwich, U.K.) to be 70% of outside concentration[316], and ozone in a Pasadena (Calif.) Gallery was half that outside[317].

Of course, looked at the other way round, this is just another illustration of how reactive ozone is, and should act as a warning against its *internal* production, as from electrostatic air filters (p. 135) or certain office copying machines using light sources with a lot of UV radiation. Usually in the latter case the ozone and nitrogen oxides can be smelt near the machine. Naturally such machines should be allowed nowhere near any exhibits.

Activated carbon filters will deal with ozone even more effectively than with sulphur dioxide.

The reader may be puzzled by reports that the stratospheric ozone layer which protects us from short-wavelength UV could be destroyed by nitrogen oxides from the exhausts of high flying supersonic jets. We have just been convincing ourselves that ozone is produced, not destroyed, by a similar kind of emission from cars. But chemists have long known that all reactions can go two ways, either backwards or forwards, and the direction depends on the relative concentrations of reactants and products. Indeed, though there is agreement that in upper atmosphere conditions there will be a net reduction in ozone concentration, the argument remains open as to whether this will be significant or not.

COMPUTERS IN ENVIRONMENT CONTROL

Obviously computers, here as elsewhere, will find their use in the construction of data bases using off-the-shelf software. No programming ability is required for this.

Some of the calculations in this book, however, will be greatly aided by a computer, and one – the colour-rendering calculation – is impractical without one. The programs will have to be written by the user.

The algorithm for the colour-rendering calculation is set out in adequate detail in ref. 261. During the course of this calculation the correlated colour temperature is required. An algorithm for this due to Robertson is described in ref. 66.

Hygrometric calculations are covered admirably by Carp[318].

DATA LOGGING

Data logging in the museum (the regular collection of relative humidity, temperature, light and other readings) divides into two stages[319,320]. The first stage consists in the collection of readings at a chosen interval (say 1 h), their optional display and their temporary storage in computer-readable digital form on one of the regular media (magnetic or solid-state). Packages can be bought for this stage.

The second stage comprises displaying and analysing the data and finally putting it into permanent storage on some high-capacity medium. Optical discs may soon be the first choice. This second stage depends very much on the museum and will therefore require a special program.

As an example of the processing in stages 1 and 2 (though how the stages are divided will vary), data may be received in millivolts from a photocell allotted channel X. In stage 1 hourly values from channel X will be stored as millivolts. It is only due to the second-stage programming that channel X is converted to, say, the south wall of room A and millivolts are converted to lux. A list of rooms with their lux values can then be put on the screen or printed out, and finally stored.

It aids compact storage if each reading is compressed to 1 byte (8 bits, a number in the range 0–255). Thus RH can be stored to the nearest whole number, temperature to half a degree by multiplying by 2 and storing as an integer, lux by dividing by 5.

A system for logging RH and temperature as well as some light levels has been operating in the National Gallery, London, since 1980 over a portion of the Gallery involving about 80 channels[319]. An RH/temperature sensor pair is mounted in each room at picture height. A cable runs from each pair to a logger in the engineer's control room. The logger prints out the current readings at intervals and warns if readings fall outside the specification. On the hour the logger sends all current data up to a small dedicated computer in the laboratory for storage over one week. Every Monday the data are transferred to a general-purpose computer for the production of summaries, graphs and tables and finally for permanent storage in condensed form on floppy disc. The program devised for this computer also allows access to the current situation in the exhibition rooms via a room plan which appears on the screen with RH and temperature data written in each room.

A slightly different system was brought into use at the Australian National Museum in Sydney shortly afterwards[320].

It uses a single small microcomputer linked to an interface unit which serves to read all channels sequentially into the computer. RH and temperature are recorded at 10 min intervals rather than 1 h in the London system. Data are stored temporarily for one week and graphed as they are stored. At the end of the week summaries of the data are printed.

Whatever configuration is used, minimum equipment comprises sensors, cables, a microcomputer, a means of reading all channels sequentially into the computer at fixed intervals (a logger), a storage device such as a floppy-disc drive, and a printer with graphics capability, e.g. a matrix printer.

In the future, to eliminate the logger and much of the cabling, it is likely that groups of sensors will be linked by short cable runs to a micro which will digitise the data and be addressable by the main computer. A single cable loop will then serve all sensors in the building.

Future Trends in Environmental Control

Every conservator who has consulted with colleagues round the world knows the situation. A small group stands round an exhibit in bad condition and discusses treatment, or indeed whether treatment is required. Someone asks the question: 'Has it changed over the last 10, 20, 50 years; how fast is it changing now?' Members of the group then venture their opinions, but no satisfactory records are available. There may be a book or a report with a photograph. The date of publication is known, but the date of the photograph remains uncertain, and in any case the details required are not clear.

Such a situation could never happen if museums were planned with preservation in mind. But planning for preservation, totally lacking in the past, is even now tacked on as an ancillary. Even where the facilities for preservation are generous, the main objective of restorers is still supposed to be what amounts in the main to cosmetic treatment, and of scientists the provision of analytical results.

It is hardly necessary to argue, however, that the very basis of a rational approach to preservation is a thorough knowledge of what is changing, how fast it is changing, and why it is changing. Thus although analysis will still be the basic activity of the museum scientist it will have an expanded aim: not only to provide data for restorers and historians but to investigate deterioration.

Methods for measuring change depend on what is to be measured, and this will usually be some visible property such as colour, cracking, corrosion or distortion. Photography can be made to record most of these visible changes with great sensitivity if properly planned, that is to say, if every aspect of the process can be repeated with sufficient accuracy and the record made sufficiently stable. For colour, reflectance spectrophotometry becomes important. In this whole field there are interesting prospects for development, but they are outside the scope of this book. The point to be emphasised is that the controls for preservation which *are* the subject of this book will depend for their advancement on very specialised and accurate record-keeping of this kind.

Museums themselves are changing, and the changes most obvious to the visitor are in display. They can be called evolutionary insofar as they present with greater truth, clarity and pleasure. Where fashion takes a part, however, – and fashion is deeply woven into the texture of all art appreciation – the conservator can be excused a little impatience. Fashion has in the past, and is still liable to, promote destruction rather than conservation. Designers very ignorant of the requirements of preservation may be assigned great responsibility, and may even avoid enquiring through fear of obstruction in their aims. Actual examples spring to mind of glaring illumination, Sahara dryness, and display techniques which involve dangerous manipulation of the exhibition material.

But conservation too has its dangers. The extremists will always be with us. His Lordship of whom it was reported in 1923, in commenting on the use of scientific investigations to preserve the colours of museum specimens, that 'he believed in using the good things of the earth as the Almighty had given them to us and not sacrificing too much to posterity'[20], is matched by the enthusiast who proposed deforestation, soil removal and tarring the entire Ajanta cave area[321] in order to protect the cave frescoes.

There are those who would build a mass of machinery into the museum, while others would struggle to keep it a haven from the technological excesses of the outside world. This book is, of course, about machinery and instruments, which I must confess to enjoy. But the aim must always be that the machinery of preservation should be unobtrusive, never standing in the way of direct enjoyment, aiding it where it can.

The difference between lighting a building meant only for people and lighting sensitive exhibits is that in the former we need only model the light so that people feel that it is right, whereas in the latter we are also fighting against its destructive energy which is measurable and must be measured. So, as we sally forth with light meter and sling hygrometer, we must be prepared for such pleasantries as greeted the architect who was diagnosed as suffering from 'foot-and-candle' disease.

In actual fact hygrometers are becoming more necessary than ever as we enter a period when the major museums, though claiming to be air-conditioned, may actually be running wildly outside their specifications. Maintenance staff soon learn to sum up the situation. Are they required to keep the first class standards of air-conditioning appropriate to a delicate collection, or merely the standards acceptable in an office block where nothing gets adjusted or checked until people start complaining of being too hot or cold? The only way of avoiding this danger is to run a regular independent check, and to allot this task to an officer of the museum officially responsible for the environment. Whether or not air-conditioning is installed, such an officer will always be needed.

Nothing should be installed which cannot be maintained, and seen to be maintained.

There is something inelegant in the mass of energy-consuming machinery needed at present to maintain constant RH and illuminance, something inappropriate in an expense which is beyond most of the world's museums. Thus the trend must be towards simplicity, reliability and cheapness. We cannot, of course, prophesy what will be developed, but I should guess that it will include means for stabilising the RH in showcases without machinery, use of solar energy for RH control in the tropics, improved building construction to reduce energy losses, and extensive electronic monitoring.

APPENDIX — SUMMARY OF SPECIFICATIONS

Specifications for Class 1 and for Class 2 environments follow.

Class 1 is appropriate for major national museums, old or new, and also for all important new museum buildings.

The aim of Class 2 is to avoid the major dangers while keeping cost and alteration to a minimum. For example, some of the most important historic houses and churches may have to be limited to Class 2 operations. Neither ducted air conditioning nor automatic light control are assumed for Class 2.

CLASS 1

LIGHT

Limitation of illuminance:

Moderately sensitive material (e.g. oil paintings)		
Daylight or artificial light at	200 ± 50 lux at eye level	(pp. 23 and 33)
Alternatively an annual exposure of not more than	650 kilolux-h	(p. 30)
Sensitive material (e.g. textiles):		
Artificial light at	50 lux	(pp. 23 and 33)
Alternatively an annual exposure of not more than	200 kilolux-h	(p. 30)
Picture galleries: highest to lowest illuminance over the whole picture-hanging area to be within a 2:1 ratio		(p. 33)
Restoration:	2000 lux	(p. 47)
Photography and television: tungsten lamps	1000 lux	(p. 46)
metal-halide lamps	2500 lux	(p. 46)
Flash exposure:	f22 at 100 ASA (20 lux-min)	(p. 48)
Colour rendering of museum light sources:	R_a about 90 or better Worst R about 80 or better Crawford Class A, B or C	(p. 54)

UV RADIATION

Proportion of UV in the light source (UV content) not to exceed 75 μW/lm (p. 20)

(A specification for UV-absorbing filters is given on p. 17.)

RELATIVE HUMIDITY

Day and night throughout the year	50 or 55 ± 5% RH	
The level may be fixed higher or lower, but for mixed collections should be in the range 45 $\cong$ 60%. Special exhibits may require special conditions		(pp.88 and 116)

TEMPERATURE

Winter	19 ± 1 °C
Summer up to	24 ± 1 °C

Note 1: Temperature must be controlled to control RH, but the level is dictated by human comfort. For fuel economy different winter and summer levels are suggested

Note 2: In storage areas or buildings closed to the public in winter, temperature can be allowed to fall, but not to the point where condensation may occur on cold or unventilated surfaces. A lower limit of 10 °C is suggested (p. 119)

AIR POLLUTION

Particulates removed to	80% efficiency on Eurovent 4/5	(p. 134)
Sulphur dioxide and nitrogen dioxide each removed to	below 10 $\mu g/m^3$	
Ozone to	below 2 $\mu g/m^3$	(p. 158)
Air flow rate: ceiling height under 3 m	8 air changes/h	
under 4–5 m	6	
under 6–8 m	4	
(can be reduced to half during non-occupancy) (see ref. 272)		

Noise:

Noise levels in exhibition rooms due to traffic and plant should not exceed (Noise Rating) NR 35 (p. 162)

CLASS 2

LIGHT

Moderately sensitive material:
Absolutely no direct sunlight
Illuminance kept within the low hundreds by use of the light meter and manual control of blinds

Sensitive material:
As for Class 1

UV RADIATION

As for Class 1

RELATIVE HUMIDITY

To be kept within the danger limits 40% and 70% by the use of room humidifiers/dehumidifiers (p. 86)

TEMPERATURE

Should be reasonably constant to stabilise RH

AIR POLLUTION

Special areas and cases can be kept free of air pollution by the use of room or case devices (p. 254)

References

Abbreviations

IIC = International Institute for Conservation of Historic & Artistic Works (6 Buckingham St., London WC2N 6BA).
ICCROM = International Centre for the Study of the Preservation and the Restoration of Cultural Property (13 Via San Michele, 00153 Rome).
ICOM = International Council of Museums (pubs. available from ICCROM).
SIC = *Studies in Conservation* (journal of IIC).
Atm. Env. = *Atmospheric Environment*

1 ZIMAN, J., Review of a book on de Broglie by A. George, *Nature* 2 Aug. (1974) 445.

2 Victoria & Albert Museum Conservation Department. *Technical Notes on the Care of Art Objects*, No. 1 – 'The deterioration of art objects by light'. (1966).

3 PADFIELD, T. and LANDI, S., 'Light-fastness of the natural dyes', *SIC* 11 (1966) 181–96.

4 LEENE, J.E., DEMENY, L., ELEMA, R.J., de GRAAF, A.J. and SURTEL, J.J., 4 Interim Reports on 'Artificial ageing of yarns . . .' to the ICOM Conservation Committee at meetings in Washington (1965), Brussels (1967), Amsterdam (1969) and Madrid (1972). Final Report, Venice (1975).

5 EGERTON, G.S. and MORGAN , A.G., 'The photochemistry of dyes', Parts 1 and 2. *J. Soc. Dyers Colour.* (1970) 79–83 and 242–49.

6 GILES, C.H. and McKAY, R.B., 'The lightfastness of dyes – a review', *Text. Res. J.*, 33 (1963) 527–77.

7 KÜHN, H., 'The effect of oxygen, relative humidity and temperature on the fading of water colours', *1967 London Conference on Museum Climatology* (ed. G. Thomson), IIC. Revised Ed. (1968) 79–88.

8 McLAREN, K., 'The spectral regions of daylight which cause fading', *J. Soc. Dyers Colour.*, 72 (1956) 89.

9 LITTLE, A.H., 'Deterioration of textile materials', *Delft Conference on the Conservation of Textiles*. IIC (2nd Ed. 1965). 67–68.

10 BOWMAN, J.G. and REAGAN, B.M., 'Filtered and unfiltered lights and their effects on selected dyed textiles', *SIC* 28 (1983) 36–44.

11 DUFF, D.G., SINCLAIR, R.S. and STIRLING, D., 'Light-induced colour changes of natural dyes', *SIC* 22 (1977) 161–69.

12 LEVISON, H.W., *Artists' Pigments – Light Fastness Tests and Ratings*, Colorlab, Florida (1976).

13 van BEEK, H.C.A. and HEERTJES, P.M., 'Fading by light of organic dyes on textiles and other materials', *SIC* 11 (1966) 123–32.

14 DEMENY, L., 'Degradation of cotton yarns by light from fluorescent lamps', *1967 London Conference on Museum Climatology* (ed. G. Thomson). IIC. Revised Ed. (1968) 53–64.

15 HARRIS, M., 'Photochemical decomposition of silk', *Am. Dyestuff Reptr.* 23 (1934) 403–5.

16 HARRIS, M. and SMITH, A.L., 'Photochemical reactions of wool', *J. Res. Nat. Bur. Stand* 20 (1938) 563–9.

17 PADFIELD, T., 'Deterioration of cellulose – a literature review', *Problems of Conservation in Museums*. ICOM (1969) 119–64.
18 FELLER, R.L., CURRAN, M. and BAILIE, B., 'Identification of traditional organic colorants employed in Japanese prints and determination of their rates of fading', *Japanese Woodblock Prints: a Catalogue of the Mary A. Ainsworth Collection*, Oberlin, Ohio (1984).
19 THOMSON, G, 'Oxygen starvation in paint and other films', *Nat. Gallery Technical Bull.* 2 (1978), 66–70.
20 BROMMELLE, N.S., 'The Russell and Abney report on the action of light on water colours', *SIC* 9 (1964) 140–52.
21 JOHNSTON-FELLER, R., FELLER, R.F., BAILIE, C.W. and CURRAN, M., 'The kinetics of fading: opaque paint films pigmented with alizarin lake and titanium oxide', *J. American Institute for Conservation* 23 (1984), 114–129.
22 JOHNSTON-FELLER, R., FELLER, R.L. and BAILIE, C.W., 'Uses of instrumental color-matching techniques for studying pigment fading', *Bull. Institute for the Conservation of Cultural Material (Australia)* 10 (1984), 71–91.
23 KECK, S., 'Mechanical alteration of the paint film', *SIC* 14 (1969) 17, Fig. 14.
24 STOLOW, N., 'Application of science to cleaning methods – solvent action studies on pigmented and unpigmented linseed oil films', *Recent Advances in Conservation* (ed. G. Thomson), Butterworths, London (1963) 84–8.
25 HOFFMANN, E. and SARACZ, A., 'Weathering of paint films', Parts 1–5. *J. Oil Colour Chem. Ass.*, (Feb. 1969) 113, (Dec 1969) 1130, (May 1971) 450, (Feb 1972) 101, (Dec 1972) 1079.
26 RITTER, H.S., 'Durability of paint films from erosion rate studies', *J. Paint Tech*, 43 (1971) 74–82.
27 FYNN, P.J., 'Cellulose behaviour with filtered sunlight', *Text. Res. J.*, 18 (1948) 350–7.
28 THOMSON, G. and STAINFORTH, S., 'Conservation and Museum Lighting'. *Information Sheet 6* from Museums Association, 34 Bloomsbury Way, London WC1. Contains lists of UV-absorbing filters, light-meters, fluorescent lamps, and grey glass available in U.K. (4th Ed. 1985).
29 BLYTHE, A.R., 'Anti-static treatment of Perspex for use in picture frames', *SIC* 19 (1974) 102–4.
30 Chartered Institute of Building Services (UK) (222 Balham High Road, London SW12 9BS), Publication No. 14, *CIBS Lighting Guide: Museums and Art Galleries* (1980)
31 ICOM, *La Lumière et la Protection des Objets et Specimens exposès dans les Musées et Galeries d'Art.* L'Association Française de l'Eclairage (no date). Also in *Lux* 63 (1971) 235–64.
32 ICCROM (13 Via di San Michele, 00153 Rome), *Catalogues of Technical Exhibitions, No. 1 – Lighting.* (1975).
33 CROLLAU, E.K. and KNORING, G.M., 'Standards of Artificial Light in museums of the USSR', *ICOM Committee for Conservation*, Venice (1975), 75/19/6.

34 LAFONTAINE, R.H. and MacLEOD, K.J., 'A statistical survey of lighting conditions and the use of ultraviolet filters in Canadian museums, archives & galleries', *J. Can. Conserv. Inst.,* **1** (1976) 41–4.

35 MACLEOD, K.J., 'Museum Lighting', *Technical Bull. No. 2*, Canadian Conservation Institute, National Museums of Canada, Ottawa K1A OM8 (repr. 1978).

36 THOMSON, G., 'A new look at colour rendering, level of illumination, and protection from ultraviolet radiation in museum lighting', *SIC* **6** (1961) 49–70.

37 BROMMELLE, N.S. and HARRIS, J.B., 'Museum Lighting', Parts 1 to 4. *Mus. J.*, **61** (1961) 169–76 and 259–67, **62** (1962) 337–46 and 176–86.

38 BROMMELLE, N.S., 'Conservation of museum objects in the tropics', *1967 London Conference on Museum Climatology.* (ed. G. Thomson). IIC. Revised Ed. (1968) 139–49.

39 BROMMELLE, N.S., 'Lighting, air-conditioning, exhibition, storage, handling & packing', *The Conservation of Cultural Property.* UNESCO Museums & Monuments series 11 (1968) 291–301.

40 HARRIS, J.B., 'Lighting problems in museums and art galleries'. *Elect. Rev.,* **10** (1964) 175.

41 HARRIS, J.B., 'Art gallery and museum lighting', *Int. Ltg. Rev.* **15** (1964) 159.

42 HARRIS, J.B., 'Some aspects of art gallery lighting', *Int. Ltg. Rev.* **15** (1964) 170–9.

43 HARRIS, J.B., 'Practical aspects of lighting as related to conservation' *1967 London Conference on Museum Climatology.* (ed. G. Thomson). IIC. Revised Ed. (1968) 133–8.

44 BRAWNE, M., *The New Museum*, Architectural Press, London (1965) 170; and *The Museum Interior*, Thames and Hudson, London (1982) 102.

45 ALLEN, W.A., 'The new museum in Lisbon for the Gulbenkian Collection. A new approach to illumination', *Mus. J.*, **71** (1971) 54–8.

46 THOMPSON, C., 'Daylight in art galleries', *Mus. J.*, **71** (1971) 59–62.

47 KIMMEL, P.S. and MAVES, M.J., 'Public reaction to museum interiors', *Mus. News,* Sept. (1972) 17–19.

48 LOE, D.L., ROWLANDS, E. and WATSON, N.F., 'Preferred lighting conditions for the display of oil and watercolour paintings', *Lighting Research and Technology* 14 (1982) 173–192.

49 Chartered Institute of Building Services (UK), *CIBS Code for Interior Lighting* (1984).

50 GLOAG, H.L. and GOLD, M.J., 'Museum and art gallery design – a short history of the daylighting of galleries'. Building Research Station (UK), *Misc. Papers No. 1* (1965).

51 THOMSON, G. 'Colour under some new fluorescent lamps', *National Gallery Technical Bulletin* (London) 9 (1985)

52 CHURCH, D.E., 'Picture gallery lighting.' *DOE Construction.* **16** HMSO (1975) 2–7.

53 EGERTON, G.S., 'Action of light on dyes in polymer materials', *Br. Polymer J.*, 3 (1971) 63–7.

54 WOLTERS, C. and KUEHN, H., 'Behaviour of painted wood panels under strong illumination', *SIC* 7 (1962) 1–9.
55 FELLER, R.L., 'Control of deteriorating effects of light on museum objects – heating effects of illumination by incandescent lamps', *Mus. News – Tech. Suppl. No. 22.* (1968) 39–47.
56 PADFIELD, T., BURKE, M. and ERHARDT, D., 'A cooled display case for George Washington's Commission', Preprints, ICOM Committee for Conservation, Copenhagen (1984) 84.17.38.
57 NIZETTE, M. and LYALL, J., '*Conservation of Photographic Materials in Australia*', Institute for the Conservation of Cultural Material (Australia), 8 (1982), 3/4, 18.
58 REILLY, J.M., SEVERSON, D.G. and McCABE, C., 'Image deterioration in albumen photographic prints', Preprints, IIC Washington Congress (1982) 61–65.
59 SWAN, A., *The Care and Conservation of Photographic Material*, Crafts Council, London (1981).
60 HANLAN, J.F., 'Effects of electronic photographic lamps on the materials or works of art', *Mus. News* **48** No. 10 (1970) 33–41.
61 ICOM Committee for Conservation – Lighting Group, *ICOM News* Dec (1969) 54.
62 STANIFORTH, S., 'Retouching and colour matching: the restorer and metamerism', *SIC* **30** (1985) 101–11.
63 McLAREN, K., 'The impact of modern lighting on the dyer', *J. Soc. Dyers Colour.* (1976) 407–16.
64 HUNT, R.W.G. and WINTER, L.M., 'Colour adaptation in picture-viewing situations', *J. Photogr. Sci.*, 23 (1975) 112–5.
65 PADGHAM, C.A. and SAUNDERS, J.E., *The Perception of Light and Colour*, Bell, London (1985).
66 WYSZECKI, G. and STILES, W.S., *Color Science*, Wiley 2nd Ed. (1982).
67 BILLMEYER, F.W. and SALTZMANN, M., *Principles of Color Technology*, Wiley, 2nd ed. (1981).
68 LYTHGOE, J.N. and NORTHMORE, D.P.M., 'Colours underwater', *Colour 73* (Abs. of papers, 2nd Congress of the Int. Colour Assoc. (1973). Hilger (London) 77–98.
69 JUDD, D.B., MACADAM, D.L. and WYSZECKI, G.W., 'Spectral distribution of typical daylight as a function of correlated color temperature', *J. Opt. Soc. Am.*, **54** (1964) 1031.
70 *Munsell Book of Color* (2 vols.). Munsell Color Co., 2441 North Calvert St., Baltimore 18, Maryland.
71 Illuminating Engineering Society (USA) *Lighting Handbook*, 3rd Ed. (1962).
72 MARR, D., *Vision*, Freeman, San Francisco (1982).
73 LAND, E.H. and McCANN, J.J., 'Lightness and retinex theory', *J. Opt. Soc. Am.*, **61** (1971) 1–11.
74 GREGORY, R.L., *Eye and Brain*, 2nd Ed. Wiedenfeld & Nicholson (1972) 204.
75 WEALE, R., 'The truth and something besides', *New Scient.*, 4 Jan (1973) 13–15.
76 BROMMELLE, N.S., 'Visual performance with limited illuminance', *Paper presented to the ICOM Committee for Conservation, Madrid 1972.* (Available from ICC, 13 Via San Michele, 00153 Rome.)
77 CRAWFORD, B.H., 'Just perceptible colour differences in relation to level of illumination', *SIC* **18** (1973) 159–66.

78 CRAWFORD, B.H., Private communication.
79 Meteorological Office. *Hygrometric Tables.* Pt. 3. (Celsius aspirated values). 2nd Ed. Met.O.265c. HMSO, London (1964).
80 RAMER, B., 'The use of colour change relative humidity indicator cards', *Conservation News* (UK Institute for Conservation 16 (Nov. 1981) 10–11.
81 PLENDERLEITH, H.J. and WERNER, A.E.A., *Conservation of Antiquities & Works of Art* 2nd Ed. OUP, London (1974).
82 LEENE, J.E. (Ed.), *Textile Conservation* Butterworths, London (1972)
83 UNESCO, *The Conservation of Cultural Property*, Museums & Monuments Series No. 11. Paris (1968).
84 BUCK, R.D., 'Some applications of mechanics to the treatment of panel paintings', *Recent Advances in Conservation.* (Ed. G. Thomson). Butterworths, London (1963) 156–62.
85 TIEMANN, H.D., *Wood Technology*, Pitman, N.Y. 2nd Ed. (1944) 158.
86 BERGER, G.A. and RUSSELL, W.H., 'Investigations into the reactions of plastic materials to environmental changes. Part I: The mechanics of the decay of paint films', *SIC* **31** (1956) 49–64.
87 CORNELIUS, F. du PONT., 'Movement of wood and canvas for paintings in response to high and low RH cycles', *SIC* **12** (1967) 76–80.
88 BERGER, G.A., 'The role of tension in the preservation of canvas paintings: a study of panoramas', Preprints, *ICOM Committee for Conservation*, Ottawa (1981), 81.2.3.
89 ORGAN, R.M., 'Examination and treatment of bronze antiquities', *Recent Advances in Conservation* (Ed. G. Thomson), Butterworths, London (1963) 104–10.
90 BUCK, R.D., 'Specifications for museum air conditioning', *Mus. News – Technical Supplement No. 6*, Dec (1964) 53–7.
91 MAJEWSKI, L., 'On conservation', *Mus. News* **51**, Sept. (1972) 8.
92 ROGERS, G. de W., 'Particular aspects of silver tarnishing', *Canadian Group Bull.* **1**, No. 1 (1976).
93 WRANGLEN, G., *Colln Czech. chem. Commun. Engl. Edn*, **36** (1971) 625–37.
94 BROMMELLE, N.S., Private communication.
95 WERNER, A.E., 'Care of glass in museums', *Mus. News – Technical Supplement*, June (1966) 45–9.
96 BRILL, R.H., 'Incipient crizzling in some early glasses', *Bull. Am. Gp. IIC.*, **12**, No. 2 (1972) 46–7.
97 RIXON, A.E., *Fossil Animal Remains – their Preservation and Conservation.* Athlone, London (1976) 139–52.
98 HOWIE, F., 'Storage environment and the conservation of geological material', *The Conservator (Journal of the United Kingdom Group of IIC)*, (1978).
99 HUECK, D.J., 'Textile pests and their control', *Textile Conservation,* (Ed. J.E. Leene) Butterworths, London (1972) 76–97.
100 NAIR, S.M., 'Biodeterioration of museum materials in tropical countries', *Conservation in the Tropics,* (Ed. O.P. Agrawal), ICC, Rome (1972) 150–7.
101 SZENT-IVANY, J.J.H., 'Identification and control of insect pests', *Conservation of Cultural Property.* UNESCO Museums & Monuments No. 11, (1968) 53–70.

102 BANKS, P.N., (book review), *SIC* 28 (1983), 47.
103 Meteorological Office, MO617a. *Tables of Temperature, Relative Humidity and Precipitation for the World,* 6 Parts (1961).
104 STOLOW, N., 'Conservation of exhibitions in the museum of the future', *Conference Proc. for 2001 – The Museum and the Canadian Public.* Canadian Museums Association (1977) 24–31.
105 BRIGGS, J.R. and SMITH, P., 'Engineering systems for galleries', *Studio International,* May/June (1975) 220–2.
106 REES-JONES, S.G., Private communication.
107 MICHALSKI, S., 'A control module for relative humidity in display cases, Preprints, *IIC Congress*, Washington (1982), footnote p. 28.
108 Electricity Council (UK), *Packaged Air Conditioning*, (1970) Fig. 36.
109 ORGAN, R.M., 'Humidification of galleries for a temporary exhibition', *1967 London Conference on Museum Climatology*, 2nd Ed., IIC, (1968) 1–13.
110 PADFIELD, T., Private communication.
111 National Institute for Occupational Safety and Health (USA), *Hazard Evaluations and Technical Assistance Branch*, report HETA 83–020–1351 (1983)
112 PADFIELD, T., 'Control of relative humidity and air pollution in show-cases and picture frames', *SIC* 11 (1966) 8–30.
113 TOISHI, K., 'Humidity control in a closed package', *SIC* 4 (1959) 81–7
114 STOLOW, N., *Controlled Environment for Works of Art in Transit.* Butterworths, London (1966).
115 STOLOW, N., 'Fundamental case design for humidity sensitive museum collections', *Mus. News,* Feb. (1966) 45–52.
116 STOLOW, N., 'Standards for the care of works of art in transit', *1967 London Conference on Museum Climatology,* 2nd Ed, IIC (1968) 271–84.
117 STOLOW, N., *Conservation and Exhibitions: Packing, Transport and Storage Considerations,* Butterworths, London (1985).
118 THOMSON, G., 'Relative humidity – variations with temperature in a case containing wood', *SIC* 9 (1964) 153–69.
119 STANIFORTH, S., 'The testing of packing cases for the transport of paintings', Preprints, *ICOM Committee for Conservation*, Copenhagen (1984), 84.12.7.
120 GREEN, T. and HACKNEY, S., 'The evaluation of a packing case for paintings', Preprints, *ICOM Committee for Conservation*, Copenhagen (1984), 84.12.1.
121 TOISHI, K., 'Relative humidity in a closed package', *Recent Advances in Conservation,* (Ed. G. Thomson), Butterworths, London (1963) 13–15.
122 ROGERS, G. de W., 'Study of factors influencing the internal environment of packing cases in transit', *ICOM Committee for Conservation,* 4th Triennial Meeting, Venice (1975) (Unpublished).
123 THOMSON, G., 'RH stabilisation in exhibition cases – hygrometric half-time', *SIC*, 22 (1977) 85–102.
124 BRIMBLECOMBE, P. and RAMER, B., Museum display cases and the exchange of water vapour', *SIC* 28 (1983), 179–188.
125 DE GUICHEN, G., 'Controle du climat autour de 197 instruments de musique', Preprints, *ICOM Committee for Conservation*, Copenhagen (1984), 84.17.19.
126 WEINTRAUB, S., Private communication.

127 SACK, S.P., 'A case study of humidity control', *Brooklyn Mus. Art. Rep.*, 5 (1963–4) 99–103.
128 CURSITER, S., 'Control of air in cases and frames', *Tech. Stud. in the Field of the Fine Arts,* 5 (1936–7) 109–16.
129 MICHALSKI, S. ref. 107 pp. 28–31.
130 LAFONTAINE, R.H., and MICHALSKI, S., 'The control of relative humidity: recent developments', Preprints, *ICOM Committee for Conservation*, Copenhagen (1984) 84.17.34
131 PADFIELD, T. and ERHARDT, D., 'Clear weather precipitation inside a museum: a case study in micrometeorology', *ICOM Committee for Conservation*, Ottawa (1981) 81.18.9.
132 ROGERS, G. de W., 'The ideal of the ideal environment', J. IIC Canadian Group, 2 No. 1 (1976) 34–39.
133 AMDUR, E.J., 'Humidity control - isolated area plan', *Mus. News – Technical Supplement*, No. 6, Dec. (1964) 58–60.
134 FEILDEN, B.M., *Conservation of Historic Buildings*, Butterworths, London (1982).
135 BUCK, R.D., 'Use of moisture barriers on panel paintings', *SIC*, 6 (1961) 9–20.
136 THOMSON, G., 'Impermanence – some chemical and physical aspects', *Mus. J.*, 64 (1964) 16–36.
137 BECK, W. and KOLLER, M., 'Problems of heating within historic buildings of Austria', Preprints, *IIC Vienna Congress* (1980) 22–29.
138 BRIGGS, J.R., 'Environmental control for old buildings', Preprints, *IIC Vienna Congress* (1980), 9–14.
139 LACY, R.E., 'Note on the climate inside a mediaeval chapel', *SIC,* **15** (1970) 65–80.
140 MASSARI, G., *L'Umidita nei Monumenti.* (1969). (Eng. version 1970). ICC, 13 via di San Michele, 00153 Roma.
141 MASSARI, G., 'Dampness and the preservation of rural paintings', & 'Etude sur les dommages causés par l'humidité aux oeuvres d'art des églises Venitiennes', *1967 London Conference on Museum Climatology*, 2nd Ed. IIC, (1968). 191–9, 285–96.
142 VOS, B.H., 'Suction of groundwater', *SIC,* **16** (1971) 129–44.
143 VOS, B.H., 'Moisture in monuments', *Application of Science in Examination of Works of Art,* (Ed. W.J. Young), Museum of Fine Arts, Boston (1973) 147–53.
144 BOEKWIJT, W.O. and VOS, B.H., 'Measuring methods for determining moisture . . . in monuments', *SIC,* **15** (1970) 81–93.
145 WINKLER, E.M., 'Salt action in urban buildings', *Application of Science in Examination of Works of Art,* (Ed. W.J. Young), Museum of Fine Arts, Boston (1973) 139–46.
146 WINKLER, E.M., *Stone – Properties, Durability in Man's Environment.* Springer, Wien (1975).
147 VAN ASPEREN de BOER, J.R.J., 'Humidity in walls in relation to the preservation of works of art', *1967 London Conference on Museum Climatology,* 2nd Ed. IIC (1968) 109–17.
148 BORDASS, W., *Heating Your Church*, CIO, London (1984).
149 KECK, C., 'On conservation: Emergency treatment of water damage', *Mus. News,* **50** June (1972) 13.
150 ORGAN, R.M. and McMILLAN, E., 'Aid to a hurricane-damaged museum', *Bul. Am. Grp – IIC,* **110** No. 1 (1969) 31–9.

151 WALSTON, S., 'Emergency conservation following the Darwin cyclone', *Inst. Conserv. Cultural Material (Australian) Bull* **2**. No. 1 (1976) 21–5. (For address see ref. 49a.)

152 ARAI, H., 'Antimicrobial factors found in virgin tumuli', Preprints, *6th Int. Biodeterioration Symposium, Washington* (1984) and 'Microbiological studies on the conservation of mural paintings in tumuli', *Int. Symposium: Conservation and Restoration of Mural Paintings (I)*, Tokyo Nat. Res. Inst. of Cultural Properties (1983) 117–124.

153 DOWMAN, E.A., *Conservation in Field Archaeology*, Methuen, London (1970).

154 WEAVER, M.E., 'Use of an inflatable air-dome to produce controlled conditions for an archaeological site', *SIC*, **18** (1973) 88–93.

155 VOUVE, J., BRUNET, J., VIDAL, P. and MARSAL, J., 'Les oeuvres rupestre de Lascaux (Montignac, France): maintain des conditions de conservation, *SIC* **28** (1983) 107–116.

156 LEFEVRE, M., 'La "Maladie verte" de Lascaux', *SIC* **19** (1974) 126–156.

157 BRUNET, J. and VIDAL, P., 'Preservation of ornamented caves of France', *Int. Symposium: Conservation and Restoration of Mural Paintings (I)*, Tokyo Nat. Res. Inst. (1983) 135.

158 MIYAKE, S., 'Preservation facilities for the tumulus Takamatsuzuka', *Int. Symposium: Conservation and Restoration of Mural Paintings (I)*, Tokyo Nat. Res. Inst. (1983) 87–103.

159 EASTLAKE, C.L., FARADAY, M. and RUSSELL, W., *Report on the subject of the protection of the pictures in the National Gallery by glass*, House of Commons, 24 May (1850).

160 AULICIEMS, A. and BURTON, I., 'Trends in smoke concentrations before and after the Clean Air Act of 1956', *Atm. Env.* **7** (1973) 1063–70.

161 MEETHAM, A.R., *Atmospheric Pollution*, Pergamon, London (1952) 187.

162 Warren Spring Laboratory. *National Survey of Air Pollution 1961–71.* Vol. 1. HMSO, London (1972) 173.

163 LEE, Jr., R.E., CALDWELL, J., ACKLAND, G.G. and FANKHAUSER, R., 'Distribution and transport of airborne particulate matter . . . in Great Britain', *Atm. Env.* **8** (1974) 1095–1109.

164 DAVIES, C.N., 'Particles in the atmosphere – natural and man-made', *Atm. Env.*, **8** (1974) 1069–79.

165 TOISHI, K. and KENJO, T., 'Alkaline material liberated into the atmosphere from new concrete', *Paint Technol.*, **39** (1967) 152–55.

166 TOISHI, K. and KENJO, T., 'Some aspects of the conservation of works of art in buildings of new concrete', *SIC*, **20** (1975) 118–22.

167 TOISHI, K. and KENJO, T., 'Simple method of measuring the alkalinity of air in new concrete buildings', *SIC*, **13** (1968) 213–4

168 PROPERTY SERVICES AGENCY, U.K. Department of the Environment, *Technical Report (M & E) TR 70* (March 1981)

169 THOMSON, G., 'Air pollution – a review for conservation chemists', *SIC*, **10** (1965) 147–67.

170 MATTHESON, M.J., STRINGER, H.L. and BUSBEE, W.L., 'Corona discharge oxidation of sulfur dioxide', *Envir. Sci. & Tech.*, Oct. (1972) 895–901.

171 MOLLER, D., 'On the global natural sulphur emission', *Atm. Env.* 18 (1984) 36.
172 LIKENS, G., 'Acid precipitations', *Chem. Enging News*, 54, 22 Nov. (1976) 29–44.
173 STERN, A.C., BOUBEL, R.W., TURNER, D.B. and FOX, D.L., *Fundamentals of Air Pollution*, 2nd ed. (1984).
174 Environmental Resources Ltd., *Acid Rain*, Graham and Trotman, London (1983).
175 SCHWAR, M.J.R. and BALL, D.J., *Thirty Years On: a Review of Air Pollution in London*, Greater London Council (1983).
176 FOWLER, D. and CAPE, J.N., *Air Pollutants in Agriculture and Horticulture*, report of the Institute of Terrestrial Ecology, Bush Estate, Penicuik, Midlothian EH26 0QB, UK, Fig. 1.
177 HACKNEY, S., 'The distribution of gaseous air pollution within museums', *SIC* 29 (1984) 105–116.
178 SPEDDING, D.J., 'Fate of sulphur-35/sulphur dioxide released in a laboratory', *Atm. Env.*, 3 (1969) 341–6.
179 THOMSON, G. and WHITE, R., 'The pH of rain and the destruction of alkaline stone', *SIC*, 19 (1974) 190–1.
180 WINKLER, E.M., 'Decay of stone', *1970 New York Conference on Conservation of Stone and Wooden Objects*, Vol. 1, 2nd Ed., (1971) 1–14.
181 SAYRE, E.V., 'Investigation of Italian frescoes, their materials, deterioration and treatment', *Application of Science in Examination of Works of Art*, (Ed. William J. Young), Museum of Fine Arts, Boston (1973) 176–81.
182 CAMUFFO, C., DEL MONTE, M., SABBIONI, C. and VITTORI, O., 'Wetting, deterioration and visual features of stone surfaces in an urban area', *Atm. Env.* 16 (1982) 2253–2259.
183 DEL MONTE, M., SABBIONI, C. and VITTORI, O., 'Airborne carbon particles and marble deterioration', *Atm. Env.*, 645–652.
184 GAURI, K.L., POPLI, R. and SARMA, A.C., 'Effect of relative humidity and grain size on the reaction rates of marble at high concentrations of SO_2', *Durability of Building Materials* 1 (1982-3) 209–216.
185 ZERONIAN, S.H., 'Reaction of cellulosic fabrics to air contaminated with sulfur dioxide', *Text. Res. J.*, Aug. (1970) 695–8.
186 FYE, C., FLASKERUD, K. and SAVILLE, D., 'Effect of a sulphur dioxide atmosphere on the breaking strength of fabrics of different fibre content', *Am. Dyestuff Reptr.*, 58, No. 14, July (1969) 16–19.
187 LANGWELL, W.H., 'Permanence of paper', Pts. 1–3 in *Tech. Bull. Br. Pap. Bd Mkrs Ass.*, 29 (1952) No. 1, 21, No. 2, 52, 30 (1953) No. 6, 2, Pts. 4 & 5 in *Proc. tech. Sect. Br. Pap. Bd Mkrs' Ass.*, 36 (1955) No. 1, 199, and 37 (1956) No. 3, 495.
188 BARROW, W.J., *Permanence/Durability of the Book*, 3 vols, W.J. Barrow Res. Lab., Richmond, Virginia, (1963).
189 BAYNES-COPE, A.D., 'The non-aqueous deacification of documents', *J. Pres. Library Archival Mater.*, 1, No. 1 (1969) 2–9.
190 FINDLAY, W.P.K., *Timber – Properties and Uses*, Crosby Lockwood Staples, London (1975).
191 COLLINGS, T.J. and YOUNG, F.J., 'Improvements in some tests and techniques in photograph conservation', *SIC*, 21 (1976) 79–84.

192 PARKER, A., *Destructive Effects of Air Pollution on Materials,* National Smoke Abatement Society (UK), (1955).

193 WATERER, J.W., *Guide to the Conservation and Restoration of . . . Leather,* Bell, London (1972).

194 HEMPHILL, J.E., NORTON, J.E., OFJORD, O.A. and STONE, R.L., 'Colorfastness to light and atmospheric contaminants', *Text. Chem. Color.,* 8 (1976) 60–62.

195 JELLINEK, H.H.G., 'Chain scission of polymers by small concentrations (1 to 5 ppm) of sulfur dioxide and nitrogen dioxide . . .', *J. Air Pollut. Control Ass.,* 20 (1970) 672–4.

196 JELLINEK, H.H.G., FLAJSMAN, F. and KRYMAN, F.J., 'Reaction of SO_2 and NO_2 with polymers', *J. Appl. Polym. Sci.,* **13** (1969) 107–16.

197 CAMPBELL, G.G., SCHURR, G.G., SLAWIKOWSKI, D.E. and SPENCE, E.J.W., 'Assessing air pollution damage', *Paint. Technol.,* **46**, June (1974) 59–71.

198 EVANS, U.R., *Introduction to Metallic Corrosion,* Arnold, London (1951) 85.

199 ODDY, W.A. and HUGHES, M.J., 'Stabilisation of active bronze and iron antiquities by the use of sodium sesquicarbonate', *SIC* **15** (1970) 183–9.

200 GETTENS, R.J., 'Mineral alteration products on ancient metal objects', *Recent Advances in Conservation,* (Ed. G. Thomson), Butterworths, London (1963) 89–92.

201 RIEDERER, J., 'Conservation of bronze sculptures', *Maltechnik–Restauro,* **1** (1972) 40–1.

202 RIEDERER, J., 'Korrosionsschaeden an Bronzen der Muenchner Residenz', *Deutsche Kunst & Denkmalpflege,* (1972) 49–56.

203 JUNGE, C.E., *Air Chemistry and Radioactivity,* Academic Press (1963).

204 VERNON, W.H.J., *Trans. Faraday Soc.,* **23** (1927) 113–83.

205 NEWTON, R.G., 'Bibliography of studies on the deterioration of stained glass', *Art Arch. Tech. Abstr.* **10**, No. 2 (1973) 132–9.

206 SCHROEDER, H., 'Susceptibility of glass to attack . . .', *Glastech. Ber.* **26**, No. 4 (1953) 91–7.

207 GILBERT, L.O., 'A biological scale for the estimation of sulphur dioxide pollution', *New Phytol.,* **69** (1970) 629–34.

208 DERWENT, R.G. and STEWART, H.N.M., 'Elevated ozone levels in the air of central London', *Nature,* 2 Feb. (1973) 342.

209 BRINER, E., 'Accelerating action of ozone in the autoxidation process', *Advances in Chemistry Series. Ozone . . .* American Chemical Society, March (1959) 184–94.

210 SCOTT, G., *Atmospheric Oxidation and Antioxidants,* Elsevier (1965) 97.

211 GRAEDEL, T.E., FRANLEY, J.P. and KAMMLOTT, G.W., 'Ozone- and photon-enhanced sulfidation of copper', *Science* (11 May 1984) 599–601.

212 LEIGHTON, P.A., *Photochemistry of Air Pollution.* Academic Press (1961).

213 GILES, C.H., 'The fading of colouring matters', *1964 Delft Conference on the Conservation of Textiles.* IIC, 2nd Ed. (1965) 8–26.

214 JAFFE, L.S., 'Effects of photochemical oxidants on materials', *J. Air Pollut. Control Ass.,* **17** (1967) 375–8.

215 MOORE, P.D., 'Ozone over Britain', *Nature,* 14 Aug (1975) 537.

216 DERWENT, R.G. and STEWART, H.N.M., 'Air pollution from the oxides of nitrogen in the United Kingdom', *Atm. Env.,* 7 (1973) 385–401.
217 SABERSKY, R.H., SINEMA, D.A. and SHAIR, F.H., 'Concentrations, decay rates, and removal of ozone', *Envir. Sci. Tech.,* 7 (1973) 347–53.
218 MUELLER, F.X., LOEB, L. and MAPES, W.H., 'Decomposition rates of ozone in living areas', *Envir. Sci. Tech.,* 7 (1973) 342–6.
219 DERWENT, R.G., *Ozone measurement in an art gallery*, Warren Spring Laboratory LR428 (AP), (1973) (Unpublished).
220 OKITA, T., KANEDA, K., YANAKA, T. and SUGAI, R., 'Determination of gaseous and particulate chloride and fluoride in the atmosphere', *Atm. Env.* 8 (1974) 927–36.
221 CROZAT, G., DOMERGUE, J.L. and BOGUI, V., 'Etude de l'aerosol atmospherique en Cote d'Ivoire et dans le Golfe de Guinée', *Atm. Env.* 7 (1973) 1103–16.
222 ODDY, W.A., 'An unsuspected danger in display', *Mus. J.,* **73** (1973) 27–8.
223 PADFIELD, T., ERHARDT, D. and HOPWOOD, W., 'Trouble in store', Preprints, *IIC Congress, Washington* (1982), 26.
224 BRADLEY, S.M., 'Safe fabrics for permanent exhibitions', Conservation News, *UK Inst. for Conservation* 23 (March 84) 16–18.
225 KIMBERLY, A.E. and EMLEY, A.L., 'A study of the removal of sulphur dioxide from library air', *Misc. Pub.,* **142**, National Bureau of Standards (1933).
226 MARSHALL, D., (Machine Control Ltd., Horsham, England), Private communication.
227 National Bureau of Standards, 'Air quality criteria for storage of paper-based archival material', *NBSIR 83-2795* (1983), Table 3, 22.
228 *New Scientist*, 'Shattering Ella', 23/30 Dec. (1976) 693.
229 American District Telegraph Company (155 Sixth Avenue, N.Y. 10013), *Ultrasonic alarm system effect on paintings,* 10 May (1967) Unpublished.
230 ABT, H.A., 'The companions of sunlike stars', *Scient. Am.,* **236** (1977) 103.
231 NASA. 'Solar electromagnetic radiation'. *NASA Space Vehicle Design Criteria.* Monograph SP–8005.
232 HENDERSON, S.T., *Daylight and its Spectrum,* Hilger, London (1970).
233 PORTER, G. and ARCHER, M.D., 'In vitro photosynthesis', *Interdis. Sci. Revs.,* **1** (1976) 121, Fig. 2.
234 HUNT, D.R.G., *Availability of Daylight*, UK Building Research Establishment (1979) Fig. 6.3.
235 Pilkington Brothers Ltd., *Windows and Environment*, (1969) 3–22.
236 HENDERSON, S.T. and MARSDEN, A.M. (Eds), *Lamps and Lighting*, Arnold, London (1972).
237 FECHNER, G.T., *Elemente der Psychophysik* (1860).
238 STEVENS, S.S., 'To honor Fechner and repeal his law', *Science*, 133 (1961) 80–86.
239 HOPKINSON, R.G., 'Light energy and brightness sensation', *Nature*, 10 Nov (1956) 1065.
240 RAWLAND, O., 'Fading of the British dyed-wool light-fastness standards in the U.K. -- some energy measurements', *J. Soc. Dyers Colour.*, **79** (1963) 697–701.

241 TENNENT, N., TOWNSEND, J.H. and DAVIS, A., 'A simple integrating dosimeter for ultraviolet light', Preprints, *IIC Congress*, Washington (1982) 32–38.
242 FELLER, R.L., 'Studies on photochemical deterioration', *ICOM Committee for Conservation*, Venice (1975) 75/19/4.
243 HARRISON, L.S., *Report on the deteriorating effects of modern light sources*, Metropolitan Museum of Art, New York (1953).
244 FELLER, R.L., 'Control of deteriorating effects of light upon museum objects', *Museum, UNESCO*, **17**, No. 2 (1964).
245 FELLER, R.L., 'The deteriorating effect of light on museum objects', *Mus. News – Tech. Suppl. No. 3*, June (1964) i–viii.
246 FELLER, R.L., 'Speeding up photochemical deterioration', *Inst. Royal du Patrimoine Artistique, Brussels, Bulletin*, **15** (1975) 135–50.
247 BELCHAMBERS, H.E. and GODBY, A.C., 'Illumination, colour rendering and visual clarity', *Ltg Res. Tech.*, 4 (1972) 104–6.
248 THOMSON, G., 'Topics in the conservation chemistry of surfaces', *Application of Science in Examination of Works of Art*. Proceedings of the Seminar, Museum of Fine Arts, Boston, Mass. (1965) 78–85.
249 STAMM, A.J., 'Wood deterioration and its prevention', *Conservation of Wooden Objects*. IIC New York Conference (1970) 11, Fig. 1 (Activation energies derived from this ref.).
250 SMITH, R.D., 'Maps – their deterioration and preservation', *Special Libraries*, Feb. (1972) 61, Table 2. (Activation energies derived from this ref.).
251 VALLENTYNE, J.R., *Geochim. cosmochim. Acta.*, **28** (1964) 157–88.
252 NORMAN, R.O.C., *Principles of Organic Synthesis*, Methuen, London (1968) 97.
253 FELLER, R.L., 'Induction time and the autoxidation of organic compounds', *Bull. Am. Inst. Conserv.*, **14**, No. 2 (1974) 142–51.
254 FELLER, R.L., 'Problems in the investigation of picture varnishes', *Conservation and Restoration of Pictorial Art*. (N.S. Brommelle and P. Smith Eds), Butterworths, London (1976) 137–44.
255 MILLS, J.M. and WHITE, R., *Organic Chemistry of Museum Materials*, Butterworths, London (1986).
256 ALLEN, N.S. and McKELLAR, J.F. (Eds), *Photochemistry of Dyed and Pigmented Polymers*, Applied Science Publishers, London (1980).
257 MILLS, J.S., 'The gas chromatographic examination of paint media. Pt. 1. Fatty acid composition and identification of dried oil films', *SIC*, **11** (1966) 92–107.
258 FELLER, R.L., 'The deterioration of organic substances and the analysis of paints and varnishes', *Proc. of the N. American International Regional Conference, Williamsburg and Philadelphia, 1972* Smithsonian Institution Press (1976) 287–99.
259 BROMMELLE, N.S., 'Colour and Conservation', *SIC*, **2** (1955) 76–86.
260 McLAREN, K., *The Colour Science of Dyes and Pigments*, Hilger, London (1983).
261 C.I.E. (Commission International de l'Eclairage). *Method of measuring and specifying colour rendering properties of light sources*. Publication No. 13.2 (TC-3.2) (1974).
262 CRAWFORD, B.H., 'Measurement of colour rendering tolerances', *J. Opt. Soc. Am.*, **49** (1959) 1147–56.
263 CRAWFORD, B.H., 'Colour rendition and museum lighting', *SIC*, **5** (1960) 41–51.

264 KELLY, K.L., 'Lines of constant correlated color temperature . . .', *J. Opt. Soc. Am.*, 53 (1963) 999.

265 CRAWFORD, B.H. and PALMER, D.A., 'Further investigations of colour rendering, and the classification of light sources', *SIC*, 6 (1961) 71–82.

266 HELSON, H., JUDD, D.B. and WARREN, R., 'Object color changes from daylight to incandescent filament illumination', *Illum. Engng*, 47 (1952) 221.

267 THORNTON, W.A., 'Lamps for assessing metamerism', *J. Illum. Eng. Soc.*, Oct. (1974) 11–18.

268 THORNTON, W.A., CORTH, R. and EVANS, G.S., 'Fluorescent light sources', *Ltg Des. App.*, Oct. (1975) 6–14.

269 THORNTON, W.A., 'The high visual efficiency of prime color lamps', *Ltg Des. App.*, Nov. (1975) 35–41.

270 HITCHCOCK, A. and JACOBY, G.C., 'Measurement of relative humidity in museums at high altitude', *SIC* 25 (1980) 78–86.

271 READING, A., 'A control philosophy for the economical air conditioning of museums and galleries', *Building Services Engineering Res. and Tech.* 4 (1983) 97–105.

272 Property Services Agency, MAGG Group (UK), *Guide for Museum and Gallery Air Conditioning, internal report* ref. 14/678/574A (1984).

273 Chartered Institute of Building Services (UK), *CIBS Guide A2*, 'Weather' (1975).

274 BUCK, R.D., 'Note on the effect of age on the hygroscopic behaviour of wood', *SIC*, 1 (1952–4) 39–44.

275 MacLEOD, K.J., 'Relative Humidity, its importance, measurement and control in museums', *Can. Conserv. Inst. Tech. Bull.*, No. 1, April (1975) 7.

276 STEVENS, W.C. and PRATT, G.H., *Kiln Operator's Handbook*, HMSO, London (1961).

277 SIEBEL, E., *Handbuch der Werkstoff Pruefung*. Vol. 4. Springer (1953) 126.

278 LAFONTAINE, R.H. and WOOD, P.A., 'The stabilization of ivory against relative humidity fluctuations, *SIC* 27 (1982), 109–117.

279 GANS, D.M., 'Water soak expansion of free paint films', *Paint. Technol.*, 44 (1972) 97–100.

280 TSOUMIS, G., 'Untersuchungen ueber die Schwankungen des Feuchtigkeitsgehaltes von lufttrockenem Holz', *Holz Roh- Werkstoff*, 18 (1960) 415–22.

281 STEVENS, W.C., 'Rates of change in the dimensions and moisture contents of wooden panels . . .', *SIC*, 6 (1961) 21–5.

282 JONES, W.P., *Air Conditioning Engineering*, 2nd Ed., Arnold, London (1973) 195.

283 *Documenta Geigy Scientific Tables*, 6th Ed., 539.

284 ASHLEY-SMITH, J. and MONCRIEFF, A.J., 'Experience with silica gel for controlling humidity in showcases', Preprints, *ICOM Committee for Conservation*, Copenhagen (1984) 84.17.1.

285 BARRETTE, B., 'Climate control: the Egyptian galleries at the Metropolitan Museum of Art', Preprints, *ICOM Committee for Conservation*, Copenhagen (1984) 84.17.6.

286 RAMER, B., 'The design and construction of two humidity-controlled display cases', Preprints, *ICOM Committee for Conservation*, Copenhagen (1984) 84.17.46.
287 SCHWEIZER, F., 'Stabilization of RH in exhibition cases: an experimental approach', Preprints, *ICOM Committee for Conservation*, Copenhagen (1984) 84.17.50.
288 WEINTRAUB, S., 'Studies on the behavior of RH within an exhibition case. Part I: measuring the effectiveness of sorbents for use in an enclosed showcase', Preprints *ICOM Committee for Conservation*, Ottawa (1981) 81.18.4.
289 PADFIELD, T., 'Design of museum show-cases', *1967 London Conference on Museum Climatology,* 2nd Ed., IIC (1968) 119–26.
290 National Bureau of Standards. *Preservation of the Declaration of Independence and the Constitution of the U.S.A.*, Circular No. 505 (1951).
291 MIURA, S., 'Studies on the behavior of RH within an exhibition case. Part II: The static and dynamic characteristics of sorbents to control the RH of a showcase', Preprints, *ICOM Committee for Conservation*, Ottawa (1981) 81.18.5.
292 ROFF, W.J. and SCOTT, J.R., *Fibres, Films, Plastics and Rubbers*, Butterworths, London (1971) 557–8.
293 LODGE, J.P., WAGGONER, A.P., KLODT, D.T. and CRAIN, C.N., 'Non-health effects of airborne particulate matter', *Atm. Env.* **15** (1981) 432–482.
294 ESMEN, N.A. and CORN, M., 'Residence time of particles in urban air', *Atm. Env.*, **5** (1971) 571–8.
295 DAVIES, C.N., 'Size distribution of atmospheric particles', *J. Aerosol Sci.*, **5** (1974) 293–300.
296 KADOWAKI, S., 'Size distribution of atmospheric total aerosols . . . in the Nagoya area', *Atm. Env.*, **10** (1976) 39–43.
297 WHITBY, K.T., CLARK, W.E., MARPLE, V.A., SVERDRUP, G.M., SEM, G.J., WILLEKE, K., LIU, B.Y.H. and PUI, D.Y.H., 'Characterisation of California aerosols – 1. Size distributions . . .', *Atm. Env.*, **9** (1975) 463–82.
298 SVERDRUP, G.M., WHITBY, K.T. and CLARK, W.E., 'Characterisation of California aerosols – II. Aerosol size distribution measurements in the Mojave Desert', *Atm. Env.*, **9** (1975) 483–94.
299 WILLEKE, K., WHITBY, K.T., CLARK, W.E. and MARPLE, V.A., 'Size distributions of Denver aerosols . . .', *Atm. Env.*, **8** (1974) 609–33.
300 WALLER, R.E., 'Studies on the nature of urban air pollution', *1967 London Conference on Museum Climatology,* 2nd Ed, IIC (1968) 65–9.
301 LEE, R.E. (Jr), CALDWELL, J.S. and MORGAN, G.B., 'Evaluation of methods for measuring suspended particulates in air', *Atm. Env.*, **6** (1972) 593–662.

302 BEAL, S.K., 'Measurements and models of indoor size spectra', *Atm. Env.*, 8 (1974) 204–5.

303 HASSID, D.V., 'The evaluation of the Machine Controls Ltd. Emcel multiple element', *Evaluation Report No. 25 – Scrubber for NOX Removal.* Ref. DVH/BD/19783 of 13 Dec 1976 (BOC Ltd.) Unpublished.

304 MOSS, R.L., POPE, O. and GIBBENS, H.R., 'The removal of sulphur dioxide from a simulated museum atmosphere using carbon filters', Warren Spring Laboratory (UK) *Report LR 525(AP)* (1985).

305 NEWMAN, L., FORREST, J. and MANOWITZ, B., 'Application of an isotopic ratio technique to a study of the atmospheric oxidation of sulfur dioxide . . .', *Atm. Env.*, 9 (1975) 959–68.

306 ELIASSEN, A. and SALTBONES, J., 'Decay and transformation rates of SO_2 . . .', *Atm. Env.*, 9 (1975) 425–9.

307 RODHE, H., PERSSON, C. and AKESSON, O., 'An investigation into regional transport of soot and sulfate aerosols', *Atm. Env.*, 6 (1972) 675–93.

308 FREIBURG, J., 'The iron catalyzed oxidation of SO_2 . . .', *Atm. Env.*, 10 (1976) 121–30.

309 CHARLSTON, R.J., VANDERPOL, A.H., COVERT, D.S., WAGGONER, A.P. and AHLQUIST, N.C., '$H_2SO_4/(NH_4)_2SO_4$ background aerosol . . .', *Atm. Env.*,,8 (1974) 1257–67.

310 GAURI, K.L., DODERER, G.C., LIPSCOMB, N.T. and SARMA, A.C., 'Reactivity of treated and untreated marble specimens in an SO_2 atmosphere', *SIC*, 18 (1973) 25–35.

311 MALESANI, P.P. and VANNUCCI, S.A., 'Decay of Pietra Serena and Pietraforte, Florentine building stones. Petrographic observations', *SIC*, 19 (1974) 36–50.

312 SPEDDING, D.J. and ROWLANDS, R.P., 'Sorption of sulphur dioxide by indoor surfaces', *J. appl. Chem.*, 20 (1970) 143–6.

313 DURBIN, P. and HECHT, T.A., 'The photochemistry of smog formation', *Clean Air*, Feb. (1975) 8–16.

314 NIEBOER, H. and Van HAM, J., 'Peroxyacyl nitrate (PAN) in relation to ozone . . .', *Atm. Env.*, 10 (1976) 115–20.

315 COX, R.A., EGGLETON, A.E.J., DERWENT, R.G., LOVELOCK, J.E. and PACK, D.H., 'Long-range transport of photochemical ozone in north-west Europe', *Nature*, 8 May (1975) 118–21.

316 DAVIES, T.D., RAMER, B. and DELANY, A.C. 'Indoor/outdoor ozone concentrations at a contemporary art gallery', *J. Air Pollution Control Assoc.* 31 (1984) 135–137.

317 SHAVER, C.L., CASS, G.R. and DRUZIK, J.R., 'Ozone and the deterioration of works of art', *Environmental Science and Technology* 17 (1983) 748–752.

318 CARP, C., 'Calculating atmospheric humidity', *SIC* 28 (1983) 24–28.

319 THOMSON, G., 'Control of the environment for good or ill? Monitoring', *National Gallery (London) Technical Bulletin* 5 (1981) 3–13; and 'Specification and logging of the museum environment', *Int. J. of Museum Management and Curatorship* 3 (1984) 317–326.

320 HORTON-JAMES, D., 'Comparison of a manual environmental monitoring system with a computer-based system', Preprints, *ICOM Committee for Conservation*, Copenhagen (1984) 84.17.26.

321 LAL, S.B., 'Geology and preservation of archaeological material', *Conservation in the Tropics* (O.P. Agrawal Ed.), ICC, Rome (1972) 134.

Index

Note

Materials such as paper, leather, wool, etc. are not indexed in every case when they appear as mere items in a list.

Absolute
  air filters, 133
  humidity, *see* Humidity
  zero, 50
  see also Moisture content
Acetate UV-absorbing filter, 18
Acetic acid, 154, 156
Acidic pollution, 130, 136, 150, 154, 256
Accumulation particulates, 247
Acrylics, 155, 238
  deterioration, 184
  glazing, 108
  sheet, 18
Activated carbon filters, 157, 220, 254, 262
Activation energy, *see* Energy
Acuity, visual, 61, 180
Adaptation, 25–26, 46–47, 52, 182, 203, 206
Adiabatic change, 212
Aerosol, 131
Air
  circulation, 101
  changes, 233, 235
  cleaning, 133, 284
  conditioning
    ducted, 104, 214, 268,
    filters for, 133, 268
    operation, 214
    packaged units, 104
    pollution,
      specification, 268
    recirculation, 252
    reducing deterioration without, 106–115
    removal of particulates, 133
  cooling, 216
  domes for excavation, 125
  locks of excavation, 125
  moisture in, 64, 81
  saturated, 68
Ajanta Caves, 266
Albedo, 167
Alizarin, *Plate 2*
Alkali from concrete, 133
Alkylamines, 124
Alum size, 144
America, North,
America, North (*cont.*)
  climate, 90
  indoor RH, 91, 234
  pollution, 132, 256
  sulphur dioxide, 256
  see also U.S.A.
Ammonia, 256
Ammonium sulphate, 146, 256
Anaerobic conditions, 124
Angle, solid, 178
Animal fibres, 145
Anisotropic changes in wood, 223
Antistatic polish, 18
Apostilb, 178
Archaeology, humidity control in, 124
Archival material, *see* Paper
Arezzo, Italy, and sulphur dioxide, 141
Arid zone, 91
Arrhenius equation, 188
'Artificial daylight' fluorescent lamp, 16
Artificial light, 50 lux, 25–27
ASHRAE Test Dust, 134, 252
Atomising humidifiers, 95
Automobiles and pollution, 150, 248, 259

Bacteria, 124
  sulphur-metabolising, 86, 257
Barber Institute, Birmingham, 38
B.C.F. and B.T.M. fire extinguishers, 159
Bassano, *Plates 7 and 8*
Beckman hygrometer, 76
Bimodal particulate distributions, 251
Biodeterioration, 86, 124,
  and biological activity, 44
Black body, 50, 201, 206,
  sun as, 167
Bloom on varnishes, 258
Blue Wool Standards, 182, 186
Blinds, 29–33
Bombay, climate, 90
Bone
  lightning for, 23
  and moisture, 66, 82
Books
  and fire extinguishers, 159
  and RH, 87
  and sulphur dioxide, 143
  see also Paper
Botanical specimens, lightning for, 23
Breathing and works of art, 111
Brightness, 61
  and luminance, 179
British Museum, 137
British Standards
  Blue Wool, 182, 186
Bronzes,
  corrosion layers, 146
  disease, 146
  and moisture, 84
  and sulphur dioxide, 146
Brussels, pollution, 140, 147
Buffers for RH control, 106–112, 232–237

Calcareous material, *see* Calcium carbonate, Marble, Stone
Calcium
  bicarbonate, 139
  carbonate, 138, 154
    and carbon dioxide, 138
    and nitrogen dioxide, 152
    and sulphur dioxide, 138
  hydroxide, 138
Calcutta, 88
Camera and colour rendering, *Plates 13–15*
Canada, 91, 117
Canadian Conservation Institute, 117
Candela, 178
Canvas and RH, 82, 225, 226
Carbon
  activated, 157, 254
  black, 15
  dioxide, 126, 169
    in air, 138
    fire extinguishers, 159
Cars and pollution, 150, 248, 259
Case, closed, 106–115, 232–237
CCT, *see* Colour Temperature
Ceiling, translucent, 29, 36, 43
  opaque centre, 31

Cellulose, 45
  activation energy, 189
  and heat, 43
  and nitrogen dioxide, 152
  and ozone, 151
  and sulphur dioxide, 143
  see also Paper
Cement dust, 133
Centrifugal humidifiers, 95
Chalk, *see* Calcium carbonate
China, 89
Chlorides, 153
  in bronzes, 84
Chroma, Munsell, 60
Chromaticity co-ordinates, 195–198
Churches, humidity control in, 119
CIE Chromaticity Chart, 56, 195–200
CIE UV bands, 17
CIELAB, 199
Cine film, 155
Clean Air Act, 132
Climate inside and outside museums, 87, 226
  and choice of RH, 88
  and wood, 225
Climatic zones, 88–91
  and wood, 225
Coal
  dust, 248
  and sulphur dioxide, 136
Coastal museums, 153
Colorimeter, 86
Colour
  adaptation, 48
  atlas, 59
  change, 10–15
  constancy, 49
  discrimination, 61
  and electronic flash, 47
  measurement, 55–60, 195–199
  Munsell system, 59
  of reflected light, 6
  rendering, 30, 48–60, 199–207, 262, *Plates 13–16*
  specification, 268,
  temperature, 49, 54, 199
  saturated, 57
  -shift method, 199
Complementary colours, 6, 57
Compression shrinkage, 83
Computer simulation of RH cycles, 237
Computers, 262
Concrete buildings, 133
Condensation, 119,122
  caused by visitors, 232
  and dew point, 92
Cone vision, 51, 177
Contrast
  sensitivity curve, 180
Convection and diffusion, 233
Cool and warm light, 25, 27, 49, 54
'Coolbeam' lamp, 7, 46, 55, 187
Coolers, 104
Copper
  corrosion, 146
  resinate, 13, *Plates 6–8*
Correlated colour temperature, *see* Colour temperature
Cotton,
  moisture in, 232, 236
  and nitrogen dioxide, 152
  and sulphur dioxide, 143
Cosine Law, Lambert's, 176
Crawford UV Monitor, 21, 175
Crawford/NPL spectral band method, 199–205
Cross-linking, 184
Cross-section of paint, *Plate 8*
Crown-silvered lamp, 174
Cumulative distributions, 247, 251
Cyclic heating and cooling, 120

Damage
  factor, 184
  by UV and visible radiaton, 15, 16, *Plate 3*, 189
  versus wavelength, 184
  see also Deterioration
Damp
  in walls, 119, 122
Dampers, 220
Data logging, 263
Daylight, 28–33, 52
  D65 Standard, 57, 165, 168, 195, 200
  damage by, 185
  factor, 185
  heat from, 186
  vision, 177
Decibel scale, 160, 182
Declaration of Independence, 233
Defensor humidifier, 96, 102
Delhi
  iron pillar, 85
  RH, 237
Dehumidifying equipment, 97–100, 103
  desiccant, 97
  refrigerant, 99
  Rotaire, 98
  Westaire, 100
Desiccant dehumidifiers, 97
Desk case, 26
Deterioration
  effect of temperature on, 43
  rate of, 43, 188
  versus visual sensitivity, 63
Dew point, 92, 216
Dichroic reflector, 7, 46, 55, 187
Diethylamino ethanol, 106
Diffusion of light, 27
Dimensional changes and RH, 222–226
Dimethyl sulphide, 256
Dimming, 207
Direct and diffuse lighting, 28, *Plates 11 and 12*
Distemper paintings, lighting for, 23
Distribution
  coefficients (CIE), 196
  of particulates
    bimodal, 250
    cumulative, 247, 251
    number, 251
    size, 244
Dominant wavelength, 58
Drawings, 38
  lighting for, 23
Dry bulb, 71, 214
Drying, 121
  agent, 109
  rapid, 125
  speed of, 121
  walls of a building, 121
Dryness due to winter heating, 87–92, 94, 115
  extreme, 124
Dulwich Art Gallery, *Plate 10*
Durban, climate, 90
Dust, Standard Test, 134, 252
Dyes
  colour, 190
    change, 10
  and light absorption, 190
  and moisture, 85
  and nitrogen dioxide, 152
  and sulphur dioxide, 145

Efficacy, *see* Luminous
Efficiency, *see* Luminous
Efflorescence, 120, 122
Egg tempera, 14
Electrical equipment and ozone, 150
Electricity, cost, 188
Electrolytic corrosion, 146
Electron volts, 190
Electronic flash, 47
Electrostatic precipitators, 135, 150, 262
Emergency RH control, 120–124
Energy
  activation, 3, 188–195
  electrical, conversion to light, 186
  light and heat, 3, 156, 189
  see also Power
Enthalpy, 210
Equilibrium Moisture Content (EMC), 228, 232
Europe
  pollution, 132, 256
  RH, 90, 91, 94, 234
  sulphur dioxide, 137, 256
'Eurovent' filter test, 134, 252
Excited state, 189, 192, 193
Exhibition cases, *see* Case
Expansion, and temperature, 44
Exponential decay and RH, 227, 234,
Exposure to light, 22
  annual, 14, 30, 268
  reducing time of, 37–38

External Design Conditions, 221
Eye, 61
  adaptation, 25, 203
  and age, 64
  and lighting, 61
  and photons, 190
  sensitivity, 56, 164, 176, 190
  and television camera, 61
  visual pigments, 197
  yellowing, 64

Fabrics and light, 10
Fading, rate of and temperature, 44
Fans, 101, 123
Far East, climate, 89
Feathers, lighting for, 23
Fechner's Law, 181
Film, 'Daylight' and 'Artificial light' 49, *Plates 13–15*
  aluminised, 35
Filters
  air, 133–135, 252–254
    Absolute, 133
    activated carbon, 157, 254, 262
    and recirculation, 252
    specifications, 268
  glass, for tungsten–iodine lamps, 8, 172
  UV-absorbing, 16–19
  yellow, 17
Fire
  detectors, 160
  extinguishers, 159
Flash, electronic, 47
Flooding, 121
Fluorinated hydrocarbons, 99
Fluorescent lamps, 8–10, 36, 45, 165, 168, 173, 187
  and colour rendering, 51, 54
  compact, 207
  control equipment, 173
  heat from, 186
  hot and cold-cathode, 173
  mercury vapour in, 8, 166
  phosphors, 8, 166, 173
  spectral power distribution in, 165, 168
  UV in, 16, 19, 175
Flux, luminous and radiant, 176
Foot-candle, 20
Formic acid, 154
Fraunhofer lines, 167
Free radicals, 193
Freezing, 116
Frescoes
  and damp walls, 122
  and nitrogen dioxide, 152
  and sulphur dioxide, 138
Frost in walls, 116
Fuels, and sulphur in, 136, 256
  burning of, 130
Fungi, 66, 121
Fur, lighting for, 23
Furniture, 224

Gas, natural, sulphur in, 136
Gaseous pollution, 135–153
  removal of, 156–158
  see also Ozone, Nitrogen dioxide, Sulphur dioxide
Giorgione, *Plate 6*
Glare, 25, 36, 42
Glass
  'Bronze', 205
  grey (tinted), 32, 35, 200
  light through, 5, 165, 168
  and moisture, 85
  transmission of, 5
  with plastic interlayer, 18, 31
  and sulphur dioxide, 147
  for tungsten halogen, 8
  weeping, 85
Glazes, paint, *Plates 6–9*
GLS lamps, 8, 174
Glue, 15
  and moisture, 66
  and RH, 84
Gold and sulphur dioxide, 147
Gouache, lighting for, 23
Green reflectance, 6, 58
Grey
  reflectance, 6, 58
  glass, 32, 35, 200
Grit, 131
Gulbenkian Museum, 32
Gum, 15
Gypsum,
  formation, 139, 257

Hair hygrometer, *see* Hygrometers
Half-time, hygrometric, 232–237
Halide, *see* Metal halide
Halogen, *see* Tunsten halogen
Halogenated methanes ('Halons'), 159
Halophosphate, 206
Harrison damage factor, 184
Heat, 43
  and biological activity, 44
  and chemical processes, 43
  and drying, 44
  effect on rate of deterioration, 43
  and embrittlement, 43
  energy, 3–4, 188
  and expansion, 43
  and light sources, 7, 45, 46 186–188
  and physical processes, 43
  radiant, 44, 45
  and rate of fading, 44
  sensible and latent, 210, 230
  and spotlights, 44
Heating,
  by electricity, 123
  by light, 186
Heating (*cont.*)
  by oil burners, 123
  see also Winter heating
Helium in cases, 233
Histograms, 244–246
HMI lamps, 47
Hofmannophilia, 86
Horn, lighting, for, 23
Hue, Munsell, 60
Humid zone, 89, 234
Humidifiers, 95–97, 102
  atomising, 95
  Defensor, 96
  evaporative, 96, 212
  maintenance, 97, 103
  steam, 220
Humidistat, 94
Humidity, 66–127, 210–241
  Absolute, 67, 79, 93, 210
  closed case, 106–115
  control, 93–127
    in archaeology, 124
    improvisation, 120–124
    mechanical, 114
    room, 93
    with windows, 123
  indicating cards, 70, 76
  measuring, 67–77
  Relative, 68
  see also RH
Humid zone, 89
Hydrochloric acid, 153
  from fire extinguishers, 159
  from PVC, 155
  removal, 156
Hydrogen
  peroxide, 151
  sulphide, 146, 154, 256
Hygrometer
  Beckman, 72
  chart, 73
  dial, 70
  electronic, 72, 76
  hair, 68
  Negretti and Zambra, 71
  Noya Sina, 72
  paper, 70
  sling, 70
  wet-and-dry, 70
  whirling, 70
Hygrometric
  chart, 77–81, 210–212
  half-time, 232
Hygrothermographs, *see* Thermo-hygrographs

Illuminance, 10–13, 20, 166, 177
  automatic control, 33
  recommended, 23, 33, 268
  specification, 268
  sun and sky, 171
  and visual acuity and colour discrimination, 61

Illuminants, Class A etc., 204
Illumination
direct and diffuse, 27–28, 31, *Plates 11–12*
limited, 22–32, 37
at London, 171
outdoor, 167
reference, 52
variation in, 62
Incandescent lamps, *see* Tungsten lamps
India, climate, 89
Indigo and nitrogen dioxide, 152
Indo-China, climate, 89
Indonesia, climate, 89
Induction period, 194
Infrared, 4–6, 43, *Plate 1*
Ink, 15
Insects and moisture, 86
International Organisation for Standardisation (ISO), 183
Iodine in tungsten–halogen lamps, 172
IR, *see* Infrared
Iron,
and moisture, 84
in particulates, 132, 256
pyrites, 86
and sulphur dioxide, 146
Isfahan, climate, 90
ISL lamp, 8, 174
Ivory
lighting for, 23
and moisture, 225

Japanese prints, 12

KJ/mole, 189
Kelvin scale, 50
Kilolux, 30
'Kolorarc', 10
'Kolorite', 54

Lacquer, lighting, for, 23
Lakes
red and yellow, 12–13, *Plate 10*
Lambert's cosine law, 176
Lamps
and control equipment, 171–174
designations, 7, 174
see also 'Coolbeam', Fluorescent, HMI, Metal halide, Tungsten, Xenon
Lascaux, 126
Latent heat, 210, 230
Lead
and organic acids, 147, 155
and sulphur dioxide, 147
Leather
lighting for, 23
and moisture, 66, 82, 87
and sulphur dioxide, 145
Lichens and sulphur dioxide, 148
Light
angle of, 35–37
Light (*cont.*)
artificial, 7, 25–27
at 50 lux, 25–27, 38–40
at 200 lux, 28–33
colour and strength change under, 2, 10–16, 44
colour of reflected, 6
cool and warm, 25, 27, 48–55
daylight at 200 lux, 28–33
diffusion of, 27–28
energy, 3–4, 164, 188–193
fastness, 11–12, 182, 186
flux, 175–179
and heat, 43–46, 186–188
material affected and unaffected by, 2, 10–16
meter, 19–20, 177
and paint medium, 14
reference sources, 203
and sulphur dioxide, 143
units, 177–179
white, *see* White
see also Lamps, Tungsten, Fluorescent, etc,
Lignin in paper, 144
Limestone and carbon dioxide, 138
and sulphur dioxide, 138–143
Linen
moisture content, 232
and sulphur dioxide, 143–145
Linseed oil, 14
Lithium chloride dessicant, 97, 98
Littlemore Engineering UV monitor, 21, 175
Log scale in normal distributions, 245
London
fogs, 149
and ozone, 152
particulates, 252
RH, 90, 234
pollution, 130, 132, 145
Los Angeles
and ozone, 150, 153, 259
particulates, 250
Louvres, 29
Lumens, 20, 166
per radiant watt, 187
Luminance, 58, 177–179
and brightness, 181
Luminosity, 179, 181
Luminous
efficacy, 172, 173, 186, 206
efficiency, 175–177, 198
emittance (exitance), 178
flux, 176
intensity, 177
Lux, 20, 178
hours, 21, 40
minutes, 48
meter, 19
values for sun and sky, 170
Luxor, climate, 90
Madras, climate, 90
Madrid, pollution, 132
Magnesium
carbonate, 48
nitrate, 113
McLaren's diagram, 186, 191
Maintenance, 97, 103
Malaysia, climate, 90
Manganese in particulates, 256
Manuscipts, lighting for, 23
Marble and sulphur dixide, 138–143
Materials
response to RH, 82–87
which change under light, 10–15
MBI, MBIF and MCF lamps, 174
Measuring
change, 265
colour, 55–60, 199–205
humidity, 67–77
UV and visible radiation, 19–20, 174–177
Mechanical
particulates, 248
RH stabilisation, 114
Mediterranean, climate, 91
Medium, paint, 14–15
see also Paint, Oil, Tempera
Mercury vapour
in fluorescent lamps, 8, 166
and ozone, 150
Metal-halide lamps, 9–10, 47, 168, 174
Metallic corrosion and RH, 84
Metals
and chlorides, 153
and nitrogen dioxide, 152
and sulphur dioxide, 146
Metamerism, 53
Meter, *see* Light meter
Micrograms, 130
Microns, 131, 244
Micrograms, 130
Microns, 131, 244
Microwatts per lumen, 166
of UV, 175
Milan, pollution, 132
Miniatures
colour change, 12
lighting for, 23
Minimum perceptible difference, 199
Mired, 201
Moisture
in air, 66, 210
of buffers, 107, 232
content, 66, 232
equilibrium, 232
reservoir, specific, 235
Moisture-absorbent materials
best RH, 88
response to RH change, 82–86
Mole, 189
Monsoon, 94, 107, 234

Mosses and sulphur dioxide, 148
Moths and RH, 86
Moulds and fungi, 66, 120
Munsell system, 59, 206
Murals, *see* Frescoes
Musical instruments, 66, 118

Nanometre, *Plate 1*
National Gallery, London, 31, 77, 83, 108, 118, *Plates 6 and 7*
  activated-carbon filters in, 157
  air conditioning, 118, 253
  histogram from catalogue, 244
  and ozone, 261
  picture sizes, 244
  and pollution, 130, 137, 257
  water spray, 156
National Gallery of Scotland, 114
National Postal Museum, 32
Natural-history specimens, 14
  lighting for, 23
New Delhi, *see* Delhi
Nikka pellets, 232
Nitric acid, 152
Nitric oxide, 259
Nitrogen
  in cases, 233
  dioxide (peroxide), 131, 145, 152, 259
    effects, 152
    levels, 152
    measurement, 255
    specification, 268
  oxides present in air, 151, 259
    removal, 156, 254
Nitrous acid, 151
Noise Rating (NR), 161, 268
Nova Sina, 76
NPL/Crawford spectral bands, 199
Nuclei, 247
Nylon and air pollution, 145

Oak
  acid vapours from, 154
  dimensional change, 224
Oil
  burners, 123
  linseed, 15
  paintings
    colour change, 12–14
      lighting for, 23–24
  and new concrete, 133
  and sulphur dioxide, 136
  see also Paint
Opaque centre ceiling, 31
Organic
  acid vapour, 155
  compounds and ozone, 150
  material, 2
'Oroglas', 18
Ottawa, 234
Oxidant pollution, 130, 136, 149
Oxidation, 193
Oxygen
  in cases, 233
  diffusion, 194
  and fading, 12
  penetration through plastics, 237–241
  removal, 194
  singlet, 193
  see also Photo-oxidation
Ozone, 131, 149–151
  effects of organic material, 150, 261
  formation, 149, 258–262
  from electrical equipment, 135, 150, 162
  levels, 152–153
  measurement, 255
  and photochemical smog, 149
  and photo-oxidation, 261
  production, 149, 258
  removal, 156
  specification, 268

Packing
  cases, 108, 109
  and vibration, 160
Paint
  cracks, 14
  cross-section, *Plate 8*
  deterioration, 184
  medium, 14–15
  and moisture, 226
  and sulphur dioxide, 12, 145
  UV-absorbing, 18
  see also Oil paint
Paintings on damp walls, 122
Pakistan, climate, 90
PAN (per-oxy acyl nitrates), 152, 260
Panel paintings and RH, 82, 227
Paper
  and light, 15, 45, 184, 189
  and moisture, 66, 70, 82, 87, 121, 225, 227, 232
  and nitrogen dioxide, 152
  rag, 143
  and sulphur dioxide, 143
  see also Cellulose
PAR lamp, 8, 174
Paraffin, 123
Parchment, and moisture, 66, 70, 87
  and sulphur dioxide, 145
Paris, pollution, 132
Particulates, 131–135, 244–253
  accumulation range, 247–252
  alkaline, 133
  classification, 247
  concentrations, 132
  natural, 132
  primary, 247
  removal, 133–135
  secondary, 247
Particulates (*cont.*)
  sizes, 131, 244–252
Parts per million and billion (ppm and ppb), 131
Pastel painting, 155
People, moisture and heat output, 230–232
Percentage saturation, 68
Permafrost, 124
Permeability of polymers, 237–241
Peroxy radical, 193
Perspex, 238
  VE, 18
Petroleum, sulphur in, 136
pH of rain, 139
Philips, 84 and 94 lamps, 165, 168
Phosphor, 9, 206–207
Photocells, 20, 29
  readings on model, 170
Photochemical,
  change, 15, 44, 184–186, 193–195
  smog, 149
  stability, standards, 183
Photocopiers and ozone, 150
Photographs and volatile sulphides, 154
  and records, 265
Photography, artificial and daylight, *Plates 13–15*
  colour film, 49
  film sensitivity to IR, 191
  lighting for, 23, 46, 268
Photolysis, 190
Photons, 4, 169, 184, 190
Photo-oxidation and ozone, 193, 261
Photopic curve, 177
Pigments
  colour change, 11–12
  sizes, 248
  and sulphur dioxide, 12, 145
Pilkington Float Glass, 5
Planckian radiator, 50, 199
Planck's constant, 189, 193
Plant fibres, 143
'Plexiglas', 18, 238
Pollens, 248
Poly-
  butyl methacrylate, 194
  carbonate, 18, 156
  ethylene ('Polythene'), 108, 110, 237–241
  iso-amyl methacrylate, 194
  methyl methacrylate, 18, 155, 238
  propylene, 156
  styrene, 156
  urethane, 156
  vinyl
    acetate, 155
    chloride, 155, 156, 238
  vinylidene chloride, 108, 238, 241
Polymers
  permeability, 238
  photo-oxidation, 191

Postal Museum, London, 23
Power
  distribution curves, 164–168
  and energy, 6, 164
  and radiant flux, 176
ppm and ppb, 131
Preferred Noise Criteria (PNC), 161
Pre-heat, 104, 214
Primary particulates, 248
Prints, lighting for, 23
Proteinaceous material, 189
Prussian blue, *Plate 2*
Psychrometer, Psychrometric *see* Hygrometer, Hygrometric
Purple reflectance, 6, 58
Pynacker, *Plate 10*
Pyrites disease, 86

Quanta, 3, 188–190
Quartz–iodine,–halogen, *see* Tungsten–halogen
Quercitron, 14
Qutb Minar pillar, 85

Radians, 178
Radiant
  flux, 176
  heat, 5, 45
Radiation
  controlling visible, 22–27
  measuring, 19–21
  and heat, 5, 44, 186–188
Radiators, 45
Radical, free, 193
  peroxy, 193
Radio waves, *Plate 1*
Radioactive decay, 234
Rain, acid, 139, 256
Rate of reaction, 192
Rayon and sulphur dioxide, 143
Reciprocity law, 21–22, 192
Recirculation and air filtration, 252
Records
  and photography, 265
  and temperature, 101
Red lakes, 12–13, *Plate 9*
Red rot of leather, 145
Reference light sources, 203
Reflectance
  curve, 56, 164, 195
  spectrophotometer, 56
  titanium dioxide, 18
Reflections, 26, 36
Refrigerant dehumidifiers, 99–100
Refrigerated showcases, 45
Re-heat, 214
Relative
  contrast sensitivity, 180
  humidity *see* RH
Replicas, 38
Retina, 51
Resonant vibration, 160
Restoration, 61
  lighting for, 23, 47, 268
RH
  and biodeterioration, 86
  for books and documents, 87
  and chemical reactions, 84
  choice of, 88
  and climate, 88, 226
  in closed case, 106
  compromise, 116
  control, 93–127
  definition, 68
  and dimensions and temperature, 228
  effect of people on, 230
  improvisation, 120–124
  limits, 87
  maintenance, 101
  mechanical stabilisation in cases, 114
  records, 101
  response of museum mateial to, 82–86, 222
  specification, 119, 268
  and sulphur dioxide, 143
  and temperature, 228
  variation which can be tolerated, 118
  see also Humidity
Rods and cones in the eye, 51
Room air cleaners, 254
Ropes, drying of, 84
Rotaire dehumidifier, 98
Rubber
  deterioration of, 184
  and ozone, 150
  vulcanized, 156
Russell and Abney, 13
Rutile, *see* Titanium dioxide

Sahara, 90
Salt, sea, wind-blown, 146, 153, 248
Salts for RH control, 113
Sandstone, alkaline and sulphur dioxide, 138
'Saran', 108, 241
Saturated
  air, 68
  colours, 57
  salt solutions, 113
Scientific notation, 161
Scottish National Gallery, 114
Sculpture, wooden, 82, 88
Secondary particulates, 248
Selenium photocell, 20, 177
Sensible heat, 210
Sensitivity
  luminous or spectral, 164, 177
  relative contrast, 180
'Sensor-check', 72, 176
Sensors, RH, 221
  enthalpy, 219
  triple-redundancy, 221
Set point, 214
Shock, 110
Show-cases, 36
  and curtains, 37
  desk-top, 26
  see also Case
Silica gel, 98, 108, 232–237
Silk, and concrete, 133
  and sulphur dioxide, 145
Silver
  and hydrogen sulphide, 146, 154
  and sulphur dioxide, 146
Singapore, 234
Size
  alum, 144
  on canvas, 225, 226
Sky, 167–171
  illuminance, 171
  overcast, 165, 168, 175
  and UV, 175
Smog, photochemical, 149, 248
Smoke, 131
Smoking, 126, 132, 248
Sodium
  bromide, 113
  chloride, 153, *see also* Salt
Solar radiation, 170
Solid angle, 178
Solvent vapours, 105
'Sonoxcarb', 158
Soot, 248
Sound and vibration, 160
Specific
  enthalpy, 212–214
  moisture reservoir, 235
  volume of air, 210
Specifications, 268
  air filters, 134, 253
  activated carbon filters, 158
  nitrogen and sulphur oxides, ozone, 158
  lighting, 23
  RH, 119
Spectral
  band method, 199
  reflectance, 6, 18, 164, 195
  sensitivity, 164, 177
  transmission, 5, 17, 201
  curves, 164–167
  power distributions, 56, 165, 168
    Standard Daylight, 196
Spectrophotometer, reflectance, 56
Spectrum, 4–7, 190, *Plate 1*
  see also Light
Spotlights, 7, 26, 42, 44
Spray, water, 105, 156
Sprinklers, automatic, 159
Sri Lanka, 88
Standard Daylight, *see* Daylight
Steradian, 178
  Hygrometric Chart, 210
Stone,
  alkaline, and sulphur dioxide, 138

Stone (*cont.*)
  and moisture, 120, 122
Storage
  pollution through, 154
  and temperature, 45
Stratosphere and ozone, 149
Sulphur dioxide,
  and calcium carbonate, 138
  and cellulose materials, 143
  concentration, 136
  damage by, 138–147
  and dyes, pigments, paints and
    synthetic textiles, 145
  fate in atmosphere, 256–258
  formation, 136, 256
  and frescoes, 138
  and glass, 147
  indoor concentration, 138
  and lichens and mosses, 148
  and light, 143
  and limestone, 138
  and marble, 138
  measurement, 255
  and metals, 146
  over London, 137
  oxidation, 136, 256
  and paper, 143
  and proteins, 145
  removal, 156–158, 254
    specification, 268
  and RH, 143
  see also under various materials
Sulphur in fuels, 136
Sulpur-metabolising bacteria, 257
Sulphur trioxide, 136
Sulphuric acid, 136, 139, 143,
    146, 256
  and bloom, 257
Sulphurous acid, 256
Sun, 167–171
  and air pollution, 150
  colour temperature, 50
  energy from, 170, 187
  and fading, 185
  heat from, 187
  illuminance from, 171
  and ozone, 258
  temperature, 167
  and UV, 175
Sunlight and oxidant smog, 149
Supersonic boom, 161
Swelling, 66, 82

Takamatsuzuka tumulus, 127
Tamen Rasset, 90
Tapestries, lighting for, 23
Tate Gallery
  air filters, 253
  and pollution, 137
Technical examination, lighting for,
  23
Television
  camera and eye, 61
Television (*cont.*)
  lighting for, 46
  sensitivity to IR, 190
Tempera
  paintings, 14
    lighting for, 23
Temperate zone, 91
Temperature, 43–46
  and activation energy, 188
  control of, 45–46
  correlated colour, 50, 199
  effect of people on, 230
  inversion, 259
  and RH, 77
  specification, 268
  and storage, 45
  of sun, 167
Temporary exhibitions, 116
Test Dusts, 134, 252
Textiles, 2
  colour change, 10
  lighting for, 23
  and moisture, 66, 70, 125
  and storage of, 45
  and sulphur dioxide, 143
Timber, *see* Wood
Tineola, 86
Thermohygrograph, 68, 72, 232
Thorn 'Kolorarc', 10
Thornton lamp, 207
Time switch, 37, 38, 41
Titanium dioxide, 18
TL lamp, 174
Tobacco smoke, 132, 248
Tombs, 124
Translucent ceiling, 31
Transmission curves, 164
  glass, 5
  grey glass, 201
  UV-absorbing filters, 17
Triphosphor fluorescent lamps, 207
Tropics, 88
  and wood, 224
Tumuli, 124
Tungsten
  halogen (iodine), 8, 9, 165, 168,
    172
    filter for, 8, 172
    heat from, 186
    and UV, 175
  incandescent, 7–10, 46, 54, 165,
    168, 182, 206
    blues under, 53, 206
    colour temperature, 50, 51
    damage by, 185
    heat from, 45, 187
    power distribution, 165, 168
    and UV, 175
Turner, 37, *Plate 5*
Two-photon reactions, 192

Ultrasonic burglar alarm, 161
Ultraviolet, *see* UV
'Univee' filter, 134
Unsaturated organic compounds, 150
U.S.A., 116
  Declaration of Independence, 233
  indoor RH, 91
  pH of rain, 139
  see also America, North
U.S.S.R., 91
UV, 4–7, 16–21, 54, 193, 201,
  Plate 2
  -absorbing filters, 16–19
  and Blue Wool, 183
  CIE bands, 17
  control, 33
  damage by, 15–16, *Plate 3*
  from fluorescent lamps, 8, 19
  measuring, 19–21, 174–175
  monitor, 20, 175
  and ozone, 258
  proportion of, 20, 165
  quanta, 190
  specification, 268
  and sulphur dioxide attack, 144
    conversion, 256
Uvethon C, 17

Value, Munsell, 59
Van Gogh, *Plate 9*
Vanadium in particulates, 256
Varnish, 15
  bloom on, 258
  UV-absorbing, 18
Vee lambda curve, 176
Veneered furniture, 66
Venice, and sulphur dioxide, 138
Ventilation and air circulation, 88,
    101, 112, 123
Verdigris, 13
Vibration, 110, 160
Victoria and Albert Museum and
    pollution, 137
Viridian reflectance, 6, *Plate 2*
Viruses, 248
Visible radiation, 4–7, *Plates 1 and 3*
  controlling, 22
  damage by, 15–16, *Plate 3*
  measuring, 19–20, 175
Visual
  performance, 180–181
Vokes filters, 134, 157, 253

Wallpapers, lighting for, 23
Walls, damp, 119–120
Warping, 66, 82, 118
  of panels, 82
Washington, 233, 234
Water
  chemical additives, 105
  content of air, 68
  as fire extinguisher, 159
  minerals in, 97
  spray, 156, 214
  vapour penetration through
  plastics, 237–241

Watercolours, 37
  colour change, 12
  lighting for, 23
Waterlogged conditions, 124
Wavelength, 4, *Plate 1*
  and damage, 184–186
  dominant, 58
Wax on panels, 227
Weakening of textiles, 2, *Plate 3*
Weeping glass, 85
Wet-and-dry-bulb hygrometer, *see* Hygrometer
Wet bulb, 71, 214
  temperature and enthalpy, 212
White, 48, 61
  light, 6–7, 56
Windows, 6–7, 34, 39
  and glare, 26
  and humidity control, 123
  and ventilation, 123
Winter heating, 85, 89, 94, 119, 234
Wood
  anisotropic drying, 223
  and climate, 224
  green state, 223
  dimensional changes, 222–226
  moisture in, 224, 236
  seasoning, 222
  volatile acids from, 154
Wool
  Blue, 182, 186
  and nitrogen dioxide, 152
  and sulphur dioxide, 145

X-rays, *Plate 1*
Xenon, 10, 165, 168
  flash, 47

Zinc
  and organic acids, 154
  sulphate, 114
Zones, climatic, 88–92